T0413714

Fuzziness: Structural Disorder in Protein Complexes

ADVANCES IN EXPERIMENTAL MEDICINE AND BIOLOGY

Recent Volumes in this Series

Volume 717
KAINATE RECEPTORS: NOVEL SIGNALING INSIGHTS
Antonio Rodriguez-Moreno and Talvinder S. Sihra

Volume 718
BRAIN INSPIRED COGNITIVE SYSTEMS 2010
Ricardo Sanz, Jaime Gomez and Carlos Hernandez

Volume 719
HOT TOPICS IN INFECTION AND IMMUNITY IN CHILDREN VIII
Nigel Curtis

Volume 720
HUMAN CELL TRANSFORMATION
Johng S. Rhim and Richard Kremer

Volume 721
SPHINGOLIPIDS AND METABOLIC DISEASE
L. Ashley Cowart

Volume 722
RNA INFRASTRUCTURE AND NETWORKS
Lesley J. Collins

Volume 723
RETINAL DEGENERATIVE DISEASES
Matthew M. LaVail, Joe G. Hollyfield, Robert E. Anderson, Christian Grimm and John D. Ash

Volume 724
NEURODEGENERATIVE DISEASES
Shamim I. Ahmad

Volume 725
FUZZINESS: STRUCTURAL DISORDER IN PROTEIN COMPLEXES
Monika Fuxreiter and Peter Tompa

Fuzziness

Structural Disorder in Protein Complexes

Edited by

Monika Fuxreiter, PhD

Institute of Enzymology, Biological Research Center, Hungarian Academy of Sciences, Budapest, Hungary

Peter Tompa, PhD

Institute of Enzymology, Biological Research Center, Hungarian Academy of Sciences, Budapest, Hungary

Springer Science+Business Media, LLC

Landes Bioscience

Springer Science+Business Media, LLC
Landes Bioscience

Printed in the USA.

Springer Science+Business Media, LLC, 233 Spring Street, New York, New York 10013, USA
http://www.springer.com

Please address all inquiries to the publishers:
Landes Bioscience, 1806 Rio Grande, Austin, Texas 78701, USA
Phone: 512/ 637 6050; FAX: 512/ 637 6079
http://www.landesbioscience.com

The chapters in this book are available in the Madame Curie Bioscience Database.
http://www.landesbioscience.com/curie

Fuzziness: Structural Disorder in Protein Complexes, edited by Monika Fuxreiter and Peter Tompa. Landes Bioscience / Springer Science+Business Media, LLC dual imprint / Springer series: Advances in Experimental Medicine and Biology.

ISBN: 978-1-4614-0658-7

Library of Congress Cataloging-in-Publication Data

Fuzziness : structural disorder in protein complexes / edited by Monika Fuxreiter, Peter Tompa.
p. ; cm. -- (Advances in experimental medicine and biology ; no. 725)
Includes bibliographical references and index.
ISBN 978-1-4614-0658-7
1. Proteins--Conformation. 2. Protein folding. 3. Protein binding. 4. Fuzzy systems. I. Fuxreiter, Monika, 1969- II. Tompa, Peter. III. Series: Advances in experimental medicine and biology ; no. 725. 0065-2598
[DNLM: 1. Protein Conformation. 2. Protein Denaturation. QU 55.9]
QP551.F99 2011
572'.633--dc23

2011021158

DEDICATION

To our future generation

FOREWORD

For more than 40 years following the determination of the first protein structure, molecular biology was guided by two central dogmas which posited that the ordered three dimensional structure of a protein is intimately linked to its biological function and that binding of a protein to ligands or to other protein molecules is exquisitely specific. It therefore came as a surprise when, in the latter half of the 1990s, experimental work on several regulatory proteins and bioinformatics surveys performed on a genomic scale showed that regions of conformational disorder are common in eukaryotic proteins involved in cellular regulation and signaling. Such intrinsically disordered proteins (IDPs) frequently function as central hubs in protein interaction networks, binding multiple protein partners. IDPs often contain short amphipathic motifs that fold into ordered structures upon binding to their targets. However, in many cases, the IDP remains disordered even in the bound state—a phenomenon aptly named “fuzziness” by Peter Tompa and Monika Fuxreiter, the editors of this volume.

Detailed characterization of fuzzy interactions will be of central importance for understanding the diverse biological functions of intrinsically disordered proteins in complex eukaryotic signaling networks. In this volume, Peter Tompa and Monika Fuxreiter have assembled a series of papers that address the issue of fuzziness in molecular interactions. These papers provide a broad overview of the phenomenon of fuzziness and provide compelling examples of the central role played by fuzzy interactions in regulation of cellular signaling processes and in viral infectivity. These contributions summarize the current state of knowledge in this new field and will undoubtedly stimulate future research that will further advance our understanding of fuzziness and its role in biomolecular interactions.

Peter Wright, PhD
Department of Molecular Biology and
Skaggs Institute for Chemical Biology
The Scripps Research Institute
La Jolla, California, USA

PREFACE

For almost four decades proteins were thought to function as entities with well-defined structures, and to each biological task a unique conformation was assigned. A decade ago it was recognized however, that some proteins do not obey this rule and act as a heterogeneous ensemble of conformations. These proteins were termed as intrinsically unstructured or intrinsically disordered proteins (IUPs or IDPs). IDPs brought a new flash into structural biology urging to change our deterministic one sequence-one structure-one function concept to a one sequence-multiple structures-one function paradigm. Many IDPs serve in molecular recognition processes, and upon targeting different partners they often adopt a well-defined three-dimensional structure. It gives the impression that although IDPs have extensive conformational freedom in the unbound state in solution they behave as 'regular' proteins upon fulfilling their functions. This view however, is misleading. The structured image of bound IDPs only reflects experimental difficulties in characterizing conformational ensembles of complexes. Indeed, many parts of IDPs preserve their structural heterogeneity even upon interacting with other molecules. Some biochemical studies demonstrate that these parts often coincide with functionally critical regions. The two statements together signify that structural disorder in complexes is important for various biological roles. This phenomenon is termed fuzziness. Structural ambiguity in complexes expands the capacity of proteins to perform multiple functions and also provides an additional level of versatility for regulation. The existence of fuzzy complexes calls for the ultimate reassessment of the classical one structure-one function paradigm and converts it to a one sequence-multiple structures-multiple functions paradigm. This book is dedicated to this new concept and via many examples introduces a new view on protein functionality.

Monika Fuxreiter, PhD
Peter Tompa, PhD
Institute of Enzymology, Biological Research Center
Hungarian Academy of Sciences, Budapest, Hungary

ABOUT THE EDITORS...

MONIKA FUXREITER is a senior scientist at the Institute of Enzymology, Hungarian Academy of Sciences in Budapest, Hungary and a visiting scientist at the Weizmann Institute of Science, in Rehovot, Israel. Her main interest is to develop mechanistic models for proteins, especially structure-function relationships for intrinsically disordered proteins and their complexes. She also studies the role of dynamism in protein evolution. She often shares ideas with her children, Krisztina and Pal. Monika Fuxreiter received her degrees from the Eötvös Loránd University, Budapest, Hungary.

ABOUT THE EDITORS...

PETER TOMPA is a professor of protein sciences at the Institute of Enzymology of the Hungarian Academy of Sciences, Budapest, Hungary. His interest focuses on the structural disorder of proteins, a phenomenon that defies the classical structure-function paradigm of proteins that is about to change our general concepts of protein structure and function. He made basic contributions to this field, he organized the first international meeting dedicated to this topic in 2007 and wrote the first monograph on structural disorder (Structure and function of intrinsically disordered proteins, 2009, CRC Press, a Taylor and Francis group, Boca Raton, FL, USA). He has published over 100 papers in international journals and holds a DSc degree from the Hungarian Academy of Sciences. Recently he took the position of director at the VIB Department of Structural Biology, Brussels, Belgium.

PARTICIPANTS

Sarah E. Bondos
Department of Molecular
and Cellular Medicine
Texas A&M Health Science Center
College Station, Texas
USA

Jean-Luc Darlix
LaboRetro
Unité de Virologie Humaine
INSERM 758,
IFR 128, ENS de Lyon
Lyon
France

Ariele Viacava Follis
Department of Structural Biology
St. Jude Children's Research Hospital
Memphis, Tennessee
USA

Monika Fuxreiter
Institute of Enzymology
Biological Research Center
Hungarian Academy of Sciences
Budapest
Hungary

Charles A. Galea
Structural Biology Division
Walter and Eliza Hall Institute
of Medical Research
Parkville
Australia

Jeffrey C. Hansen
Department of Biochemistry
and Molecular Biology
Colorado State University
Fort Collins, Colorado
USA

Hao-Ching Hsiao
Department of Molecular
and Cellular Medicine
Texas A&M Health Science Center
College Station, Texas
USA

Roland Ivanyi-Nagy
Molecular Parasitology Group
The Weatherall Institute
of Molecular Medicine
University of Oxford
Oxford
UK

Elizabeth A. Komives
Department of Chemistry
and Biochemistry
University of California San Diego
San Diego, California
USA

Richard W. Kriwacki
Department of Structural Biology
St. Jude Children's Research Hospital
and
Department of Molecular Sciences
University of Tennessee Health
Sciences Center
Memphis, Tennessee
USA

Kevin A.W. Lee
Department of Biology
Hong Kong University of Science
and Technology
Hong Kong
China

Sonia Longhi
Architecture et Fonction des
Macromolécules Biologiques
Universités d'Aix-Marseille I et II
Marseille
France

Steven J. McBryant
Department of Biochemistry
and Molecular Biology
Colorado State University
Fort Collins, Colorado
USA

Régis Pomès
Molecular Structure and Function
Hospital for Sick Children
and
Department of Biochemistry
University of Toronto
Toronto
Canada

Sarah Rauscher
Molecular Structure and Function
Hospital for Sick Children
and
Department of Biochemistry
University of Toronto
Toronto
Canada

Alexander B. Sigalov
SignaBlok, Inc.
Shrewsbury, Massachusetts
USA

Peter Tompa
Institute of Enzymology
Biological Research Center
Hungarian Academy of Sciences
Budapest
Hungary

G. Rickey Welch
Department of Biological Sciences
and Department of History
University of Maryland
Baltimore, Maryland
USA

Peter Wright, PhD
Department of Molecular Biology
and
Skaggs Institute for Chemical Biology
The Scripps Research Institute
La Jolla, California
USA

CONTENTS

CHAPTER 1

FUZZY COMPLEXES:
A More Stochastic View of Protein Function

Monika Fuxreiter and Peter Tompa
Institute of Enzymology, Biological Research Center, Hungarian Academy of Sciences, Budapest, Hungary
Emails: monika@enzim.hu; tompa@enzim.hu

Abstract: Intrinsically disordered proteins (IDPs) are widespread in eukaryotic proteomes and challenge the classical structure-function paradigm that equates a folded 3-D structure with protein function. However, IDPs often function by molecular recognition, in which they bind a partner molecule and undergo "induced folding" or "disorder-to-order transition" upon binding, which apparently suggests that in a functional context IDPs become ordered. Whereas this observation would restore the "prestige" of the classical structure-function paradigm, a closer inspection of the complexes of IDPs reveals that they do not always become fully ordered, but preserve functionally significant disorder in the complex with their binding partner(s). This phenomenon, which we termed "fuzziness", is the ultimate extension of structural disorder to the functional native state of proteins. In this introductory chapter, we outline the most important aspects of fuzziness, such as its structural categories, molecular mechanisms of function it mediates and the biological processes, in which it plays a distinguished role. As confirmed by all the other chapters of the book, we will show that new cases of fuzziness pop up at an accelerating pace, underscoring that this phenomenon presents a widespread novel paradigm of protein structure and function.

INTRODUCTION AND OVERVIEW

Our classical view of protein function is engraved in a stable three- dimensional architecture of amino acids arranged in a given order. For many decades an unambiguous relationship between sequence and a well-defined three-dimensional structure has been assumed, the latter underlying a given biological task. In case of enzymes, for example,

Fuzziness: Structural Disorder in Protein Complexes, edited by Monika Fuxreiter and Peter Tompa.

the protein provides a scaffold for the active site, where side chains participating in the chemical reaction are located and ideally oriented to interact with the substrate. To achieve high catalytic rates, the enzymatic environment must be complementary to the transition state of the reaction, which requires a given arrangement of dipoles, i.e., a given fold.[1] Although the importance of protein dynamics has been recognized in many biological processes at all levels, ranging from variations in sidechain rotamers during enzymatic catalysis[2-4] or to more substantial conformational rearrangement in allosteric proteins,[5] they still could be interpreted within the framework of the classical structure-function paradigm, as variations around an optimal structure.

Multiple, independent evidence demonstrates, however that proteins can exist simultaneously in different, yet functionally relevant conformations.[6-8] Such degree of structural diversity, i.e., interconversion of proteins between many conformations, is in apparent contradiction with our classical view of one structure-one function relationship. The number of experimentally observed protein examples that lack a well-defined three-dimensional structure increases exponentially (http://www.disprot.org).[9] Intrinsically disordered proteins or regions (IDPs/IDRs) are represented in all three kingdoms of life, with increasing propensity in more complex organisms.[10-12] More than 50% of eukaryotic proteins have at least one long (>30 aa) IDR. Partial or local disorder is observed in almost every protein, but it seems to be more abundant in certain protein classes.[13] Protein disorder enables complex regulatory functions, illustrated by high abundance of IDPs in transcription,[14] cell-cycle regulation or signal transduction.[15,16]

IDPs are distinguished in molecular recognition functions.[17] They are considered to have high specificity in target selection and fast association/dissociation rates upon interaction with the partner. Equilibrium among many conformations of the ensemble can be modulated by binding of a partner, leading to binding of multiple, unrelated ligands.[18] Binding of IDPs to a partner is often accompanied by adopting a well-defined structure.[19] This coupled folding-binding process can occur between an IDP and a globular partner, but also between two IDPs, with the resulting complexes being amenable to structural studies. This gives the impression that even though the free state of IDPs cannot be characterized by a unique structure, their functional state is conformationally unambiguous. Furthermore, some structural elements of IDPs are biased for partner recognition/binding.[20] Thus, structural data on a few dozen IDP complexes argues that the classical structure-function paradigm can also be extended to IDPs, if their bound state is considered. In this sense, IDPs represent a special class of proteins, where structural multiplicity is beneficial for partner selection, but behave as "traditional" proteins after binding.

Why would, however, nature limit the capabilities of IDPs after they attain the bound state? In other functions, like entropic springs/bristles, IDPs preserve their conformational freedom in all circumstances.[21] It would be a twisted logic not to exploit these beneficial features after interacting with the primary partner. Hence, it is possible that the complexed state of IDPs be considered in a way similar to IDPs themselves 15 years ago and look for structural properties not properly recognized/represented in their bound states. Hence we were seeking for cases, where structural disorder is a functionally essential property of an IDP complex. Indeed upon an initial survey, 26 such complexes were found, where structural polymorphism or disorder was present in the bound state with significant contribution to function.[22] Similarly to fuzzy logic in mathematics, the terminology fuzzy complex was used.

Structural disorder in fuzzy complexes, such as protein disorder in the free state comprises many different states along a wide spectrum, ranging from local to global disorder, from compact to extended states. Fuzziness was classified into four major categories. In polymorphic complexes at least one of the partners adopts a few or multiple alternative conformations. In flanking and clamp complexes the disordered segment neighbors or connects ordered binding region(s). The random complexes are extreme cases of fuzziness, where binding does not induce ordering of the interacting regions. These categories are not distinct they can also overlap as illustrated by the examples below. The phenomenon of sequence independence in protein-protein interactions was also related to fuzziness.

The recognition of the biological relevance of disordered bound states motivated the investigation of such 'unorthodox' associations. Complexes reported to be fuzzy up to date are summarized in Table 1. Accumulation of structural data did not only provide a more detailed view on how conformational heterogeneity is realized in the bound state, but also revealed those novel mechanisms, which could be mediated by fuzziness. In the following we review how polymorphism or disorder in protein complexes can be exploited in a variety of biological processes. The chapter will be organized as follows. In the following section, basic definitions of the categories of fuzziness will be given with a few illustrative examples. Additional sections discuss the types of interactions in fuzzy complexes, biological processes where fuzziness plays a distinguished role, and the importance of fuzzy complexes in molecular biology.

FIVE CATEGORIES OF FUZZINESS

Polymorphic Complexes

Alternative conformations (one/multiple) are observed in the complex that can also underlie different biological functions. Structural variability in these cases could be resolved, like upon binding nuclear localization signals (NLSs) to α-importin. NLS peptides primarily bind via two clusters of basic residues connected by a linker, which is variable in length.[23] The same NLS peptide exhibits different side-chain conformations enabling its association with different cargo proteins.

Different side-chain conformations can also lead to alternative set of interactions between two molecules, as seen upon binding T-cell factor 4 (Tcf4) transcription factor to β-catenin. Tcf4 binds to β-catenin in an extended conformation, where the middle highly acidic segment forms alternative salt bridges with the partner.[24] Diminishing any of these contacts impairs binding affinity, indicating that variability in charge-charge interactions is required for the optimal complex.

Alternative contacts may impart different morphology on larger assemblies. For example, actin polymerization is regulated by various proteins containing tandem repeats of WH2 domains (e.g., thymosine, ciboulot, Spire, Cordon bleu).[25] Binding of the regulators to the disordered subdomain-2 of actin induces their folding. The variable interactions of the WH2 C-terminal tail affects the spacing between the ordered parts of the different WH2 domains and thereby the organisation of the actin polymer.

Whereas direct structural evidence is missing in most cases, different biological outcomes might result from alternative conformations. A disordered loop of dihydropyridine receptor (DHPR) can bind to ryanodine receptor (RyR) with different affinities, resulting in activation or inhibition of RyR channel opening.[26] The opposite effects are likely due to alternative interactions between the disordered peptide and gating residues of the receptor that can block the passage of chloride ions.

Table 1. Fuzzy complexes

Model	IDP	Partner	Evidence	Ref.
Static			S	
Polymorphic	NLS	α-importin	S	23
	Msp90 MEEVD	Ppp5 TPR domain	B	77
	Tcf4 CBD	β-catenin	B	24
	RyR	DHPR	B	26
	CFTR R domain	CFTR	B	52
	Inhibitor 2	Protein phosphatase 1	B	78
	Prion	Prion amyloid	B	47
	RNase 1	RNase inhibitor	S	79
	WH2	Actin	S	25
	Myelin	Actin	S	56
Dynamic				
Clamp	Ste5	Fus3	S	27
	Oct-1 trans. fac.	Ig-k promoter	S	28
	NLS	α-importin	S	23
	Cellulase E	Cellulose	B	80
	Myosin VI	Actin filament	S	23
	L7	L12	B	68
	UPF1	UPF2	S	59
Flanking	Hsp25	α-acrystalline	S	81
	RNAPII CTD	mRNA maturation factors	S	61
	Measles virus nucloprotein	Phosphoprotein	S	71
	CREB KID	CBP KIX	S	32
	Proline rich peptides	SH3 domain	S	36
	IA_3 inhibitor	Aspartic acid protease	S	35
	SF1 splicing factor	U2AF	S	33
	FnBP	Fn3 domain	S	82
	SP1 trans. fac	Transcription preinitiation complex (PIC)	B	83
	β-catenin	APC	B	62
	p27	Cdk—cyclin		54
	Ebola virus nucleoprotein	VP35 and VP24	B	72
	Ubx homeodomain	DNA	S	34
	SSB	DNA	S	40
	Ets-1	DNA	S	84
Random	T cell receptor zeta chain	T cell receptor zeta chain	S	39
	Elastin	Elastin	S	85
	Sic1	Cdc4	S	49
	UmuD2	UmuD2	S	38
Sequence independent				
	GCN4	PIC	B	41
	Gal4	PIC	B	42
	EFP	PIC	B	43
	Linker histone H1 CTD	DFF40	B	45
	Core histone H4 NTD	nucleosome	B	46
	Ure2 prion	Ure2 prion	B	86
	Sup35 prion	Sup35 prion	B	

Clamp Complexes

Many IDPs interact with their partners in a bi-partite manner, with the linker region lacking permanent contacts and remaining disordered. The length of the linker is critical as this segment modulates the separation of the clamps and is responsible to correctly position specificity determining interactions. The scaffolding protein Ste5 contacts the fusion protein 3 (Fus3) by two distinct segments connected by an 8 residue long linker. Only simultaneous binding of the two ordered regions provides measurable affinity that requires the presence of the connecting segment that dynamically interchanges among many conformations.[27] The involvement of two such segments connecting three binding regions has been observed in the case of inhibitor-1c in complex with protein phosphatase 2B (Fig. 1A).

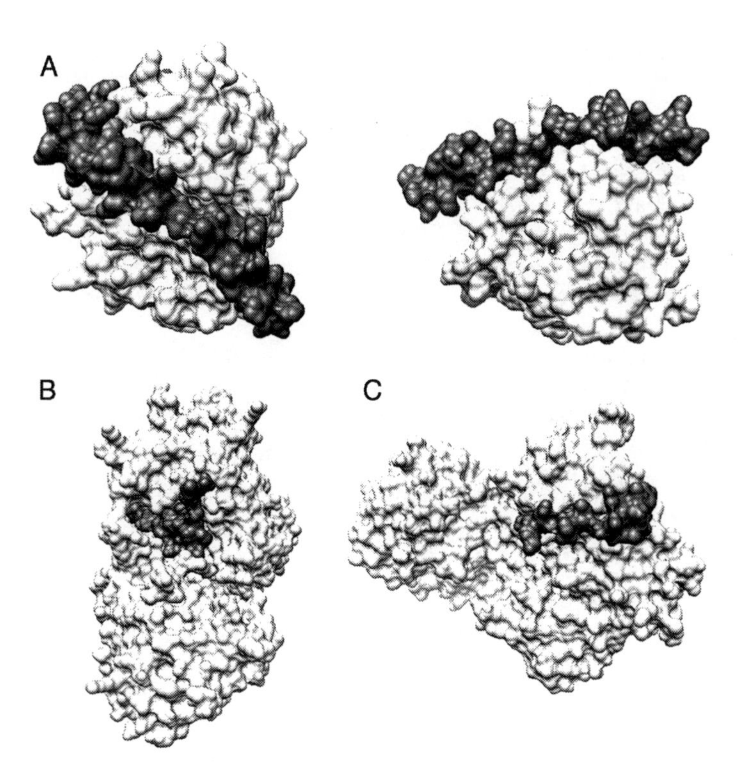

Figure 1. Fuzziness and function in protein-protein interactions. Three examples demonstrate the functional involvement of fuzzy regions of protein complexes. The figure shows the structure of three complexes, with fuzzy parts remaining disordered in the bound state. The partner (dark grey) binds and IDP (light grey), which undergoes limited induced folding. A significant part of the IDP remains disordered, and contributes to function. A) The complex of Inhibitor-1c with protein phosphatase 2B (PP2B, pdb 2O8G). The inhibitor binds its partner via three discontinuous binding segments, with the intervening regions contributing to post-translational regulatory modifications. B) C-terminal domain (CTD) of RNA polymerase II (RNAP II) bound to its partner, mRNA guanylyltransferase (pdb 1P16). Large part of the CTD, which is about 350 amino acids long, remains disordered in the complex, and mediates interactions with many other factors involved in mRNA maturation. C) WH2 domain of WASP in complex with G-actin (pdb 2A3Z). The adjacent Pro-rich domain of the IDP remains disordered, and mediates interaction with G-actin sequestering proteins (e.g. profilin) to release G-actin for actin polymerization.

Disordered linkers often have excess of charged residues, which provide transient, nonspecific electrostatic contacts with the partner. The octamer binding factor 1 (Oct1) has two separate DNA binding module, each recognizing a 4-5 base pair sequence. For optimal affinity and specificity, a 23 aa long linker region is required; mutating some of its charged residues or shortening it causes severe drop in selectivity.[28]

Flexible/disordered linkers may also enable processivity of binding, as exemplified by movements along a polymeric partner. Myosin VI has two heads, separated by a 80-residue linker, which can interact with actin polymer. In each step, the head performs a diffusive search promoted by electrostatic interactions between the disordered part of the linker region and the actin polymer.[29]

Flanking Complexes

IDPs often utilize a short recognition element to establish specific contacts with their partners. Indeed, many protein-protein interactions are mediated by short recognition elements ('linear' motifs).[30,31] The disordered environment in which the motif is embedded, not only imparts plasticity to localize the target site, but may also promote complex formation via transient, nonspecific interactions. Shortening the flanking region results in gradual decrease in binding affinity. The kinase inducible domain (KID) of CREB interacts with the KIX domain of CBP via 29 amino acid residues.[32] Doubling the size of the flanking regions results in a five-fold decrease of the dissociation constant. The binding affinity of splicing factor 1 (SF1) to U2 small nuclear RNA auxiliary factor also increases two-fold upon inclusion of noncontacting residues in the motif and five-fold for the full-length protein.[33] The interaction of Drosophila Hox protein Ultrabithorax with cognate DNA is primarily mediated by a Hox domain, whereas a long disordered N-terminal tail contributes many additional, transient regulatory interactions which either activate or inhibit the primary Hox-DNA interaction. Thus, this fuzzy region presents an intricate interplay of intra- and intermolecular interactions that fine-tune specificity of DNA recognition.[34]

Flanking regions may also induce formation of secondary structure required for binding. The yeast IA3 inhibitor has subnanomolar potency against yeast aspartic protease. The inhibitor is disordered in isolation, while 30 residues adopt an α-helical conformation upon interaction with the protease. Deletion or mutation of inhibitor residues that are not in physical contact significantly compromises binding[35] due to decreased electrostatic stabilization of the helix. Similarly, upon interaction of Src SH3 domain with a proline-rich peptide, only two prolines make direct contacts with the partner, the rest contribute to the formation of the left-handed PPII helix required for recognition.[36]

Disordered flanking regions ('tails') often play important role in target site localization of DNA-binding proteins. The N-terminal tail of homeodomains is a crucial factor in selectivity and remains partly unstructured in the complex.[37]

Random Complexes

These are the most bizarre examples of fuzziness, when the IDP interacts with the partner only via transient contacts, which apparently do not induce its (even partial) folding. By a variety of biophysical techniques, these IDPs behave similar to their free form and show no signs of an ordered structure.

The UmuD2 gene product is a dimer that can undergo autocleavage. The resulting proteins, shortened by 24 residues each, form a stable homodimer, but show a random-coil signal in CD at physiologically relevant concentrations.[38] According to various spectroscopical methods, T-cell receptor ζ chains also lack secondary structure upon homodimerization.[39] Random behavior is not limited to protein-protein interactions. Binding of single-stranded DNA binding protein to DNA does not induce disorder-to order transition of the protein, which keeps fluctuating among several alternative conformations.[40]

Sequence Independent Interactions

Some IDPs can form a full-affinity complex even if their sequence is shuffled. Most probably, the variability of specific contacts is responsible for this phenomenon, which can result in structural heterogeneity in the bound state. In most cases, structural evidence is available for the disordered state of such segments (missing coordinates from crystal structures, CD) but no detailed view of such interactions could be obtained.

The classical observation is that in transcription factors the acidic transactivator segments can be replaced by random acidic sequences without compromising transcriptional response.[41,42] In Ewing's sarcoma fusion proteins the repeat units can be freely interchanged or even reversed without impairing function.[43]

The C-terminal domains (CTD) of linker histones are highly variable in sequence, but show stable amino acid composition.[44] Binding of the linker histone H1 CTD to DFF40 is resistant to scrambling of the repeat units with respect of the globular domain.[45] The redundant function of H2 and H4 core histone N-terminal domain (NTD) in chromatin condensation could also be mediated by the multiplicity of conformations rather than a single state.[46]

INTERACTIONS IN FUZZY COMPLEXES

Alternating Interactions

Conformational heterogeneity can result in alternative permanent or transient contacts with the partner. Variations in permanent interactions can lead to different morphologies in large assemblies, like in prions, where different hydrogen bonding patterns lead to different packing of β strands.[47] Plasticity of structures might also explain why amyloid formation is not sensitive to sequence changes.[48]

Alternative contacts are often formed by electrostatic interactions that generate a conformational ensemble of the complex. The cyclin dependent kinase inhibitor Sic1 contains 9 recognition sites for the cell-division cycle protein 4 (Cdc4), which can interact with a single site of the receptor. Interactions are induced by phosphorylation that causes transient, local ordering around the motifs.[49] The whole complex, however, remains dynamic, which makes the phosphorylated sites interchangeable.

Interactions Induced by Posttranslational Modifications

Disordered regions are the preferred sites of posttranslational modifications[50] even when some parts are bound to a partner. Phosphorylation can induce formation of transient local ordering, further posttranslational modifications or contacts with other partners. The degree of phosphorylation can be utilized to fine-tune the strength of interaction in the complex via long-range electrostatic interactions as seen in the Sic1-Cdc4 complex.[51]

Multiple phosphorylation in the disordered regulatory R domain of the cystic fibrosis transmembrane-conductance regulator (CFTR) induces formation of transient α-helices.[52] These fluctuating secondary structure elements establish dynamic contacts with the nucleotide binding domains (NBD) of CFTR and thereby prohibit their dimerization. This couples the phosphorylation of R domain to the mechanism of channel opening.

Phosphorylation may induce further interactions that can can mediate a functional outcome distinct from that of the original binding. The cyclin dependent kinase inhibitor $p27^{Kip1}$ binds to cyclin A-Cdk complex and blocks the progression from G1 to S phase in the cell cycle.[53] The flexibility of the molecule allows phosphorylation of T88 causing its exclusion from the ATP binding pocket of Cdk2. This process re-activates the kinase enabling its action on T187, located in the long, disordered C-terminal tail of p27.[54] Phosphorylation of T187 initiates its interactions with the SCF/Skp2 that polyubiquitinates p27 leading to its degradation and progression to the S phase.

Interactions with Alternative Partners (Promiscuity)

Plasticity of disordered regions often enables the adaptation of the same motif to different partners or a variable arrangement of the recognition motifs, which can mediate interactions with alternative partners. The proline-rich segment of myelin basic protein (MBP) can bind to several SH3 domain-containing proteins, e.g., Yes1, PSD95, cortactin, PexD, Abl, Fyn, c-Src, Itk.[55] Besides the direct contacts with the proline residues, all these complexes are facilitated by long-range electrostatic interactions between the disordered region of myelin outside the binding context and the partner.[56]

Disordered charged tails are utilized to fine-tune binding strength of homeodomains to different DNA sequences.[34,57] Although the disordered tail primarily serves as a nonspecific anchor via transient electrostatic interactions, it can also establish a few specific contacts without adopting an ordered structure.[58]

Simultaneous Interactions with Different Partners

Disordered bound segments may contain recognition sites for other proteins that can be targeted even in the complex. Nonsense-mediated decay factors UPF1 and UPF2 interact in a bipartite manner, leaving the connecting segment disordered.[59] This binding mode is not only required for full affinity, but also promotes association with the translation termination factor eRF3 and initiates complex formation with the ribosome and the exon junction complex (EJC).

A considerable fraction of the general transcription factor TFIIF remains unstructured upon binding to RNAP II.[60] These regions are highly charged, sensitive to proteolysis (exposed) and also interact with the Mediator, thereby contributing to the assembly of the preinitiation complex (PIC). The C-terminal domain of RNAP II is highly disordered and serves as a scaffold for complexes participating transcription termination and mRNA maturation:[61] most of the CTD remains disordered in any of the complexes with its multitude of partners (Fig. 1B), thereby mediating further regulatory interactions.

BIOLOGICAL PROCESSES DISTINGUISHED BY FUZZINESS

Structural disorder correlates in general with signaling and regulation and contributes to the function of these IDPs in molecular recognition. Fuzziness is most often seen in these physiological processes and when protein function fails, also in pathological processes.

Signaling

Efficient information flow requires large sensitivity for environmental signals and rapid response. This is normally achieved via transient contacts among molecules or conformational rearrangements that alter the interaction pattern. Ligand-dependent conformational equilibrium and dynamically fluctuating interactions of fuzzy complexes offer efficient means for molecular communication. Many fuzzy examples are associated with signalling pathways. For example, β-catenin is a transcription co-activator that also plays role in Wnt signalling pathway. It is bound to a multi-protein complex including the Adenomatus Polyposis Coli (APC) protein, axin and protein phosphatase 2. Phosphorylation of APC is responsible for turning the Wnt signals on and off by changing the affinity of the β-catenin binding. As β-catenin remains partly disordered in the complex it can also interact with the ubiquitination machinery without dissociating from APC.[62] A similar fuzzy signaling complex has been observed between the WH2 domain of Wiskott-Aldrich syndrome protein (WASP) and G-actin (Fig. 1C), in which the Prorich region of WASP remains disordered and available for further interactions, with actin-sequestering profilin, for example.

Transcription and Translation Regulation

Transcription regulation involves various multi-protein complexes that act in a given spatial and temporal order. Many of these complexes interact via disordered regions that may remain disordered even in the bound state.[14] For example, C-terminal tails of linker histones and of core histones are critical for organizing chromatin assemblies and are invisible in the nucleosome complex.[63] Their macromolecular interactions (e.g., with chromatin remodelling complexes) are modulated by posttranslational modifications, which alter local structure and dynamics.[64] Function of histone tails is also insensitive to sequence shuffling, indicating a variability of interactions.[45,46] Hub-behavior of architectural transcription factors is also linked to their highly disordered state in complex that facilitates simultaneous interactions with many partners.[65] Low-resolution structural data reflect significant structural disorder in the pre-initiation complex in co-activators (e.g., Mediator complex[66]), in some RNAP bound GTFs (e.g., TFIIF[67] and TFIIB) and also in CTD of RNAPII. The function of RNAPII CTD depends on the degree of phosphorylation modulating the shape and interactions of this molecular scaffold.[61]

mRNA maturation involves various proteins with long disordered regions, the dynamics of which in the complex mediate further interactions (e.g., UPF2 in complex with UPF1 in nonsense-mediated decay, NMD[59]). The length of interdomain region in ribosomal proteins (L7/L12) was also shown to be critical for function.[68]

Cell-Cycle Regulation

Tuning the cell-cycle requires ultrasensitive regulatory devices. Interactions between cyclin-dependent kinases and their inhibitors can be influenced by different regions of the inhibitor and also by posttranslational modifications, both linked to structural disorder in the complex. The interaction of cyclin dependent kinase inhibitor Sic1 with Cdc4 is gradually tuned by phosphorylation. Although only one out of several phosphorylated Cic1 sites can interact with the target receptor, phosphorylation of the rest interferes with local ordering and dynamics of the complex.[49,51] In addition, long-range electrostatic interactions also contribute to affinity. In the case of $p27^{Kip1}$, phosphorylation of a distant site of the inhibitor is utilized as a switch for G1 to S transition. The phosphorylation occurs in a probabilistic manner depending on the exclusion of the primary p27 contact site (T88) from the active pocket of Cdk.[69]

Viral Assemblies

Viral nucleoproteins are responsible for encapsidating the viral genome such as single stranded RNA and influence its replication via complex interactions. The measles virus nucleoprotein utilizes a conserved region (Box2) to interact with the phosphoprotein, which is affected by the C-terminal segment that does not establish direct contacts with the phosphoprotein. While Box2 adopts an α-helical structure in the bound form, the C-terminal region remains disordered in the complex.[70,71] The Ebola virus nucleoprotein also preserves disordered features upon associating with other virus-like particles.[72]

Amyloid Formation

In amyloid formation proteins are converted from their normal physiological state to a highly ordered fibrillar aggregate, with physiological (e.g., physiological prions) or pathological (systemic and tissue-specific disorders) consequences.[73] The 3D structure of several amyloids is recently solved at residue-level resolution by solid-state NMR and EPR spectroscopy.[74,75] The defining and unifying feature of these amyloids is a highly-ordered parallel β-sheet core, in which individual polypeptide chains are stacked in register, running perpendicular to the major axis of the fibril. Invariably, however, part of the polypeptide chain in the amyloid remains intrinsically disordered and thus it is excluded from the ordered core.[76] These fuzzy regions (e.g., in the amyloids of α-synuclein, Aβ peptide, the yeast prion HET-s and Ure2p), have characteristic amino acid composition and higher predicted disordered value than segments incorporated into the cross-β core. Because these region serve different functions, such as the accommodation of destabilizing residues and the mediation of secondary interactions between protofibrils, they represent a special case of fuzziness.

CONCLUSION AND FUTURE IMPLICATIONS

The concept of fuzzy complexes has only been recently established,[22] yet the number of supporting examples is growing rapidly. Whereas detailed structural data is still sparse, the idea of conformational heterogeneity in the bound state is in sharp contrast with the classical view of one structure-one protein function. As briefly illustrated in this chapter, fuzziness extends the functional capacity of proteins by enabling diverse, yet very specific interactions. Posttranslational modifications in the complexed state offer an additional and versatile device for regulation. With the recognition that many IDPs function by molecular recognition, in which they undergo induced folding, it has been temporarily thought that IDPs can be brought under the umbrella of the classical structure-function paradigm. Now, with the recognition of fuzziness, the basic idea is clear: the deterministic relationship between sequence and function has to be replaced by a more probabilistic view. The underlying structural and functional details however, await to be elucidated. This book collects representative and functionally interesting cases from diverse biological processes. Based on the abundance of IDRs in complex systems and the benefits of fuzziness, we anticipate this type of interaction to be very common in eukaryotic proteomes. The highlighted examples demonstrate that even if high-resolution structural studies are not feasible, low-resolution techniques and/or biochemical approaches may provide valuable information. In the near future, more data has to be collected to obtain a comprehensive picture of how fuzzy complexes function. If we succeed to develop a more stochastic formalism between sequence and function, it can be used to reinterpret the functional readout of protein-protein interactions. This will ultimately have a thorough impact on our ability to describe all cellular processes at the molecular level.

ACKNOWLEDGEMENTS

This work was supported by the FP7 Infrastructures grant BioNMR (no. 261863) and Human Mobility grant MB08-A-80125 (Marie Curie Actions) from the European Commission

REFERENCES

1. Warshel A, Sharma PK, Kato M et al. Electrostatic basis for enzyme catalysis. Chem Rev 2006; 106:3210-3235.
2. Henzler-Wildman KA, Thai V, Lei M et al. Intrinsic motions along an enzymatic reaction trajectory. Nature 2007; 450:838-44.
3. Todd AE, Orengo CA, Thornton JM. Plasticity of enzyme active sites. Trends Biochem Sci 2002; 27:419-26.
4. Freedman SJ, Sun ZY, Kung AL et al. Structural basis for negative regulation of hypoxia-inducible factor-1alpha by CITED2. Nat Struct Biol 2003; 10:504-12.
5. del Sol A, Tsai CJ, Ma B et al. The origin of allosteric functional modulation: multiple pre-existing pathways. Structure 2009; 17:1042-50.
6. Romero P, Obradovic Z, Kissinger CR et al. Thousands of proteins likely to have long disordered regions. Pac Symp Biocomputing 1998; 3:437-48.
7. Tompa P. Intrinsically unstructured proteins. Trends Biochem Sci 2002; 27:527-33.
8. Wright PE, Dyson HJ. Intrinsically unstructured proteins: re-assessing the protein structure-function paradigm. J Mol Biol 1999; 293:321-31.
9. Sickmeier M, Hamilton JA, LeGall T et al. DisProt: the Database of Disordered Proteins. Nucleic Acids Res 2007; 35:D786-D793.
10. Dunker AK, Obradovic Z, Romero P et al. Intrinsic protein disorder in complete genomes. Genome Inform Ser Workshop Genome Inform 2000; 11:161-71.
11. Tompa P, Dosztanyi Z, Simon I. Prevalent structural disorder in E. coli and S. cerevisiae proteomes. J Proteome Res 2006; 5:1996-2000.
12. Ward JJ, Sodhi JS, McGuffin LJ et al. Prediction and functional analysis of native disorder in proteins from the three kingdoms of life. J Mol Biol 2004; 337:635-45.
13. Xie H, Vucetic S, Iakoucheva LM et al. Functional anthology of intrinsic disorder. 1. Biological processes and functions of proteins with long disordered regions. J Proteome Res 2007; 6:1882-98.
14. Fuxreiter M, Tompa P, Simon I et al. Malleable machines take shape in eukaryotic transcriptional regulation. Nat Chem Biol 2008; 4:728-37.
15. Dunker AK, Brown CJ, Lawson JD et al. Intrinsic disorder and protein function. Biochemistry 2002; 41:6573-82.

16. Uversky VN, Oldfield CJ, Dunker AK. Showing your ID: intrinsic disorder as an ID for recognition, regulation and cell signaling. J Mol Recognit 2005; 18:343-84.
17. Tompa P. The interplay between structure and function in intrinsically unstructured proteins. FEBS Lett 2005; 579:3346-54.
18. James LC, Roversi P, Tawfik DS. Antibody multispecificity mediated by conformational diversity. Science 2003; 299:1362-67.
19. Wright PE, Dyson HJ. Linking folding and binding. Curr Opin Struct Biol 2009; 19:31-38.
20. Fuxreiter M, Simon I, Friedrich P et al. Preformed structural elements feature in partner recognition by intrinsically unstructured proteins. J Mol Biol 2004; 338:1015-26.
21. Patel SS, Belmont BJ, Sante JM et al. Natively unfolded nucleoporins gate protein diffusion across the nuclear pore complex. Cell 2007; 129:83-96.
22. Tompa P, Fuxreiter M. Fuzzy complexes: polymorphism and structural disorder in protein-protein interactions. Trends Biochem Sci 2008; 33:2-8.
23. Fontes MR, Teh T, Toth G et al. Role of flanking sequences and phosphorylation in the recognition of the simian-virus-40 large T-antigen nuclear localization sequences by importin-alpha. Biochem J 2003; 375:339-49.
24. Graham TA, Ferkey DM, Mao F et al. Tcf4 can specifically recognize beta-catenin using alternative conformations. Nat Struct Biol 2001; 8:1048-52.
25. Renault L, Bugyi B, Carlier MF. Spire and Cordon-bleu: multifunctional regulators of actin dynamics. Trends Cell Biol 2008; 18:494-504.
26. Haarmann CS, Green D, Casarotto MG et al. The random-coil 'C' fragment of the dihydropyridine receptor II-III loop can activate or inhibit native skeletal ryanodine receptors. Biochem J 2003; 372:305-16.
27. Bhattacharyya RP, Remenyi A, Good MC et al. The Ste5 scaffold allosterically modulates signaling output of the yeast mating pathway. Science 2006; 311:822-26.
28. van Leeuwen HC, Strating MJ, Rensen M et al. Linker length and composition influence the flexibility of Oct-1 DNA binding. C. EMBO J 1997; 16:2043-53.
29. Rock RS, Ramamurthy B, Dunn AR et al. A flexible domain is essential for the large step size and processivity of myosin VI. Mol Cell 2005; 17:603-09.
30. Neduva V, Russell RB. Linear motifs: evolutionary interaction switches. FEBS Lett 2005; 579:3342-45.
31. Fuxreiter M, Tompa P, Simon I. Local structural disorder imparts plasticity on linear motifs. Bioinformatics 2007; 23:950-56.
32. Zor T, Mayr BM, Dyson HJ et al. Roles of phosphorylation and helix propensity in the binding of the KIX domain of CREB-binding protein by constitutive (c-Myb) and inducible (CREB) activators. J Biol Chem 2002; 277:42241-48.
33. Selenko P, Gregorovic G, Sprangers R et al. Structural basis for the molecular recognition between human splicing factors U2AF65 and SF1/mBBP. Mol Cell 2003; 11:965-76.
34. Liu Y, Matthews KS, Bondos SE. Multiple intrinsically disordered sequences alter DNA binding by the homeodomain of the Drosophila hox protein ultrabithorax. J Biol Chem 2008; 283:20874-87.
35. Phylip LH, Lees WE, Brownsey BG et al. The potency and specificity of the interaction between the IA3 inhibitor and its target aspartic proteinase from Saccharomyces cerevisiae. J Biol Chem 2001; 276:2023-30.
36. Yu H, Chen JK, Feng S et al. Structural basis for the binding of proline-rich peptides to SH3 domains. Cell 1994; 76:933-45.
37. Billeter M, Qian YQ, Otting G et al. Determination of the nuclear magnetic resonance solution structure of an Antennapedia homeodomain-DNA complex. J Mol Biol 1993; 234:1084-93.
38. Simon SM, Sousa FJ, Mohana-Borges R et al. Regulation of Escherichia coli SOS mutagenesis by dimeric intrinsically disordered umuD gene products. Proc Natl Acad Sci USA 2008; 105:1152-57.
39. Sigalov A, Aivazian D, Stern L. Homooligomerization of the cytoplasmic domain of the T cell receptor zeta chain and of other proteins containing the immunoreceptor tyrosine-based activation motif. Biochemistry 2004; 43:2049-61.
40. Savvides SN, Raghunathan S, Futterer K et al. The C-terminal domain of full-length E. coli SSB is disordered even when bound to DNA. Protein Sci 2004; 13:1942-47.
41. Hope IA, Mahadevan S, Struhl K. Structural and functional characterization of the short acidic transcriptional activation region of yeast GCN4 protein. Nature 1988; 333:635-40.
42. Sigler PB. Transcriptional activation. Acid blobs and negative noodles. Nature 1988; 333:210-12.
43. Ng KP, Potikyan G, Savene RO et al. Multiple aromatic side chains within a disordered structure are critical for transcription and transforming activity of EWS family oncoproteins. Proc Natl Acad Sci USA 2007; 104:479-84.
44. Hansen JC, Lu X, Ross ED et al. Intrinsic protein disorder, amino acid composition, and histone terminal domains. J Biol Chem 2006; 281:1853-56.

45. Lu X, Hamkalo B, Parseghian MH et al. Chromatin condensing functions of the linker histone C-terminal domain are mediated by specific amino acid composition and intrinsic protein disorder. Biochemistry 2009; 48:164-72.
46. McBryant SJ, Klonoski J, Sorensen TC et al. Determinants of histone H4 N-terminal domain function during nucleosomal array oligomerization: roles of amino acid sequence, domain length, and charge density. J Biol Chem 2009; 284:16716-22.
47. Krishnan R, Lindquist SL. Structural insights into a yeast prion illuminate nucleation and strain diversity. Nature 2005; 435:765-72.
48. Ross ED, Edskes HK, Terry MJ et al. Primary sequence independence for prion formation. Proc Natl Acad Sci USA 2005; 102:12825-30.
49. Mittag T, Orlicky S, Choy WY et al. Dynamic equilibrium engagement of a polyvalent ligand with a single-site receptor. Proc Natl Acad Sci USA 2008; 105:17772-77.
50. Iakoucheva LM, Radivojac P, Brown CJ et al. The importance of intrinsic disorder for protein phosphorylation. Nucleic Acids Res 2004; 32:1037-49.
51. Borg M, Mittag T, Pawson T et al. Polyelectrostatic interactions of disordered ligands suggest a physical basis for ultrasensitivity. Proc Natl Acad Sci USA 2007; 104:9650-55.
52. Baker JM, Hudson RP, Kanelis V et al. CFTR regulatory region interacts with NBD1 predominantly via multiple transient helices. Nat Struct Mol Biol 2007; 14:738-45.
53. Lacy ER, Filippov I, Lewis WS et al. p27 binds cyclin-CDK complexes through a sequential mechanism involving binding-induced protein folding. Nat Struct Mol Biol 2004; 11:358-64.
54. Galea CA, Nourse A, Wang Y et al. Role of intrinsic flexibility in signal transduction mediated by the cell cycle regulator, p27 Kip1. J Mol Biol 2008; 376:827-38.
55. Polverini E, Rangaraj G, Libich DS et al. Binding of the proline-rich segment of myelin basic protein to SH3 domains: spectroscopic, microarray, and modeling studies of ligand conformation and effects of posttranslational modifications. Biochemistry 2008; 47:267-82.
56. Ahmed MA, Bamm VV, Shi L et al. Induced secondary structure and polymorphism in an intrinsically disordered structural linker of the CNS: solid-state NMR and FTIR spectroscopy of myelin basic protein bound to actin. Biophys J 2009; 96:180-91.
57. Damante G, Pellizzari L, Esposito G et al. A molecular code dictates sequence-specific DNA recognition by homeodomains. EMBO J 1996; 15:4992-5000.
58. Toth-Petroczy A, Simon I, Fuxreiter M et al. Disordered tails of homeodomains facilitate DNA recognition by providing a trade-off between folding and specific binding. J Am Chem Soc 2009; 131:15084-85.
59. Clerici M, Mourao A, Gutsche I et al. Unusual bipartite mode of interaction between the nonsense-mediated decay factors, UPF1 and UPF2. EMBO J 2009; 28:2293-306.
60. Corden JL. Tails of RNA polymerase II. Trends Biochem Sci 1990; 15:383-87.
61. Proudfoot NJ, Furger A, Dye MJ. Integrating mRNA processing with transcription. Cell 2002; 108:501-12.
62. Ha NC, Tonozuka T, Stamos JL et al. Mechanism of phosphorylation-dependent binding of APC to beta-catenin and its role in beta-catenin degradation. Mol Cell 2004; 15:511-21.
63. Luger K, Mader AW, Richmond RK et al. J. Crystal structure of the nucleosome core particle at 2.8 A resolution. Nature 1997; 389:251-60.
64. Jenuwein T, Allis CD. Translating the histone code. Science 2001; 293:1074-1080.
65. Reeves R. Molecular biology of HMGA proteins: hubs of nuclear function. Gene 2001; 277:63-81.
66. Taatjes DJ, Marr MT, Tjian R. Regulatory diversity among metazoan co-activator complexes. Nat Rev Mol Cell Biol 2004; 5:403-10.
67. Chung WH, Craighead JL, Chang WH et al. RNA polymerase II/TFIIF structure and conserved organization of the initiation complex. Mol Cell 2003; 12:1003-13.
68. Bubunenko MG, Chuikov SV, Gudkov AT. The length of the interdomain region of the L7/L12 protein is important for its function. FEBS Lett 1992; 313:232-34.
69. Galea CA, Wang Y, Sivakolundu SG et al. Regulation of cell division by intrinsically unstructured proteins: intrinsic flexibility, modularity, and signaling conduits. Biochemistry 2008; 47:7598-609.
70. Bernard C, Gely S, Bourhis JM et al. Interaction between the C-terminal domains of N and P proteins of measles virus investigated by NMR. FEBS Lett 2009; 583:1084-89.
71. Bourhis JM, Canard B, Longhi S. Structural disorder within the replicative complex of measles virus: functional implications. Virology 2006; 344:94-110.
72. Shi W, Huang Y, Sutton-Smith M et al. A filovirus-unique region of Ebola virus nucleoprotein confers aberrant migration and mediates its incorporation into virions. J Virol 2008; 82:6190-99.
73. Chiti F, Dobson CM. Protein misfolding, functional amyloid, and human disease. Annu Rev Biochem 2006; 75:333-66.
74. Chen M, Margittai M, Chen J et al. Investigation of alpha-synuclein fibril structure by site-directed spin labeling. J Biol Chem 2007; 282(34):24970-79.

75. Margittai M, Langen R. Fibrils with parallel in-register structure constitute a major class of amyloid fibrils: molecular insights from electron paramagnetic resonance spectroscopy. Q Rev Biophys 2008; 41:265-97.
76. Tompa P. Structural disorder in amyloid fibrils: its implication in dynamic interactions of proteins. FEBS J 2009; 276:5406-15.
77. Cliff MJ, Harris R, Barford D et al. Conformational diversity in the TPR domain-mediated interaction of protein phosphatase 5 with Hsp90. Structure 2006; 14:415-26.
78. Hurley TD, Yang J, Zhang L et al. Structural basis for regulation of protein phosphatase 1 by inhibitor-2. J Biol Chem 2007; 282:28874-83.
79. Kover KE, Bruix M, Santoro J et al. The solution structure and dynamics of human pancreatic ribonuclease determined by NMR spectroscopy provide insight into its remarkable biological activities and inhibition. J Mol Biol 2008; 379:953-65.
80. von Ossowski I, Eaton JT, Czjzek M et al. Protein disorder: conformational distribution of the flexible linker in a chimeric double cellulase. Biophys J 2005; 88:2823-32.
81. Lindner RA, Carver JA, Ehrnsperger M et al. Mouse Hsp25, a small shock protein. The role of its C-terminal extension in oligomerization and chaperone action. Eur J Biochem 2000; 267:1923-32.
82. Schwarz-Linek U, Pilka ES, Pickford AR et al. High affinity streptococcal binding to human fibronectin requires specific recognition of sequential F1 modules. J Biol Chem 2004; 279:39017-25.
83. Gill G, Pascal E, Tseng ZH et al. A glutamine-rich hydrophobic patch in transcription factor Sp1 contacts the dTAFII110 component of the Drosophila TFIID complex and mediates transcriptional activation. Proc Natl Acad Sci USA 1994; 91(1):192-96.
84. Lee GM, Pufall MA, Meeker CA et al. The affinity of Ets-1 for DNA is modulated by phosphorylation through transient interactions of an unstructured region. J Mol Biol 2008; 382:1014-30.
85. Pometun MS, Chekmenev EY, Wittebort RJ. Quantitative observation of backbone disorder in native elastin. J Biol Chem 2004; 279(9):7982-87.
86. Ross ED, Baxa U, Wickner RB. Scrambled prion domains form prions and amyloid. Mol Cell Biol 2004; 24(16):7206-13.

CHAPTER 2

DYNAMIC FUZZINESS DURING LINKER HISTONE ACTION

Steven J. McBryant and Jeffrey C. Hansen*
Department of Biochemistry and Molecular Biology, Colorado State University, Fort Collins, Colorado, USA
**Corresponding Author: Jeffrey C. Hansen—Email: jeffrey.c.hansen@colostate.edu*

Abstract: Linker histones are multi-domain nucleosome binding proteins that stabilize higher order chromatin structures and engage in specific protein-protein interactions. Here we emphasize the structural and functional properties of the linker histone C-terminal domain (CTD), focusing on its intrinsic disorder, interaction-induced secondary structure formation and dynamic fuzziness. We argue that the fuzziness inherent in the CTD is a primary molecular mechanism underlying linker histone function in the nucleus.

INTRODUCTION

Linker histones (e.g., H1, H5) comprise a large family of chromatin associated proteins. All eukaryotes contain at least one gene encoding a linker histone. Humans have 11 histone H1 isoforms that are developmentally regulated.[1-3] Linker histones are small proteins (~21 kDa) that have three distinct domains. The N-terminal ~35 residues lack secondary structure, the middle ~65 residues fold into a well defined structure and the C-terminal ~100 residues also are unstructured (Fig. 1). The unique molecular properties of the linker histone C-terminal domain (CTD) are the focus of this chapter. Specifically, we discuss evidence that the linker histone CTD is intrinsically disordered and dynamically fuzzy. Moreover, we speculate that the fuzziness and plasticity inherent in CTD interactions with other macromolecules allow H1 to be a multifunctional "hub" protein in the nucleus.

The chapter first briefly reviews chromatin fiber dynamics to provide the background needed to understand linker histone function. We then present evidence that the linker histone CTD is intrinsically disordered and fuzzy when it binds to DNA. The notion of

Fuzziness: Structural Disorder in Protein Complexes, edited by Monika Fuxreiter and Peter Tompa.

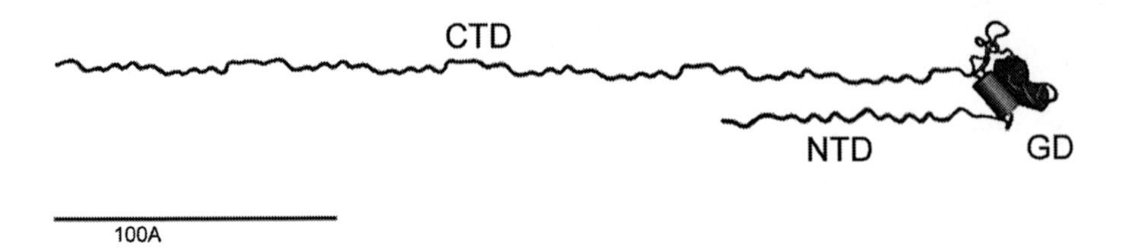

Figure 1. Illustration of linker histone tripartite domain organization drawn to scale. The image was constructed by starting with the PDB file of the H1 GD (1GHC). Amino acid residues from core histone H2B (from the crystal structure of the nucleosome[92]) were concatamerized using COOT,[93] resulting in an extended polypeptide chain (3.5 Å per residue). 30 and 100 residues were then added to the N- and C-terminus of the GD, respectively, to model the length of the intrinsically disordered NTD and CTD of the H1 protein.

CTD fuzziness is reinforced by our concluding discussions of a well characterized linker histone CTD-protein interaction and the effects of phosphorylation on CTD function.

CHROMATIN FIBER STRUCTURE AND DYNAMICS

As organisms have evolved and become more complex, their genomes have also become more complex. The complete genome of a typical eukaryotic cell is nearly two meters in length yet is constrained within a ~10 micrometer nucleus. Despite being in this highly condensed state, the genomic DNA must be available to be used for functional processes such as DNA replication, transcription and damage repair. To achieve a high level of chromosomal compaction and couple compaction with closely controlled genomic access, metazoans have evolved a specific set of chromosomal proteins termed the core and linker histones.

The bulk of genomic DNA is bound to octamers made up of two each of the core histone proteins, H2A, H2B, H3 and H4.[4] 147 bp of genomic DNA is wrapped in ~1.75 superhelical turns around the histone octamer in each nucleosome, forming the first level of DNA condensation. Nucleosomes are spaced between 10-60 bp apart along the chromosomal DNA, forming nucleosomal arrays. The intervening DNA between nucleosomes is called linker DNA. Nucleosomal arrays containing bound linker histones are called chromatin fibers. Nucleosomal arrays and chromatin fibers make up the next levels of DNA compaction. In higher eukaryotes, linker histones are present in nearly stoichiometric amounts with nucleosomes.[5] The primary function of linker histones has long been believed to influence the higher order structural transitions of condensed structures of chromatin fibers.

Chromatin fibers are conformationally dynamic. Typically, there are two defining salt-dependent condensation transitions seen in vitro: folding and oligomerization (for reviews see refs. 6-11). In the absence of salts, chromatin fibers are in extended, beads on a string-like conformation, with no or few nucleosome-nucleosome interactions. However, under more physiological conditions (such as in 1-2 mM $Mg2^{+}$ or 100-200 mM NaCl) chromatin fibers fold into a thickened 30 nm diameter fiber. Folding involves short-range, intrafiber, nucleosome-nucleosome interaction.[12-18] Salt also induces chromatin fibers to reversibly self-associate into large (>100S), soluble, oligomeric assemblages.

Oligomerization of short chromatin fiber fragments in vitro is thought to be reflective of longer ranger fiber-fiber interactions that occur in long chromosomal fibers.[6,11]

EFFECTS OF LINKER HISTONES ON CHROMATIN DYNAMICS

Linker histones affect chromatin dynamics in several ways. The first recognized benchmark of linker histone binding to chromatin fibers was that 168 bp of DNA was protected from nuclease digestion in an H1-bound fiber,[19,20] compared to the 147 bp of protection afforded by the nucleosome alone. This suggested that linker histones (1) bind near the entry-exit sites of DNA on the nucleosome and (2) interact with 20 bp of linker DNA and protect it from nuclease accessibility. Later studies using nucleosomal arrays and chromatin fibers confirmed that linker histones decrease the angle of the DNA as it enters and exits the nucleosome, forming what is known as the apposed stem motif.[21-23] Apposition causes the two DNA strands (one entering and one leaving the nucleosome) to come in close physical contact. The results from many studies over the years are consistent with the conclusion that a major function of linker histones is to bind to and deform linker DNA (see for example, refs. 5,24-29).

In terms of higher order chromatin fiber dynamics, several important early papers defined a critical role for linker histones in the formation and maintenance of folded chromatin fibers.[30-36] Specifically, at physiological salt concentrations, endogenous H1-containing chromatin folded into 30-50 nm thick super-helical fibers with a 10 nm pitch and between 6-10 nucleosomes per helical turn. Nucleosomal arrays alone condensed, but not to the same degree as H1-containing chromatin fibers. It was thus proposed that H1 functions to induce the formation of the folded ~30 nm fiber. Subsequent studies of defined nucleosomal array and chromatin fiber model systems performed over a 20 year period have firmly documented that nucleosomal arrays alone can fold in the absence of bound H1, but that linker histones are required for stability of the 30 nm fiber.[5,8,37,38] Another characteristic of linker histone action is that the salt concentrations needed to induce condensation decrease significantly when linker histones are bound to nucleosomal arrays.[6,25,39] This presumably results in part from electrostatic screening of the DNA during linker histone binding, which in turn facilitates more stable nucleosome-nucleosome interactions.

There has been much debate regarding the higher-order structures of linker histone-stabilized chromatin, specifically the architecture of the canonical 30 nm fiber (for reviews see refs. 8,38,40). Two main models have been proposed. In the first, the chromatin fiber forms a twisted ribbon-like structure based on a two start helix.[5,41-43] Different linker DNA lengths are accommodated by varying the diameter of the fiber. The second model is built upon a coiling of successive nucleosomes along the fiber, forming a one-start solenoid six nucleosomes in circumference with bent linker DNA.[37,44] In addition, in the solenoid model the linker DNA must be long enough to accommodate H1 binding and the subsequent alteration of the linker DNA path.

Regardless of the specific structure(s) that equate to the 30 nm fiber, there is widespread agreement that linker histones perform at least three key functions during chromatin condensation: They bind to nucleosomal particles, they bind to and alter linker DNA structure and they stabilize the folded and oligomeric conformations of chromatin fibers. The remainder of the chapter focuses on the protein chemistry of linker histones and the fuzziness of the intrinsically disordered linker histone CTD during linker histone action.

LINKER HISTONE DOMAIN PROPERTIES

To appreciate how linker histones influence chromatin fiber structure and stability, it is important to understand the unique features and functions of the linker histone domains. As mentioned previously (Fig. 1), linker histones have a tripartite domain organization, with long unstructured N- and C-terminal "tails" and a central folded globular domain (GD).[45] Below we discuss key structural features of these three domains.

The Linker Histone Globular Domain Binds Nucleosomes

The 3-dimensional structures for the GDs of histone H1 and H5 have been determined.[46-48] The GD folds into a classical winged-helix motif (Figs. 1,2), with two distinct DNA binding regions.[49] Nucleosome binding by the GD is asymmetric, in that binding occurs between the final superhelical turn of the DNA and the histone octamer.[50-53] It has been proposed that the shorter linker histone NTD protects 0-5 bp on one side and the longer CTD protects 15-20 bp on the opposite side, of the nucleosome.[54-57] The globular domain is primarily responsible for the increase in nuclease protection from 146 to 168 bp[19,20] and has been shown to be sufficient for binding to nucleosomes.[51] However, the GD alone is unable to stabilize condensed chromatin, a function that resides in the linker histone CTD[24,39,58] (see below).

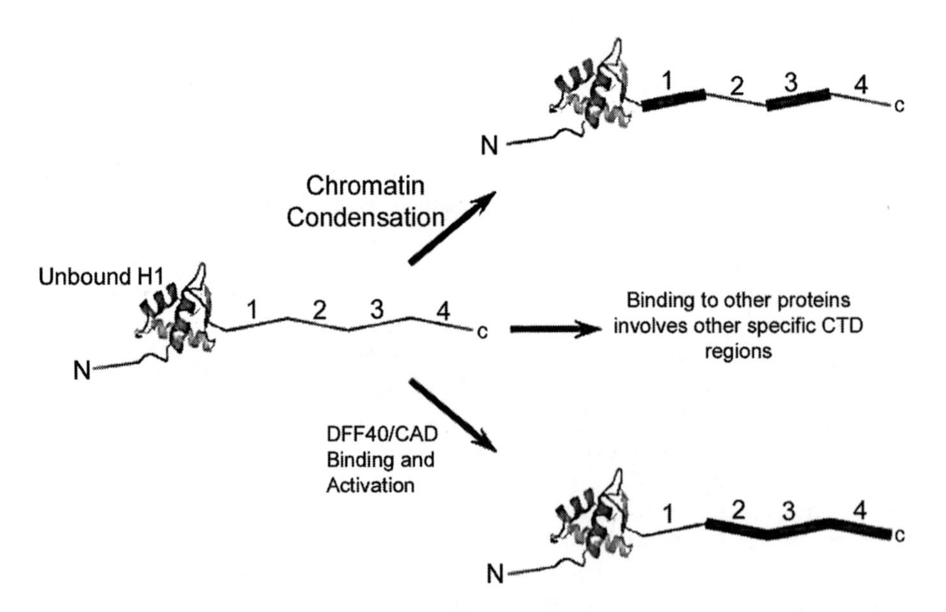

Figure 2. Model of linker histone CTD fuzziness. Shown schematically on the left is an unbound linker histone, in which the CTD is in a fully disordered state. Numbers refer to the CTD regions that were sequentially deleted in the experiments of Lu and Hansen 2004 and Widlak et al 2005. Thickened lines indicate the region of the CTD that are functional for the processes indicated.

The Linker Histone NTD and CTD Have Unique Amino Acid Compositions and Undergo DNA-Dependent Disorder to Order Transitions

The linker histone NTD is typically ~35 amino acids in length. No specific role in chromatin condensation has been elucidated for the NTD, though it has been proposed to assist in locating and anchoring the GD to the nucleosome.[24] The extensive posttranslational modification of the NTD suggests it also may serve as a binding platform for other proteins.[1] The NTD has been shown to be extensively disordered in aqueous solution. However, the NTDs of two H1 isoforms (H1o and H1e) form α-helix in the presence of long DNA and the solvent triflouroethanol (TFE).[59,60] The NTD is rich in basic residues (21% arginine and lysine in mouse H1o). The overall amino acid complexity is low; besides the arginine and lysine, only glycine, alanine, proline and serine/threonine are well represented. Thus, the linker histone NTD has the properties of an intrinsically disordered domain.

The linker histone CTD is ~100 residues long, binds to linker DNA and stabilizes folded and oligomeric chromatin fibers.[24,39] Data indicate that the linker histone CTD also is intrinsically disordered. The CTD is devoid of secondary structure in aqueous solution.[61-63] The addition of TFE or perchlorate ions (NaClO4) effectively increases the helical content within full length peptides or shorter peptides derived from the linker histone CTD.[61,64] Perchlorate ions are thought to mimic the charge on the phosphates of the DNA backbone.[62] A short CTD peptide corresponding to the N-terminal most residues that abut the GD (aa 99-121) was shown to contain up to 45% α-helix in 90% TFE, using a combination of CD and proton NMR.[64] The resulting helix spans residues 99-117 and is amphipathic, with six positively charged residues on one face and three valines on the opposing face. When this same short peptide was bound to DNA, both helical and turn components were more prevalent, relative to the free peptide.[65] A recent study of full-length CTD peptides derived from H1o (somatic isoform) and H1t (germ line specific isoform) used CD and IR spectroscopy to study the peptide structure when free in aqueous solution and when bound to DNA.[66] In solution, CD showed that both peptides were essentially random coil, with a component of 'turns', which was suggested to correlate with a nascent helix.[67] However, increases in all types of secondary structure were observed for both the H1o and H1t CTDs when bound to DNA.[66] Interestingly, shifting the peptides from low (10 mM) to a more physiological NaCl concentration (140 mM) eliminated any contribution to the CTD-DNA spectrum from random coil, with concomitant increases in all secondary structures. Thus, when bound to DNA, the linker histone CTD becomes most ordered under ionic conditions where chromatin is stably folded into 30 nm fibers.

The CTDs of the H1 isoforms have a unique amino acid composition, in that they are highly enriched in lysine (~40%), alanine (~20-30%) and proline (~13%) and contain only a few other types of amino acids (e.g., threonine/serine, glycine and valine). Noticeably absent from the isoform CTDs are substantive amounts of 13 of the common amino acids, particularly the acidic and hydrophobic residues (reviewed in ref. 68). There is no sequence conservation among the various H1 isoform CTDs,[69-71] yet the unique amino acid composition signature is retained. Moreover, in most isoform CTDs the lysine residues are almost uniformly distributed along the length of the peptide,[72] which would lead to formation of induced secondary structures with nearly uniform positive charge density.

THE LINKER HISTONE C-TERMINAL DOMAIN IS DYNAMICALLY FUZZY

The concept of "fuzziness" in macromolecular interactions has recently been discussed in detail.[73,74] Two classes of fuzziness have been proposed. In "static fuzziness" a region of a protein exists in more than one stable conformation and hence these regions tend to be absent from atomic structures. The second class is "dynamic fuzziness", where a region of a protein exists in an ensemble of equilibrating conformations and disorder may remain present in the bound state. As discussed below, all available evidence suggests that the CTD forms dynamic fuzzy complexes while functioning to both stabilize condensed chromatin fibers and mediate specific protein-protein interactions. Furthermore, the fuzziness of the linker histone CTD appears to be modulated by posttranslational modifications such as phosphorylation.

Linker Histone CTD-DNA Interactions

An in vivo study using fluorescence recovery after photobleaching showed that the CTD is nearly as important as the GD for stabilizing the binding of the linker histone to chromatin in vivo.[75] Early studies of native chromatin fibers bound to trypsin-truncated linker histones showed that the CTD is absolutely required for stabilization of folded chromatin structures.[24,29,58] The importance of the CTD was recently confirmed using chromatin fiber model systems assembled with truncated recombinant H1o.[39] The recombinant studies went even further, making uniform ~24 residue truncations beginning from the extreme C-terminus. This effectively divided the H1o CTD into four regions (Fig. 2) having approximately equal charge density and amino acid composition. These constructs were then used to test whether the entire CTD was needed for H1o function. The rather surprising answer is that it was not, as region 4 could be deleted without affecting chromatin condensation. Deletions of regions 3 and 4 led to a partial disruption of folding, indicating that region 3 participates in stabilization of the folded fiber. A truncated H1o protein that had regions 2, 3 and 4 removed behaved the same as the mutant with only regions 3 and 4 removed, indicating no condensing functions resided in region 2. Finally, all of the remaining functions of the CTD relating to condensation were lost when all four regions were deleted, leaving only the NTD fused to the GD. Thus, even though the H1o CTD is 97 residues long, only the discontinuous 24-residue regions 1 and 3 were necessary to mediate CTD function during chromatin condensation (Fig. 2).

An even more unexpected result was obtained from experiments that shuffled the order of CTD regions 1-4 identified in the earlier studies;[76] when functional region 1 (Fig. 2; residues 96-121) was replaced by either "nonfunctional" region 2 or 4, wild-type ability to condense chromatin fibers was maintained. Additionally, scrambling the primary sequence of region 1 also had no affect on function. Finally, all mouse somatic H1 isoforms could condense chromatin fibers equally well in vitro.[76] The conclusion from these experiments was that the functional CTD regions were defined less by primary sequence than by amino acid composition and position relative to the winged helix GD. These results further suggest that strategic portions of the CTD are engaged with linker DNA in strategic locations while stabilizing folded and oligomeric chromatin fibers. Presumably, the unique amino acid composition of the CTD allows formation of secondary structure in those regions bound to linker DNA, while the nonfunctional regions remain disordered and unbound. Thus, the linker histone CTD appears to display dynamic fuzzy behavior while binding to linker DNA and stabilizing condensed chromatin fibers.

Linker Histone CTD-Protein Interactions

The chromatin condensation results raise an intriguing question. Why are linker histone CTD lengths conserved at around 100 residues if only 75 residues are needed to stabilize condensed chromatin fibers? At least part of the answer is that the terminal residues of the CTD in some cases are required for H1 to recognize with protein surfaces.[77,78] Linker histones are increasingly being shown to interact with many different cellular proteins,[79] although in general the H1 domains responsible for mediating the protein-protein interaction have not been determined. Two exceptions are the chromosomal protein, BAF1 and the apoptotic nuclease, DFF40/CAD. Using pull-down approaches in combination with truncation mutants, BAF1 was shown to bind specifically and with micromolar affinity to the entire CTD of the H1.1 isoform.[77] The interaction of DFF40/CAD with the H1 CTD has been examined in great detail using CTD peptides and the same H1o truncation mutants used to study chromatin condensation.[39] The truncation studies indicated that successive deletion of CTD regions 4, 3 and 2 led to successive disruption of DFF40/CAD activation by H1. However, deletion of region 1 had little further affect (Fig. 2). Thus, for the H1 CTD-DFF40/CAD interaction, region 4 was functional while region 1 was not. This is opposite of the result obtained for H1 CTD function in chromatin condensation. Consistent with the truncation results, peptide studies indicated that residues 146-193 (i.e., the most C-terminal 47 residues) were as effective as the full CTD (residues 97-193) or the full-length protein. Interestingly, a CTD peptide that was only 24 residues in length failed to activate DFF40/CAD, indicating the minimal length needed for function was between 27 and 47 residues. These results demonstrate that the linker histone CTD can recognize and stably bind specific protein surfaces. They also showed that the region(s) needed to bind DFF40/CAD overlap with, but are distinct from, the CTD regions involved in chromatin condensation. Of note, all six mouse somatic isoforms activated the enzyme equally.[78] Thus, amino acid composition, position of key CTD regions relative to the GD and length of the functional CTD regions also are key molecular factors in the recognition of protein surfaces by the H1 CTD.

Fuzzy Interactions and CTD Posttranslational Modifications

The fuzziness of the linker histone CTD can be modulated by posttranslational modifications. Certain specific CTD residues can be phosphorylated, acetylated, methylated, ubiquitinated and formylated (see refs. 1,3 for excellent reviews). Phosphorylation is the best characterized CTD modification. Phosphorylation of H1 by one or more cyclin-dependent kinases occurs at the conserved KS/TPXK or KS/TPK motifs found exclusively in the unstructured NTD and CTD.[80] The number and utilization of the phosphorylation motifs is isoform specific. For example, the human linker histone present in differentiated cells (H1o) lacks these consensus sites, while the somatic H1.1-H1.5 isoforms have up to eight sites in their CTDs.[3] Interestingly, in the case of the CTD, most phosphorylation sites are clustered in the middle of the domain,[1] i.e., corresponding to regions 2 and 3 above (Fig. 2).

At the molecular level, phosphorylation of even a single CTD residue can have several significant effects on the properties of the domain. For example, phosphorylation converts Ser/Thr into highly negatively charged residues, which are almost completely absent from the CTDs of all the linker histone isoforms.[68] Thus, in the vicinity of a phosphorylation site the amino acid composition changes significantly and the uniform positive charge density of the CTD will be disrupted. These changes are likely to affect the local fuzziness of the CTD. Indeed, model studies have demonstrated that phosphorylation alters the secondary structure content of DNA-bound CTD peptides, but not peptides that are free in solution.[66] Moreover, the extent of secondary structure alteration was dependent on the isoform CTD used. The unphosphorylated linker histone binds DNA more tightly than the highly phosphorylated form.[81,82] Thus, the effects of phosphorylation are coupled to both the CTD disorder to order transition(s) and macromolecular recognition.

CTD phosphorylation clearly is an important regulator of linker histone function in vivo. H1 phosphorylation is very high in rapidly proliferating cells and almost undetectable in quiescent cells, suggesting a role in regulating cell-cycle transitions (for review see refs. 83). Consistent with these results, linker histones are phosphorylated in a cell-cycle dependent manner, with highest levels detected in S and G2 phases and lowest levels at the end of mitosis.[84-87] Recently, Thiriet and Hayes[88] showed that DNA replication is strongly dependent on linker histone phosphorylation and phosphorylation correlates with H1 eviction from chromatin and more rapid progression through replication. Using fluorescence recovery after photobleaching (FRAP), phosphorylation has been suggested to decrease the residency time of H1 on chromatin.[89] These results are consistent with a role for linker histone phosphorylation in loosening protein-DNA interactions during interphase. In seeming contradiction, the stability of metaphase chromosomes requires H1 hyperphosphorylation.[90] One possible explanation for these results is that hyperphosphorylation causes adoption of a DNA-induced secondary structure that stabilizes metaphase chromosomes. An alternative hypothesis is that hyperphosphorylation helps dissociate the H1 CTD from DNA, thereby creating a chromatin fiber structure that can be more easily molded into a supercondensed metaphase chromosome by other macromolecules.

Importantly, phosphorylation of any one of the four phosphate acceptor residues on the H1b CTD disrupts interaction of H1 with heterochromatin protein 1 (HP1).[91] This implies that the unmodified CTD is involved in the HP1-H1 interaction and that regulation of CTD fuzziness by phosphorylation extends to H1-protein interactions.

CONCLUSION

The linker histone CTD is an intrinsically disordered protein domain that can recognize both DNA and protein surfaces. Molecular determinants of function appear to be amino acid composition, the position of strategic CTD region(s) with respect to the GD and region length. Although the CTD has been maintained throughout evolution at ~100 residues, only portions of the complete domain are required for macromolecular recognition. In the cases examined thus far, the regions of the CTD that function in DNA recognition are different than those that function in protein recognition. These properties indicate that the linker histone CTD is a dynamically fuzzy protein domain. By all accounts the fuzziness of the CTD allows linker histones to be multifunctional regulatory hubs in the nucleus, not just single-function proteins involved in chromatin architecture.

ACKNOWLEDGEMENTS

J.C.H. acknowledges National Institutes of Health grant GM45916 for supporting his research on linker histone structure and function.

REFERENCES

1. Godde JS, Ura K. Cracking the enigmatic linker histone code. Journal of Biochemistry 2008; 143(3):287-293.
2. Khochbin S, Wolffe AP. Developmentally regulated expression of linker-histone variants in vertebrates. Eur J Biochem 1994; 225(2):501-510.
3. Happel N, Doenecke D. Histone H1 and its isoforms: contribution to chromatin structure and function. Gene 2009; 431(1-2):1-12.
4. Klug A, Rhodes D, Smith J et al. A low resolution structure for the histone core of the nucleosome. Nature 1980; 287(5782):509-516.
5. Woodcock CL, Skoultchi AI, Fan Y. Role of linker histone in chromatin structure and function: H1 stoichiometry and nucleosome repeat length. Chromosome Res 2006; 14(1):17-25.
6. Hansen JC. Conformational dynamics of the chromatin fiber in solution: determinants, mechanisms and functions. Annu Rev Biophys Biomol Struct 2002; 31:361-392.
7. Horowitz-Scherer RA, Woodcock CL. Organization of interphase chromatin. Chromosoma 2006; 115(1):1-14.
8. Tremethick DJ. Higher-order structures of chromatin: the elusive 30 nm fiber. Cell 2007; 128(4):651-654.
9. Woodcock CL, Dimitrov S. Higher-order structure of chromatin and chromosomes. Curr Opin Genet Dev 2001; 11(2):130-135.
10. Zheng C, Hayes JJ. Structures and interactions of the core histone tail domains. Biopolymers 2003; 68(4):539-546.
11. Lu X, Klonoski JM, Resch MG et al. In vitro chromatin self-association and its relevance to genome architecture. Biochem Cell Biol 2006; 84(4):411-417.
12. Dorigo B, Schalch T, Bystricky K et al. Chromatin fiber folding: requirement for the histone H4 N-terminal tail. J Mol Biol 2003; 327(1):85-96.
13. Kan PY, Caterino TL, Hayes JJ. The H4 tail domain participates in intra- and internucleosome interactions with protein and DNA during folding and oligomerization of nucleosome arrays. Mol Cell Biol 2009; 29(2):538-546.
14. Kan PY, Lu X, Hansen JC et al. The H3 tail domain participates in multiple interactions during folding and self-association of nucleosome arrays. Mol Cell Biol 2007; 27(6):2084-2091.
15. Shogren-Knaak M, Ishii H, Sun JM et al. Histone H4-K16 acetylation controls chromatin structure and protein interactions. [see comment]. Science 2006; 311(5762):844-847.
16. Tse C, Sera T, Wolffe AP et al. Disruption of higher-order folding by core histone acetylation dramatically enhances transcription of nucleosomal arrays by RNA polymerase III. Mol Cell Biol 1998; 18(8):4629-4638.
17. Wang X, Hayes JJ. Acetylation mimics within individual core histone tail domains indicate distinct roles in regulating the stability of higher-order chromatin structure. Mol Cell Biol 2008; 28(1):227-236.
18. Zheng C, Lu X, Hansen JC et al. Salt-dependent intra- and internucleosomal interactions of the H3 tail domain in a model oligonucleosomal array. J Biol Chem 2005; 280(39):33552-33557.
19. Noll M, Kornberg RD. Action of micrococcal nuclease on chromatin and the location of histone H1. J Mol Biol 1977; 109(3):393-404.
20. Simpson RT. Structure of the chromatosome, a chromatin particle containing 160 base pairs of DNA and all the histones. Biochemistry 1978; 17(25):5524-5531.
21. Furrer P, Bednar J, Dubochet J et al. DNA at the entry-exit of the nucleosome observed by cryoelectron microscopy. J Struct Biol 1995; 114(3):177-183.
22. Bednar J, Horowitz RA, Dubochet J et al. Chromatin conformation and salt-induced compaction: three-dimensional structural information from cryoelectron microscopy. J Cell Biol 1995; 131(6 Pt 1):1365-1376.
23. Bednar J, Horowitz RA, Grigoryev SA et al. Nucleosomes, linker DNA and linker histone form a unique structural motif that directs the higher-order folding and compaction of chromatin. Proc Natl Acad Sci USA 1998; 95(24):14173-14178.
24. Allan J, Mitchell T, Harborne N et al. Roles of H1 domains in determining higher order chromatin structure and H1 location. J Mol Biol 1986; 187(4):591-601.
25. Carruthers LM, Bednar J, Woodcock CL et al. Linker histones stabilize the intrinsic salt-dependent folding of nucleosomal arrays: mechanistic ramifications for higher-order chromatin folding. Biochemistry 1998; 37(42):14776-14787.
26. Hamiche A, Schultz P, Ramakrishnan V et al. Linker histone-dependent DNA structure in linear mononucleosomes. J Mol Biol 1996; 257(1):30-42.
27. Krylov D, Leuba S, van Holde K et al. Histones H1 and H5 interact preferentially with crossovers of double-helical DNA. Proc Natl Acad Sci USA 1993; 90(11):5052-5056.
28. Renz M, Nehls P, Hozier J. Involvement of histone H1 in the organization of the chromosome fiber. Proc Natl Acad Sci USA 1977; 74(5):1879-1883.
29. Thomas JO. Histone H1: location and role. Curr Opin Cell Biol 1999; 11(3):312-317.

30. Allan J, Staynov DZ, Gould H. Reversible dissociation of linker histone from chromatin with preservation of internucleosomal repeat. Proc Natl Acad Sci USA 1980; 77(2):885-889.
31. Modak SP, Lawrence JJ, Gorka C. Selective removal of histone H1 from nucleosomes at low ionic strength. Mol Biol Rep 1980; 6(4):235-243.
32. Osipova TN, Pospelov VA, Svetlikova SB et al. The role of histone H1 in compaction of nucleosomes. Sedimentation behaviour of oligonucleosomes in solution. Eur J Biochem 1980; 113(1):183-188.
33. Finch JT, Klug A. Solenoidal model for superstructure in chromatin. Proc Natl Acad Sci USA 1976; 73(6):1897-1901.
34. Renz M, Nehls P, Hozier J. Involvement of histone H1 in the organization of the chromosome fiber. Proc Natl Acad Sci USA 1977; 74(5):1879-1883.
35. Thoma F, Koller T. Influence of histone H1 on chromatin structure. Cell 1977; 12(1):101-107.
36. Thoma F, Koller T, Klug A. Involvement of histone H1 in the organization of the nucleosome and of the salt-dependent superstructures of chromatin. J Cell Biol 1979; 83(2 Pt 1):403-427.
37. Routh A, Sandin S, Rhodes D. Nucleosome repeat length and linker histone stoichiometry determine chromatin fiber structure. Proc Natl Acad Sci USA 2008; 105(26):8872-8877.
38. Robinson PJ, Rhodes D. Structure of the '30 nm' chromatin fibre: a key role for the linker histone. Curr Opin Struct Biol 2006; 16(3):336-343.
39. Lu X, Hansen JC. Identification of specific functional subdomains within the linker histone H10 C-terminal domain. J Biol Chem 2004; 279(10):8701-8707.
40. Staynov DZ. The controversial 30 nm chromatin fibre. Bioessays 2008; 30(10):1003-1009.
41. Dorigo B, Schalch T, Kulangara A et al. Nucleosome arrays reveal the two-start organization of the chromatin fiber. [see comment]. Science 2004; 306(5701):1571-1573.
42. Woodcock CL. Chromatin architecture. Curr Opin Struct Biol 2006; 16(2):213-220.
43. Schalch T, Duda S, Sargent DF et al. X-ray structure of a tetranucleosome and its implications for the chromatin fibre. Nature 2005; 436(7047):138-141.
44. Kruithof M, Chien FT, Routh A et al. Single-molecule force spectroscopy reveals a highly compliant helical folding for the 30-nm chromatin fiber. Nat Struct Mol Biol 2009; 16(5):534-540.
45. Hartman PG, Chapman GE, Moss T et al. Studies on the role and mode of operation of the very-lysine-rich histone H1 in eukaryote chromatin. The three structural regions of the histone H1 molecule. Eur J Biochem 1977; 77(1):45-51.
46. Zarbock J, Clore GM, Gronenborn AM. Nuclear magnetic resonance study of the globular domain of chicken histone H5: resonance assignment and secondary structure. Proc Natl Acad Sci USA 1986; 83(20):7628-7632.
47. Ramakrishnan V, Finch JT, Graziano V et al. Crystal structure of globular domain of histone H5 and its implications for nucleosome binding. Nature 1993; 362(6417):219-223.
48. Cerf C, Lippens G, Ramakrishnan V et al. Homo- and heteronuclear two-dimensional NMR studies of the globular domain of histone H1: full assignment, tertiary structure and comparison with the globular domain of histone H5. Biochemistry 1994; 33(37):11079-11086.
49. Goytisolo FA, Gerchman SE, Yu X et al. Identification of two DNA-binding sites on the globular domain of histone H5. EMBO J 1996; 15(13):3421-3429.
50. Zhou YB, Gerchman SE, Ramakrishnan V et al. Position and orientation of the globular domain of linker histone H5 on the nucleosome. Nature 1998; 395(6700):402-405.
51. Pruss D, Bartholomew B, Persinger J et al. An asymmetric model for the nucleosome: a binding site for linker histones inside the DNA gyres. Science 1996; 274(5287):614-617.
52. Hayes JJ, Kaplan R, Ura K et al. A putative DNA binding surface in the globular domain of a linker histone is not essential for specific binding to the nucleosome. J Biol Chem 1996; 271(42):25817-25822.
53. Hayes JJ. Site-directed cleavage of DNA by a linker histone—Fe(II) EDTA conjugate: localization of a globular domain binding site within a nucleosome. Biochemistry 1996; 35(37):11931-11937.
54. Hayes JJ, Pruss D, Wolffe AP. Contacts of the globular domain of histone H5 and core histones with DNA in a "chromatosome". Proc Natl Acad Sci USA 1994; 91(16):7817-7821.
55. Hayes JJ, Wolffe AP. Preferential and asymmetric interaction of linker histones with 5S DNA in the nucleosome. Proc Natl Acad Sci USA 1993; 90(14):6415-6419.
56. An W, Leuba SH, van Holde K et al. Linker histone protects linker DNA on only one side of the core particle and in a sequence-dependent manner. Proc Natl Acad Sci USA 1998; 95(7):3396-3401.
57. An W, van Holde K, Zlatanova J. Linker histone protection of chromatosomes reconstituted on 5S rDNA from Xenopus borealis: a reinvestigation. Nucleic Acids Res 1998; 26(17):4042-4046.
58. Allan J, Hartman PG, Crane-Robinson C et al. The structure of histone H1 and its location in chromatin. Nature 1980; 288(5792):675-679.
59. Vila R, Ponte I, Jimenez MA et al. An inducible helix-Gly-Gly-helix motif in the N-terminal domain of histone H1e: a CD and NMR study. Protein Sci 2002; 11(2):214-220.

60. Vila R, Ponte I, Collado M et al. DNA-induced alpha-helical structure in the NH2-terminal domain of histone H1. J Biol Chem 2001; 276(49):46429-46435.
61. Hill CS, Martin SR, Thomas JO. A stable alpha-helical element in the carboxy-terminal domain of free and chromatin-bound histone H1 from sea urchin sperm. EMBO J 1989; 8(9):2591-2599.
62. Clark DJ, Hill CS, Martin SR et al. Alpha-helix in the carboxy-terminal domains of histones H1 and H5. EMBO Journal 1988; 7(1):69-75.
63. Bradbury EM, Cary PD, Chapman GE et al. Studies on the role and mode of operation of the very-lysine-rich histone H1 (F1) in eukaryote chromatin. The conformation of histone H1. Eur J Biochem 1975; 52(3):605-613.
64. Vila R, Ponte I, Jimenez MA et al. A helix-turn motif in the C-terminal domain of histone H1. Protein Science 2000; 9(4):627-636.
65. Vila R, Ponte I, Collado M et al. Induction of secondary structure in a COOH-terminal peptide of histone H1 by interaction with the DNA: an infrared spectroscopy study. J Biol Chem 2001; 276(33):30898-30903.
66. Roque A, Ilboro I, Arrondo et al. DNA-induced secondary structure of the carboxyl-terminal domain of histone H1. J Biol Chem 2005; 280(37):32141-32147.
67. Dyson HJ, Rance M, Houghten RA et al. Folding of immunogenic peptide fragments of proteins in water solution. I. Sequence requirements for the formation of a reverse turn. J Mol Biol 1988; 201(1):161-200.
68. Hansen JC, Lu X, Ross ED et al. Intrinsic protein disorder, amino acid composition and histone terminal domains. J Biol Chem 2006; 281(4):1853-1856.
69. Ponte I, Vila R, Suau P. Sequence complexity of histone H1 subtypes. Mol Biol Evol 2003; 20(3):371-380.
70. Van Holde K. Chromatin. New York: Springer-Verlag, 1988.
71. Wolffe A. Chromatin: structure and function. 3rd edition. San Diego: Academic Press; 1998.
72. Subirana JA. Analysis of the charge distribution in the C-terminal region of histone H1 as related to its interaction with DNA. Biopolymers 1990; 29(10-11):1351-1357.
73. Tompa P. Structure and funtion of intrisically dissordered proteins. Chapman and Hall: CRC Press, 2010.
74. Tompa P, Fuxreiter M. Fuzzy complexes: polymorphism and structural disorder in protein-protein interactions. Trends Biochem Sci 2008; 33(1):2-8.
75. Misteli T, Gunjan A, Hock R et al. Dynamic binding of histone H1 to chromatin in living cells. Nature 2000; 408(6814):877-881.
76. Lu X, Hamkalo B, Parseghian MH et al. Chromatin condensing functions of the linker histone C-terminal domain are mediated by specific amino acid composition and intrinsic protein disorder. Biochemistry 2009; 48(1):164-172.
77. Montes de Oca R, Lee KK, Wilson KL. Binding of barrier to autointegration factor (BAF) to histone H3 and selected linker histones including H1.1. Journal of Biological Chemistry 2005; 280(51):42252-42262.
78. Widlak P, Kalinowska M, Parseghian MH et al. The histone H1 C-terminal domain binds to the apoptotic nuclease, DNA fragmentation factor (DFF40/CAD) and stimulates DNA cleavage. Biochemistry 2005; 44(21):7871-7878.
79. McBryant SJ, Lu X, Hansen JC. Multifunctionality of the linker histones: an emerging role for protein-protein interactions. Cell Res 2010;20(5):519-28.
80. Langan TA, Gautier J, Lohka M et al. Mammalian growth-associated H1 histone kinase: a homolog of CDC2+/CDC28 protein kinases controlling mitotic entry in yeast and frog cells. Mol Cell Biol 1989; 9(9):3860-3868.
81. Hendzel MJ, Lever MA, Crawford E et al. The C-terminal domain is the primary determinant of histone H1 binding to chromatin in vivo. J Biol Chem 2004; 279(19):20028-20034.
82. Green GR, Lee HJ, Poccia DL. Phosphorylation weakens DNA binding by peptides containing multiple "SPKK" sequences. J Biol Chem 1993; 268(15):11247-11255.
83. Roth SY, Allis CD. Chromatin condensation: does histone H1 dephosphorylation play a role? Trends Biochem Sci 1992; 17(3):93-98.
84. Gurley LR, D'Anna JA, Barham SS et al. Histone phosphorylation and chromatin structure during mitosis in Chinese hamster cells. Eur J Biochem 1978; 84(1):1-15.
85. Bradbury EM, Inglis RJ, Matthews HR et al. Phosphorylation of very-lysine-rich histone in Physarum polycephalum. Correlation with chromosome condensation. Eur J Biochem 1973; 33(1):131-139.
86. Hohmann P. Phosphorylation of H1 histones. Mol Cell Biochem 1983; 57(1):81-92.
87. Talasz H, Helliger W, Puschendorf B et al. In vivo phosphorylation of histone H1 variants during the cell cycle. Biochemistry 1996; 35(6):1761-1767.
88. Thiriet C, Hayes JJ. Linker histone phosphorylation regulates global timing of replication origin firing. J Biol Chem 2009; 284(5):2823-2829.
89. Lever MA, Th'ng JP, Sun X et al. Rapid exchange of histone H1.1 on chromatin in living human cells. Nature 2000; 408(6814):873-876.
90. Th'ng JP, Guo XW, Swank RA et al. Inhibition of histone phosphorylation by staurosporine leads to chromosome decondensation. J Biol Chem 1994; 269(13):9568-9573.

91. Hale TK, Contreras A, Morrison AJ et al. Phosphorylation of the linker histone H1 by CDK regulates its binding to HP1alpha. Mol Cell 2006; 22(5):693-699.
92. Luger K, Mader AW, Richmond RK et al. Crystal structure of the nucleosome core particle at 2.8 A resolution. Nature 1997; 389(6648):251-260.
93. Emsley P, Cowtan K. Coot: model-building tools for molecular graphics. Acta Crystallogr D Biol Crystallogr 2004; 60:2126-2132.

CHAPTER 3

INTRINSIC PROTEIN FLEXIBILITY IN REGULATION OF CELL PROLIFERATION:

Advantages for Signaling and Opportunities for Novel Therapeutics

Ariele Viacava Follis,[1] Charles A. Galea[2] and Richard W. Kriwacki*,[1,3]

[1]*Department of Structural Biology, St. Jude Children's Research Hospital, Memphis, Tennessee, USA;*
[2]*Structural Biology Division, Walter and Eliza Hall Institute of Medical Research, Parkville, Australia;*
[3]*Department of Molecular Sciences, University of Tennessee Health Sciences Center, Memphis, Tennessee, USA*
**Corresponding Author: Richard W. Kriwacki—Email: richard.kriwacki@stjude.org*

Abstract: It is now widely recognized that intrinsically disordered (or unstructured) proteins (IDPs, or IUPs) are found in organisms from all kingdoms of life. In eukaryotes, IDPs are highly abundant and perform a wide range of biological functions, including regulation and signaling. Despite increased interest in understanding the structural biology of IDPs, questions remain regarding the mechanisms through which disordered proteins perform their biological function(s). In other words, what are the relationships between disorder and function for IDPs? Several excellent reviews have recently been published that discuss the structural properties of IDPs.[1-3] Here, we discuss two IDP systems which illustrate features of dynamic complexes. In the first section, we discuss two IDPs, p21 and p27, which regulate the mammalian cell division cycle by inhibiting cyclin-dependent kinases (Cdks). In the second section, we discuss recent results from Follis, Hammoudeh, Metallo and coworkers demonstrating that the IDP Myc can be bound and inhibited by small molecules through formation of dynamic complexes. Previous studies have shown that polypeptide segments of p21 and p27 are partially folded in isolation and fold further upon binding their biological targets. Interestingly, some portions of p27 which bind to and inhibit Cdk2/cyclin A remain flexible in the bound complex. This residual flexibility allows otherwise buried tyrosine residues within p27 to be phosphorylated by nonreceptor tyrosine kinases (NRTKs). Tyrosine phosphorylation relieves kinase inhibition, triggering Cdk2-mediated phosphorylation of a threonine residue within the flexible C-terminus of p27. This, in turn, marks p27 for ubiquitination and proteasomal degradation, unleashing full Cdk2 activity which drives cell cycle

Fuzziness: Structural Disorder in Protein Complexes, edited by Monika Fuxreiter and Peter Tompa.

progression. p27, thus, constitutes a *conduit* for transmission of proliferative signals via posttranslational modifications. Importantly, activation of the p27 signaling conduit by oncogenic NRTKs contributes to tumorigenesis in some human cancers, including chronic myelogenous leukemia (CML)[9] and breast cancer.[10] Another IDP with important roles in human cancer is the proto-oncoprotein, Myc. Myc is a DNA binding transcription factor which critically drives cell proliferation in many cell types and is often deregulated in cancer. Myc is intrinsically disordered in isolation and folds upon binding another IDP, Max and DNA. Follis, Hammoudeh, Metallo and coworkers identified small molecules which bind disordered regions of Myc and inhibit its heterodimerization with Max. Importantly, these small molecules—through formation of dynamic complexes with Myc—have been shown to inhibit Myc function in vitro and in cellular assays, opening the door to IDP-targeted therapeutics in the future. The p21/p27 and Myc systems illustrate, from different perspectives, the role of dynamics in IDP function. Dynamic fluctuations are critical for p21/p27 signaling while the dynamic free state of Myc may represent a therapeutically approachable anticancer target. Herein we review the current state of knowledge related to these two topics in IDP research.

INTRODUCTION: INTRINSICALLY DISORDERED (OR UNSTRUCTURED) PROTEINS

Many proteins, which play a wide range of biological roles, either entirely lack secondary and/or tertiary structure, or possess long segments that lack secondary and/or tertiary structure, under physiological conditions.[3,12-14] These are commonly termed intrinsically disordered (or unstructured) proteins, abbreviated IDPs (or IUPs). Bioinformatics analyses of whole genome sequences using disorder predictors[15,16] indicated that 6-33% of proteins in bacteria and 35-51% of proteins in eukaryotes contain disordered regions of 40 or more consecutive residues.[15,17] The greater abundance of IDPs in the latter was proposed to be due to the greater need for protein-mediated signaling, regulation and control in eukaryotes.[17] It is now widely recognized that IDPs play broad biological roles in all kingdoms of life.[17,18]

Functions of Disordered Proteins

IDPs are involved in many cellular functions, including regulation of cell division, transcription and translation, signal transduction, protein phosphorylation, storage of small molecules, chaperone action, transport and regulation of the assembly or disassembly of large multi-protein complexes, amongst many others.[19,20] Indeed, the majority of transcription factors[21] and proteins involved in signal transduction[22] in eukaryotes are predicted to be disordered or contain long disordered segments. In general, it has been shown that the expression of IDPs is exquisitely tightly regulated at many stages, including transcription, translation and degradation.[23] It is thought that such tight regulation is required to maintain regulatory IDPs at relatively low levels to prevent deleterious interactions with their diverse biological targets. Disruption of this tight regulatory system can have dire consequences. For example, 79% of human cancer-associated proteins (HCAPs) have been classified as IDPs, compared to 47% of all eukaryotic proteins in the SWISS-PROT database.[22] Further, a large fraction of human oncoproteins known to exert their tumorigenic effects through over-expression, including Myc (discussed later), are enriched in disordered segments which mediate promiscuous signaling interactions.[24] The latter observations highlight the importance of disorder in the function of proteins that regulate processes often dysregulated in cancer such as cell proliferation, apoptosis and DNA repair.

Although many IDPs function by folding into an ordered conformation upon binding their biological targets, for many others, disordered conformations mediate biological function. For example, disordered segments serve as linkers in many IDPs (e.g., between folded domains in multi-domain proteins such as p53).[25] In other cases, IDPs function as entropic bristles (e.g., NF-M and NF-H which serve a repulsive spacers in neurofilaments),[26] springs (e.g., titin, which induces passive tension in muscle filaments)[27] and semi-permeable barriers (FG-domains within nucleoporins of the nuclear pore complex).[28] In these examples, disordered polypeptide segments often occur in conjunction with other folded or partially folded domains.

Folding-upon-Binding

While IDPs are disordered in isolation under physiological conditions, they often perform their biological functions by binding specifically to other biomolecules through the process of folding-upon-binding. In general, folding-upon-binding reactions are enthalpically driven to overcome the accompanying large and unfavorable entropies of binding, as shown for protein-DNA interactions[29] and protein-protein interactions.[30,31] Due to the extended nature of many IDPs which fold upon binding their targets, the magnitudes of both the favorable enthalpy change for binding (ΔH) and unfavorable entropy change for binding (ΔS) are approximately proportional to the length of the disordered polypeptide segment involved in binding.[31] This allows a range of different size binding sites to be targeted by IDPs through evolutionary tuning of the binding favorability and structural complementarity of IDPs and the protein surfaces they target. While the loss of conformational freedom due to folding upon binding (ΔS_{conf}) is entropically unfavorable, it is partially compensated by the entropically favorable release of bound water molecules (ΔS_{HE}) upon binding of an IDP to a protein surface (the hydrophobic effect). While some segments of the polypeptide backbone of IDPs involved in specific protein-protein interactions may become rigid after folding upon binding, other segments may remain dynamic within complexes,[32] mitigating to some extent the unfavorable ΔS_{conf}. Further, the methyl groups of either IDPs and/or their binding targets, that mediate intermolecular hydrophobic interactions, may experience motional restriction to different extents upon binding, providing an additional mechanism for modulating ΔS of binding.[33] These two mechanisms allow tuning of the affinity of interactions (ΔG) through evolutionary variation of the associated entropy changes. Consequently, the values of dissociation constants (K_d) observed for IDPs binding their biological targets span a wide range, from low nanomolar values (tight binding; e.g., p27 binding to Cdk2/cyclin A)[31] to high micromolar values (weak binding; e.g., WASP binding to Cdc42).[34] As a general rule, weak interactions involving IDPs involve relatively small amounts of buried surface area and tight interactions involve the burial of very large surfaces.

Functional Advantages of Disorder

The inherent flexibility of IDPs is thought to confer certain functional advantages over more highly structured proteins. First, some IDPs bind specifically to more than one biological target and thus exhibit diverse biological functions, often with involvement in signaling and regulation.[35] For instance, p21 binds and regulates the catalytic activity of several cyclin-dependent kinase (Cdk)/cyclin complexes, an early example of "binding promiscuity".[36] p21 and p27 also bind additional partners in both the cell nucleus and cytoplasm, extending their functions to include regulation of apoptosis, cell motility and transcription (37 and references therein). Another example is the p53 tumor suppressor protein. While the DNA binding and tetramerization domains are folded, the N- (residues 1-97) and C-terminal domains (residues 363-393) of p53 are intrinsically unstructured in isolation and mediate interactions with numerous binding partners that modulate p53 activity in diverse ways (38 and references therein). Promiscuous binding activities allow p53 to regulate diverse cellular processes such as cell division, apoptosis and DNA repair.[38] Finally, it has been suggested that IDPs are specialized to function as hubs in protein interaction networks due to their propensity for promiscuous interactions;[39] however, the generality[40] and validity[41] of this general concept has been challenged.

Second, because a large fraction of residues within IDPs are solvent exposed, even within multi-protein assemblies, these sites are accessible for posttranslational modification (PTM), allowing control of protein function, localization and turnover. For example, the majority of known phosphorylation, acetylation and ubiquitination sites in p53 occur within the disordered N- and C-terminal domains and modification of these sites alters the function, localization and turnover of p53.[42] PTM sites are often clustered within disordered polypeptide segments, affording accessibility not only to modification enzymes but also to other proteins that interact specifically with the modified sites to transduce biological signals. An example of this is the phosphorylation/ubiquitination cascade that regulates p27 function.[43]

Third, disordered polypeptide segments within proteins, which are often highly susceptible to proteolytic cleavage in vitro, may influence the rate of IUP degradation in cells. However, a recent study by Tompa and coworkers of >3,000 yeast proteins showed that protein disorder was a poor predictor of the in vivo rate of protein turnover; hence, while it is intuitively obvious that polypeptide disorder is associated with proteolytic susceptibility, protein degradation in vivo is highly regulated and influenced by many other factors.[44] For example, Shaul and coworkers have shown that p53 is degraded by the 20S proteasome via a ubiquitination-independent "default" pathway. These authors proposed that disordered segments of p53 and other proteins,[45] are signals for 20S proteasome-mediated degradation and that the formation of multi-protein assemblies masks these signals and guards against degradation.[46] A class of proteins referred to by Shaul and coworkers as "nannys" are thought to interact with un-complexed IDPs and mediate their direct degradation by the 20S proteasome.[47] This may represent a mechanism for sensing imbalances in the levels of subunits within multi-subunit assemblies, allowing subunits present in excess to be degraded by default.[45,47] Thus, the physical properties of disordered polypeptide segments allow proteins to be extensively regulated by PTM and provide the opportunity for rapid turnover and possibly quality control during assembly of multi-protein complexes.

Finally, the noncompact nature of IDPs may facilitate biomolecular interactions by increasing intermolecular association rates. Wolynes and coworkers[48] postulated that disordered proteins have a greater "capture radius" than compact, folded proteins. According to their so called "fly-casting" mechanism, a segment of an extended, unfolded protein first binds relatively weakly to the surface of a target, followed by folding to reel in the target. By being extended, IDPs sample larger solution volumes, in a sense reducing the dimensionality of the search for their partners.[48] For example, p27 binds via a sequential mechanism to Cdk2/cyclin A, with an extended domain at the N-terminus binding first to a compact surface on cyclin A, followed by extensive folding of p27 and remodeling of Cdk2 as the inhibited p27/Cdk2/cyclin A complex is fully assembled.[31]

This fly-casting-like mechanism may facilitate assembly of Cdk/cyclin complexes under the low concentration conditions found in cells. Another example of this mechanism was revealed recently by Wright and coworkers[49] in studies of the phosphorylated KID (pKID) domain of CREB binding to the KIX domain of the CREB binding protein (CBP). Intrinsically unstructured pKID was shown to initially dock nonspecifically on the surface of KIX, allowing rapid searching of the KIX surface for the specific binding site, followed finally by folding into the specific, high-affinity complex.[49] However the benefits of the fly-casting mechanism were shown in a recent computational study to be at least partially counter-balanced by the effect of extended IDP conformations on diffusion rates in solution;[50] these considerations would affect the initial rate of encounters between IDPs and their targets but not that of the subsequent, reduced dimensionality search for a specific binding site through surface scanning. In addition to a role for disordered domains in recognition events, Hilser and Thompson have proposed that linking ligand binding with disordered domain folding provides a mechanism for optimizing allosteric coupling in multi-domain proteins.[51]

INTRINSICALLY DISORDERED PROTEINS IN MAMMALIAN CELL CYCLE REGULATION

In eukaryotes, cyclin-dependent kinases (Cdks) are the master time keepers of cell division.[52] Many proteins regulate the Cdks, both directly and indirectly and, in turn, the catalytic activity of the Cdks regulates the activity of myriad downstream targets.[53,54] While many of these regulatory proteins are folded, many others are intrinsically unstructured. In this chapter, we focus on a small subset of these IDPs: the cyclin-dependent kinase regulators (CKRs) p21, p27 and p57^{Kip2} (p57)[53] that regulate cell division through direct interactions with Cdk/cyclin complexes (Fig. 1A). Through binding promiscuity,[36] the CKRs regulate Cdk/cyclin complexes that control 1) entry into G_1 phase (Cdk4 and Cdk6 paired with D-type cyclins) and 2) progression from G_1 to S phase (Cdk2 paired with A- and E-type cyclins).[53] Further, p21 and p27 exhibit functional diversity by having seemingly opposite effects on these different Cdk/cyclin complexes, promoting the assembly and catalytic activity of some (e.g., Cdk4 paired with D-type cyclins) and potently inhibiting others (e.g., Cdk2 paired with A- and E-type cyclins).[53] In the following sections, we discuss results from our laboratory and others on the structural and dynamic features of the CKRs and the relationship of these features to their diverse biological functions.

Domain Structure of CKRs

The CKRs p21, p27 and p57 contain a conserved, 60 residue-long N-terminal kinase inhibitory domain (KID, residues 28-90 in p27) (Fig. 1B) and several nuclear localization signals (NLSs)[55,56] within their C-terminal domains (Fig. 1C). p21 and p57 also contain a PCNA-binding domain within their C-termini that, when bound, inhibits the ability of PCNA to stimulate DNA synthesis.[57,58] Further, p27 and p57 possess a C-terminal QT domain that contains a critical threonine residue (T187 in p27 and T310 in p57) that, when phosphorylated by Cdk2, triggers ubiquitination of p27[59,60] and p57[61] by SCF/Skp2. Human and mouse p57 have an additional domain comprised of multiple Pro-Ala repeats and mouse p57 contains a segment rich in acidic residues; these domains were proposed to mediate protein-protein interactions.[62]

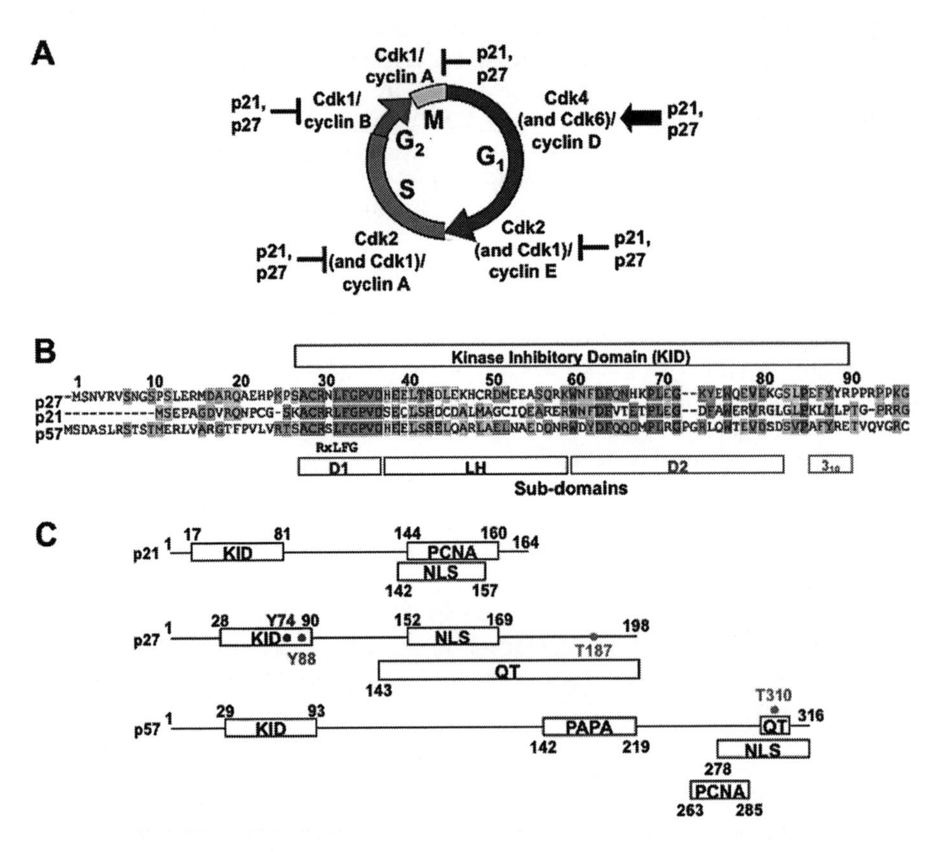

Figure 1. Regulation of the eukaryotic cell division cycle. A) Illustration of various stages of the cell division cycle and Cdk/cyclin complexes that play key roles in regulating progression through the different stages are indicated. Initiation of cell division in G_1 phase requires the activity of Cdk4/cyclin D and Cdk6/cyclin D and progression into S phase (when DNA synthesis or replication occurs, "S") requires Cdk2/cyclin E and Cdk2/cyclin A (and similar complexes with Cdk1). The activity of Cdk1/cyclin B and Cdk1/cyclin A are required for entry into mitosis ("M"). While initially thought to be universal inhibitors of these Cdk/cyclin complexes, p21 and p27 have been show to activate Cdk4/cyclin D and Cdk6/cyclin D under certain circumstances (indicated by arrow). B) Alignment of sequences of the kinase inhibitory domains (KID) of p27, p21 and p57. The boundaries of sub-domains D1 (blue), LH (black), D2 (red) and 3_{10} (green) are indicated, as is the "RxLF" motif which is recognized by cyclin A. C) Illustration of the domain structure of p21, p27 and p57. PCNA, PCNA binding domain; NLS, nuclear localization signal; QT, QT domain which contains one or more QT motifs that are either known or putative phosphorylation sites. PAPA, domain with multiple repeats of PAPA motif. The locations of known phosphorylation sites, Y74, Y88 and T187 in p27 and T310 in p57, are also indicated. Reproduced from Galea CA et al. Biochemistry 2008; 47(29):7598-609, with permission from the American Chemical Society. A color version of this image is available at www.landesbioscience.com/curie.

The Structure of p27 Bound to Cdk2/Cyclin A

The crystal structure of the p27/Cdk2/cyclin A ternary compex showed that the KID of p27 binds in a highly extended conformation to both proteins of the Cdk2/cyclin A complex (Fig. 2), burying over 2,000 Å^2 of solvent exposed surface.[63] Several sub-domains of the disordered p27-KID, which possess many residues conserved among p21, p27 and p57

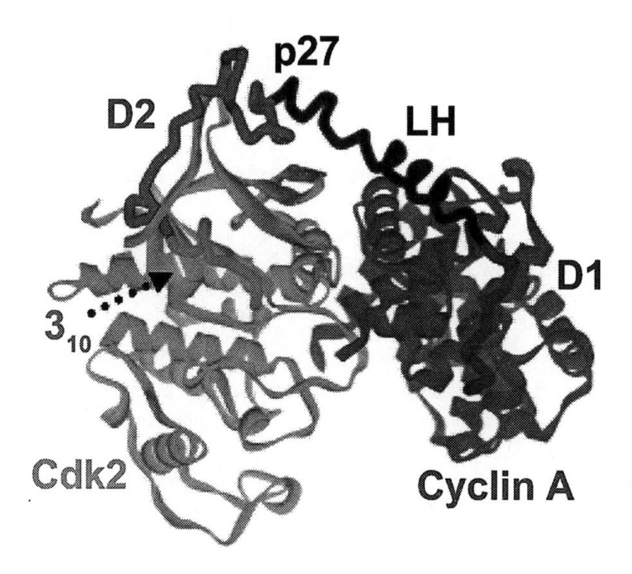

Figure 2. Structure of p27-KID bound to Cdk2/cyclin A. Crystal structure of p27-KID/Cdk2/cyclin A determined in 1996 by Pavletich and coworkers.[57] The sub-domains of p27, D1, LH, D2 and 3_{10} are indicated using the color scheme of Figure 1. Reproduced from Galea CA et al. Biochemistry 2008; 47(29):7598-609, with permission from the American Chemical Society. A color version of this image is available at www.landesbioscience.com/curie.

(Fig. 1B), adopt secondary structure in the Cdk2/cyclin A-bound state (Fig. 2): sub-domain D1, containing the conserved "RxL" motif,[64] binds in an extended conformation on the surface of cyclin A; sub-domain LH forms a 22 residue-long α-helix that spans the nearly 40 Å gap between Cdk2 and cyclin A; sub-domain D2 forms a β-hairpin and an intermolecular β-sheet (with Cdk2) on the surface of the N-terminal lobe of Cdk2; and sub-domain 3_{10} forms a 3_{10}-helix that inserts into the ATP binding pocket of Cdk2. p27 inhibits Cdk2 1) by inserting sub-domain 3_{10} into the active site and blocking access to ATP and 2) by remodeling the catalytic cleft through displacement of a β-strand of Cdk2 by sub-domain D2 of p27.[63] Further, sub-domain D1 of p27 blocks the substrate binding site on cyclin A which recognizes Cdk2 substrates possessing a RxL motif.[64] It has been proposed that similar mechanisms are ultilized by all CKRs to regulate the Cdks that control the G_1/S transition during cell division.[31,65] However, due to the inherent limitations of X-ray crystallography this structural model does not provide insights into the role of the intrinsic flexibility of p27 and other CKRs in recognizing and binding to Cdk/cyclin complexes.

The CKRs are IDPs

Analysis using proteolysis, CD and NMR spectroscopy showed that p21, p27 and p57 are largely disordered, with ~15-20% α-helical content.[31,36,66,67] Secondary structure[31] and disorder prediction (using FoldIndex,[68] IUPred[69] and PONDR)[70] also indicated that these proteins are predominantly disordered. 2D ^{1}H-^{15}N HSQC NMR spectra for p21, p27 and p57 exhibited limited resonance dispersion (backbone amide protons resonate between 7.8 and 8.5 ppm), a feature typical of intrinsically unstructured proteins.[31,36,66]

However, despite being disordered, these proteins were shown to be potent inhibitors of various Cdks in vitro[71-73] and in vivo.[53] As early as 1996, structural data for p21[36] and the previously identified heat-resistant nature of p27[73] strongly suggested that polypeptide disorder was associated with the biological function of the CKRs.

The CKRs Exhibit Partially Populated Secondary Structure in Solution

Although p21, p27 and p57 can be categorized as IDPs based on their lack of tertiary structure, CD spectra indicated the presence of a small amount of α-helical secondary structure[31,36,66,67] within the KID of each protein.[36,66,67] Subsequently, we used NMR spectroscopy to localize this secondary structure within p27-KID. Analysis of secondary $^{13}C_{\alpha}$ chemical shift values ($\Delta\delta^{13}C_{\alpha}$) indicated that the small amount of α-helical secondary structure observed using CD was localized to sub-domain LH (~30% α-helical) while other sub-domains of p27-KID appeared to lack secondary structure.[31] To probe the dynamics of p27-KID, we measured $\{^{1}H\}$-^{15}N heteronuclear NOE (hetNOE) values; this NMR relaxation parameter is sensitive to fluctuations of amide groups on the high picosecond to low nanosecond timescale. The hetNOE values observed for p27-KID indicated that the polypeptide backbone experiences intermediate dynamics, being less rigid than a folded protein but more rigid than a random coil. Interestingly, partially restricted motions were observed not only for sub-domain LH, but also for most residues in sub-domains D2 and 3_{10}. In contrast, residues within sub-domain D1 (residues 27-35) exhibited negative or near zero hetNOE values, consistent with a high degree of flexibility.

More recently, we probed the structure and dynamics of p27-KID using a novel approach that utilized amide proton-amide proton ($^{1}H_{N}$-$^{1}H_{N}$) NOE data from NMR spectroscopy and molecular dynamics (MD) computations to provide insights into how structure fluctuates with time.[74] Interestingly, these studies revealed that several segments of p27-KID exhibited discreet structures which we termed intrinsically folded structural units (IFSUs). The IFSUs occurred within sub-domains LH, D2 and 3_{10}, which also exhibited positive hetNOE values. Sub-domain LH adopted helical structure, as expected from chemical shift analysis. While sub-domain D2 formed a β-hairpin and nascent helical structure and sub-domain 3_{10} formed a single turn of helix. These results indicated that p27-KID is quite rich in transient, discrete structures, in contrast to the picture which first comes to mind for an IDP. Importantly, with the exception of the nascent helical segment of sub-domain D2, these structural features resemble those observed when p27-KID is bound to Cdk2/cyclin A.

Finally, we recently characterized the solution structure of the ~100 residue-long C-terminal domain of p27 using NMR spectroscopy.[75] This domain contains several phosphorylation sites, including T187 mentioned earlier and threonine 157 (T157) within the NLS which is phosphorylated by Akt in breast cancer,[76] and several lysine residues that are likely sites of ubiquitination. Earlier results from CD and NMR suggested that this segment of p27 is highly disordered;[31] these results were confirmed by the recent NMR studies which showed that this domain lacks tertiary and secondary structure on the basis of $\Delta\delta^{13}C_{\alpha}$ and hetNOE values.[75]

The CKRs Fold Sequentially upon Binding Specific Cdk/Cyclin Complexes

In 1996, Kriwacki and Wright demonstrated that sub-domains D2 and 3_{10} of p21 folded upon binding to Cdk2 using NMR spectroscopy and, through proteolytic mapping, deduced that sub-domain D1 bound to cyclin A within the Cdk2/cyclin A complex.[36] Later in the same year, Pavletich and coworkers demonstrated that p27-KID adopted a highly extended, folded conformation when bound to Cdk2/cyclin A.[63] As noted earlier, several of the IFSUs detected in p27-KID prior to binding were also observed in the Cdk2/cyclin A bound state. For example, one part of sub-domain D2 and sub-domain 3_{10} maintained their β-hairpin and helical conformations, respectively, when bound to Cdk2. Importantly, the single turn of helix observed for apo sub-domain 3_{10} occupied the ATP binding site of Cdk2 within the ternary complex, with tyrosine 88 (Y88) bound in place of the purine ring of ATP.[63] Sub-domain LH formed a 22 residue long α-helix that linked sub-domains D2 and D1, which were bound to Cdk2 and cyclin A, respectively. Sub-domain D1 exhibited a high degree of disorder and flexibility in the free state and adopted an extended, rigid conformation upon binding to a pocket on the surface of cyclin A that is conserved in many cyclins that regulate cell division.[31,63,65] Finally, another part of sub-domain D2, which exhibited nascent helical features prior to binding, adopted an extended conformation and formed an intermolecular β-sheet upon binding to the N-terminal domain of Cdk2.

While the crystal structure defined the Cdk2/cyclin A-bound conformation of p27 and provided key insights into the molecular basis of specific recognition of Cdk/cyclin complexes,[63] it does not explain why the CKRs have evolved to be disordered or how disorder plays a role in their biological functions. Answers to these questions came from studies that probed the mechanism through which p27 binds to Cdk2/cyclin A.[31] Isothermal titration calorimetry (ITC) was used to determine values of thermodynamic parameters (ΔG, ΔH and ΔS) associated with p27 binding to Cdk2/cyclin A and to quantitatively characterize the polypeptide folding which accompanies binding.[29] Further, surface plasmon resonance (SPR) was used to analyze the kinetics of p27 association with and dissociation from Cdk2/cyclin A. Results from these two methods, coupled with our knowledge of structure and dynamics, indicated that the sub-domains of p27 bind to Cdk2/cyclin A via a sequential mechanism; first the highly flexible sub-domain D1 binds cyclin A, followed by docking of helical sub-domain LH and finally by docking and folding of sub-domains D2 and 3_{10} to Cdk2 (Fig. 3). In addition, other studies[65] suggested that sub-domain D1 rapidly scans the surfaces of protein complexes for a conserved binding pocket, as found on cyclin A and other cyclins that regulate cell division. When sub-domain D1 encounters the cyclin pocket, p27 transiently binds, providing time for other sub-domains to sequentially dock and fold into the final, inhibited ternary complex with Cdk2/cyclin A. Amino acids that comprise the cyclin A docking site for p27 are highly conserved in Cdk/cyclin complexes that directly regulate cell division and are inhibited by p21 and p27. However, these residues are not conserved in Cdk/cyclin complexes involved in other biological functions. Therefore, we proposed that the intrinsic disorder of p21 and p27 evolved to allow specific molecular recognition through sequential folding upon binding.[31,65] The extended character of p27 when bound to Cdk2/cyclin A has evolved to accommodate the large distance (40 Å) between the specificity determining site on cyclin A and the site of inhibition (ATP binding pocket) on Cdk2. It is possible that simultaneously engaging these two sites through interactions with the two ends of an extended polypeptide chain provided evolutionary advantages over an alternative scheme involving interactions mediated by multiple, folded protein domains.

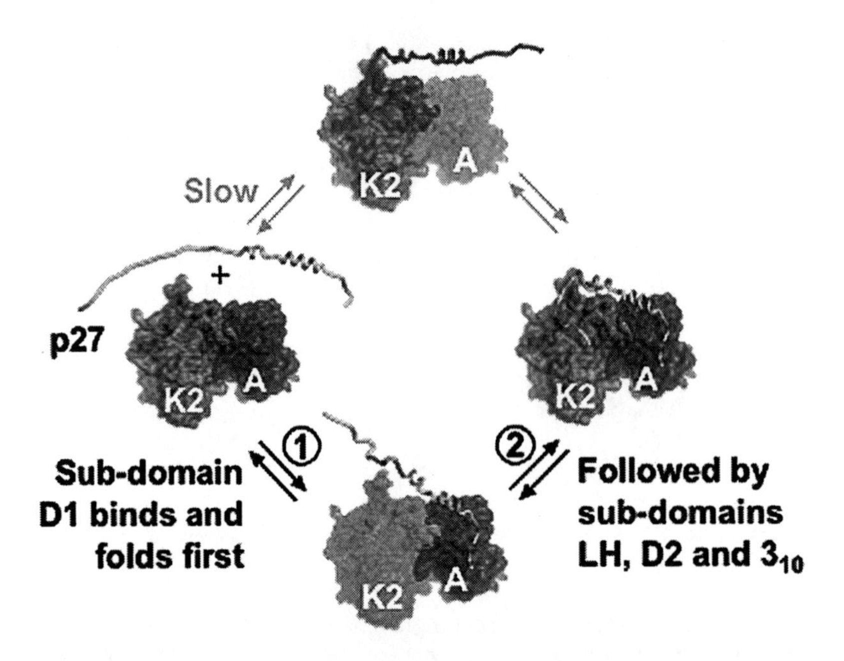

Figure 3. The sub-domains of p27 sequentially fold and bind Cdk2/cyclin A. Results from NMR, ITC and SPR support a scheme in which the "RxLF" motif within sub-domain D1 of p27 binds first to cyclin A ("1"), followed by folding and docking of sub-domain LH, followed by binding of sub-domain D2 to Cdk2 (with extensive remodeling of Cdk2, followed finally by binding of sub-domain 3_{10} in the ATP binding pocket of Cdk2 ("2"). This sequential scheme provides a mechanism for specificity toward Cdk/cyclin complexes which preserve the binding site for the "RxLF" motif within sub-domain D1 of p27. Reproduced from Galea CA et al. Biochemistry 2008; 47(29):7598-609, with permission from the American Chemical Society. A color version of this image is available at www.landesbioscience.com/curie.

Regulation of p27 Function through Posttranslational Modification: The Importance of Flexibility in Signaling

Posttranslational modifications regulate the localization, turnover and activity of p27 (77 and references therein). For example, Akt-mediated phosphorylation of T157 within p27's NLS in breast cancer cells prevents its interactions with the nuclear import machinery and leads to cytoplasmic localization. p27, normally located in the nucleus, encounters new targets in the cytoplasm and exhibits a gain of oncogenic function. In a further example, phosphorylation of p27 on Ser 10 promotes its interaction with the shuttling protein, CRM1, leading to export from the nucleus. Finally, cells entering the division cycle contain super-stoichiometric levels of p27 (with respect to Cdk/cyclin complexes) and phosphorylation-dependent ubiquitination and degradation of p27 by the 26 S proteasome is required for Cdk/cyclin complexes to be activated, allowing progression through the cell division cycle.

p27 degradation is regulated by two E3 ubiquitin ligases during cell division, KPC1 in the cytoplasm in G1 phase and SCF^{Skp2} in the nucleus in S and G2 phases (43 and references therein). While KPC1 ubiquitinates unphosphorylated and free p27, SCF^{Skp2} targets p27 that is phosphorylated on T187 and which is bound to Cdk2/cyclin E or Cdk2/cyclin A. Ubiquitinated p27 is degraded by the 26S proteasome. Active cyclin E/Cdk2 can phosphorylate cyclin/Cdk-bound p27 on T187. However, while nonp27-bound, active Cdk2/cyclin phosphorylates Cdk-bound p27 efficiently (because T187 is within the C-terminal regulatory domain of p27 that does not participate in direct interactions with Cdk2/cyclin complexes), structural and biochemical studies have demonstrated that p27-bound Cdk2 is catalytically inactive.[31] This presented an apparent paradox since elimination of p27 appeared to require the activity of the enzyme (Cdk2) that it was known to potently inhibit. However, studies by Hengst, Kriwacki and coworkers[43] demonstrated that phosphorylation of p27 at a previously uncharacterized site within the p27/Cdk2/cyclin A complex provided a mechanistic "missing link" leading to a solution to this conundrum.

A key finding was that p27 was phosphorylated on tyrosine 88 (Y88) within the kinase inhibitor domain (KID) by nonreceptor tyrosine kinases, including Abl[43], Lyn[43] and Src[10]. Phosphorylation of residue Y88 on p27 bound to Cdk2/cyclin A was enabled by local flexibility within the 3_{10}-helix containing Y88, relieved inhibition of Cdk2 and promoted Cdk2-mediated phosphorylation of T187 via a pseudo uni-molecular reaction mechanism.[43] NMR studies showed that phosphorylation of Y88 (pY88) within the Cdk2 binding domain of p27 by Abl kinase led to ejection of the inhibitory 3_{10}-helix of p27 (sub-domain 3_{10}) from the ATP binding pocket of Cdk2 while leaving other interactions between p27 and Cdk2/cyclin A unperturbed (Fig. 4, "Step 1"). Surprisingly, Cdk2 retained significant catalytic activity even though pY88-p27 remained tightly bound to the Cdk/cyclin complex.[43] Consequently, residue T187 within the intrinsically unstructured and highly dynamic C-terminal domain of p27 could then be phosphorylated by the partially reactivated Cdk2 within the same pY88-p27/Cdk2/cyclin A ternary complex[43] (Fig. 4, "Step 2"). Although Cdk2 within this phosphorylated ternary complex exhibited sub-maximal catalytic activity, tethering of p27 to cyclin A/Cdk2 strongly promoted the phosphorylation of T187 through the uni-molecular mechanism (Fig. 4B). p27 that has been phosphorylated on both Y88 and T187 (pY88/T187-p27) can be poly-ubiquitinated by the SCF^{Skp2} ubiquitin ligase and degraded, resulting in complete reactivation of Cdk2/cyclin A. The accumulation of free, active cyclin A/Cdk2 may further promote Thr187 phosphorylation of p27 within remaining p27/Cdk2/cyclin A complexes and accelerate progression from G_1 to S phase of the cell division cycle.

We propose that the intrinsic disorder and flexibility of p27 are evolutionarily advantageous by enabling the structural fluctuations (also termed "fuzziness") and posttranslational modifications associated with the phosphorylation/poly-ubiquitination cascade that regulates p27 turnover at the G_1/S boundary during cell division. First, segmental flexibility between the sub-domains D2 and 3_{10} allows Y88 within sub-domain 3_{10} to fluctuate between ATP pocket-bound and solvent exposed conformations. This so-called molecular fuzziness allows Abl and other NRTKs to access and phosphorylate otherwise occluded Y88. Notably, in addition to targeting Y88, Src also phosphorylates p27 on tyrosine 74 (Y74). This implies that the β-hairpin secondary structure harboring Y74 within p27 sub-domain D2 (Fig. 4A) also exhibits fuzziness by fluctuating between bound and solvent exposed conformations so as to provide Src access to Y74. Second, after Y88 (and sometimes Y74) has been phosphorylated and sub-domain 3_{10} has been ejected from the Cdk2 active site, the extreme flexibility of the p27 C-terminus allows T187 to approach the substrate binding site of Cdk2 and be phosphorylated. This flexibility also ensures that phosphorylated T187 is accessible for recognition by SCF^{Skp2}. Interestingly, T187 phosphorylated p27 is recognized by SCF^{Skp2} only when bound to Cdk2/cyclin A (or cyclin E). Third, the p27 C-terminus contains six lysine residues that are likely sites for poly-ubiquitination by SCF^{Skp2}. Intrinsic flexibility within this polypeptide segment will not only make these sites accessible for poly-ubiquitination, but also ensures that

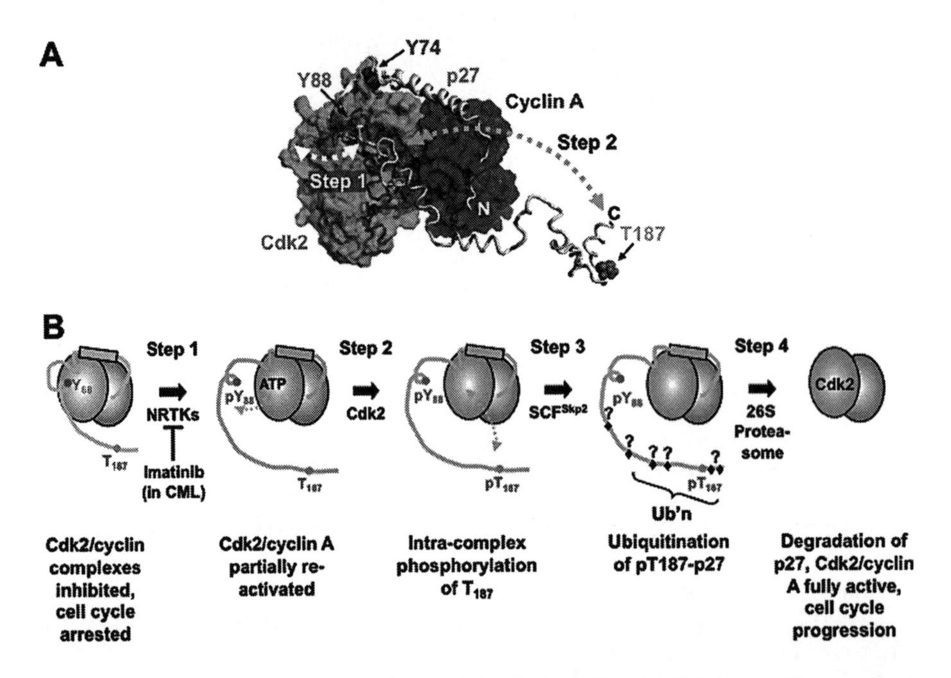

Figure 4. The p27 molecule is a signaling conduit. A) A single snap-shot from a 13 ns molecular dynamics trajectory illustrating the structure of p27 bound to Cdk2/cyclin A (cyan and magenta, respectively).[68] The 100 residue-long C-terminal domain of p27 (yellow tube), which contains T187 (orange), is intrinsically unstructured and highly dynamic in this trajectory. Also illustrated are two critical tyrosine residues, Y74 and Y88 (red and green, respectively), which are phosphorylated by nonreceptor tyrosine kinases (NRTKs). Phosphorylation of Y88 ("Step 1") and possibly Y74, ejects sub-domain 3_{10} from the ATP binding pocket of Cdk2 (indicate by white arrow), allowing T187 within the flexible C-terminal domain to encounter the Cdk2 active site ("Step 2", indicated by grey arrow) and be phosphorylated by Cdk2. B) Scheme illustrating the two-step p27 phosphorylation mechanism involving Y88 and T187 which triggers p27 poly-ubiquitination and 26S proteasomal degradation in both normal and cancer cells. The pseudo-uni-molecular nature of step 2 is illustrated. Reproduced from Galea CA et al. Biochemistry 2008; 47(29):7598-609, with permission from the American Chemical Society. A color version of this image is available at www.landesbioscience.com/curie.

covalently linked poly-ubiquitin chains are accessible for processing by the 26S proteasome and its various accessory proteins.[78] Notably, ubiquitinated p27, within Cdk2/cyclin complexes, is selectively degraded by the 26S proteasome, leading to release of fully active Cdk2/cyclin complexes in the nucleus. Thus, the intrinsic flexibility of p27 critically mediates each step of this multi-step, posttranslational modification cascade. In view of this, we have proposed that p27 acts as a molecular signaling conduit which integrates proliferative signals from NRTKs (through phosphorylation on Y74 and Y88), participates in the processing of these signals (through reactivation of Cdk2 and phosphorylation of T187) and finally transduces these signals by triggering its own poly-ubiquitination and degradation (through phospho-T187-dependent interactions with SCF^{Skp2}). While p27 exhibits modular structure within the kinase inhibitory domain when free in solution, the segments connecting these modules are highly flexible, allowing the individual modules to function independently as part of this signaling conduit. The flexibility and modularity which enable this structural independence allow signals to flow through the p27 conduit as a consequence of these sequential phosphorylation and ubiquitination modifications.

INHIBITION OF Myc-DEPENDENT CELL PROLIFERATION: TARGETING INTRINSICALLY DISORDERED REGIONS WITH SMALL MOLECULES

Myc Drives Cell Proliferation in Response to Mitogenic Signals

The protein product of the *MYC* gene, Myc, is an intrinsically disordered transcription factor of the basic helix-loop-helix leucine zipper family (bHLHZip) that broadly controls cell proliferation through protein-DNA and protein-protein interactions. Importantly, Myc function is deregulated in the majority of human cancers.[79-81] Deregulation can arise from missense point mutations,[82] insertional mutagenesis,[83,84] chromosomal translocation[85,86] and gene amplification.[87,88] Alternatively, perturbation of upstream and downstream signaling pathways can deregulate Myc function through alteration of protein expression, posttranslational modification and/or degradation.[89-92] The Myc protein, which normally has a very short half-life and is subject to tight regulation, is a potent proliferative agent, promoting or silencing the expression of a large number of genes,[93,94] leading to cell cycle progression[95] and proliferation.[96-98]

The Myc isoform c-Myc is comprised of 439 amino acids and is intrinsically disordered in the absence of its binding partners. Two critical domains within c-Myc include the N-terminal transcriptional activation domain (residues 1-143)[99,100] and a C-terminal basic, helix-loop-helix leucine zipper (bHLHZip) DNA binding domain (residues 350-439).[5,101] The transactivation domain mediates interactions with proteins involved in regulating chromatin structure and gene expression, including TRAPP, INI1, RNA polymerase II and PTEFb.[102-105] The bHLHZip domain mediates heterodimerization with a similar domain within a partner protein, Max and, when paired with Max, mediates specific binding to E-box DNA sites with the conserved sequence, 5′-CACGTG-3′.[106-108] The bHLHZip domains of c-Myc and Max are intrinsically disordered in isolation and cofold upon heterodimerization, which promotes further cofolding of basic residues within the bHLHZip domain upon specific binding to E-box DNA sites.[5,109] The HLH portion of the c-Myc/Max heterodimer (Fig. 5A) forms a four-helix bundle structure which extends at the C-termini into a left handed coiled-coil formed by the two leucine zipper motifs. Interhelical hydrophobic interactions stabilize this heterodimeric structure. In the presence of E-box DNA, the two N-terminal basic regions form α-helices which extend from the N-termini of the HLH/four-helix bundle and bind symmetric sites within the major grooves of the palindromic DNA double-helix. The DNA-bound c-Myc/Max heterodimeric complex further mediates interactions with proximal transcription factor/DNA complexes, resulting in repression of gene expression.[110]

Inhibition of c-Myc Activity through Disruption of Myc/Max Heterodimers: From Dominant-Negative Protein Domains to Small Molecules

Because c-Myc function is very commonly deregulated in human cancers, significant consideration has been given to therapeutic strategies to limit cancer cell proliferation through inhibition of c-Myc function. Several approaches toward inhibition of c-Myc function have been attempted.[8,11,111-119] Due to the dependence of c-Myc activity on heterodimerization with Max, a widely explored approach has involved disruption of c-Myc/Max dimers.[8,11,116-119] The therapeutic promise of this approach has subsequently been validated.[120-122] In particular, a landmark study was performed by Evan and coworkers using a mouse model of lung cancer in which Myc plays a role in tumorigenesis.[122] Myc

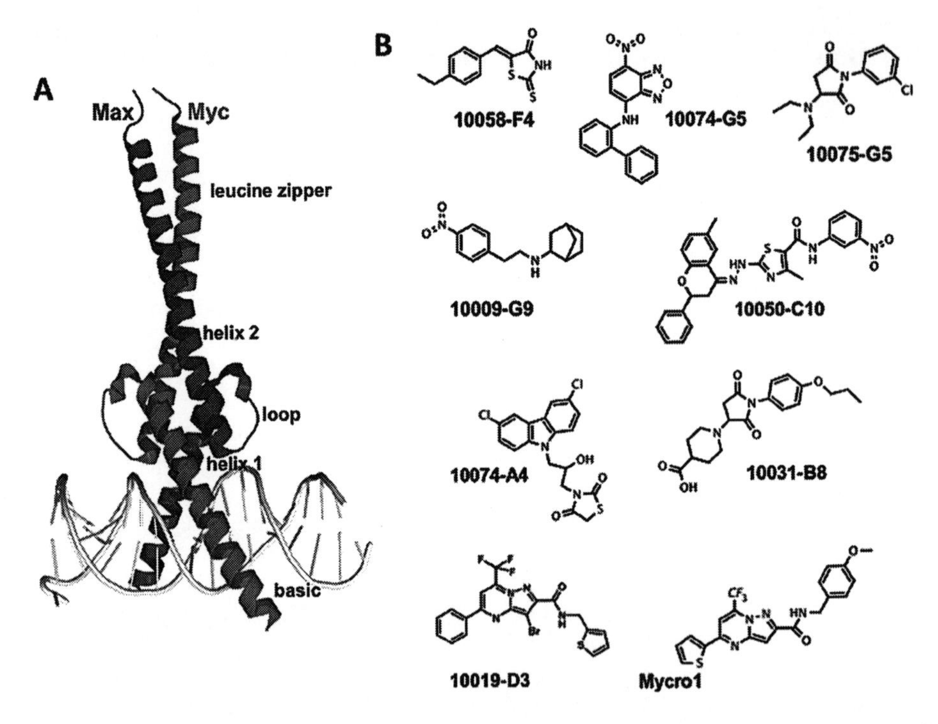

Figure 5. Inhibition of Myc function through disruption of Myc/Max dimers. A) Crystal structure of the Myc/Max heterodimer bound to target E-box DNA, containing the palindromic recognition sequence CACGTG (PDB: 1nkp[96]). B) Inhibitors of Myc/Max heterodimer formation first reported by Yin et al[116] and 'Mycro' compound reported by Kiessling et al.[114] A color version of this image is available at www.landesbioscience.com/curie.

inhibition was achieved through chromosomal insertion of a doxycycline-inducible gene for a dominant-negative, modified Myc bHLHZip domain (termed Omomyc) which binds all Myc isoforms (c-Myc, n-Myc and L-Myc), prevents heterodimerization with Max and inhibits Myc function.[123] While tumors developed in the lungs of untreated mice, tumors did not develop or regressed from lungs of mice in which Myc function was systemically inhibited through doxycycline-induced, Omomyc expression.[122] Importantly, Myc inhibition was associated with relatively mild and reversible side-effects in highly proliferative tissue types.[122]

In contrast to many current drug targets, Myc is not a receptor or enzyme and therefore does not possess a rigidly defined active site that can be occupied by inhibitory small molecules. Further, Myc lacks secondary and tertiary structure when not complexed with one of its biological targets[124] and therefore does not exhibit well-defined surface features that potentially could be targeted by small molecules to inhibit protein-protein or protein-DNA interactions. Thus, Myc, as an intrinsically disordered protein, represents a novel type of drug target. Despite the fundamental challenges associated with "drugging" an intrinsically disordered protein, several groups performed screens to identify small molecules that inhibit c-Myc/Max heterodimerization using in vitro, fluorescence-based assays[11,117-119] and a yeast two-hybrid assay.[8] These screens of combinatorial or diversity oriented libraries of low molecular weight compounds identified numerous molecules which exhibited K_i values in the low micromolar range. Interestingly, the so-called "hit" compounds from different studies exhibited low structural similarity. These initial successes motivated studies into the mechanism(s) by which small molecules inhibited Myc-Max dimerization.

"Hit" Molecules Bind to the Intrinsically Disordered c-Myc Monomer

An initial hypothesis was that inhibitory small molecules bound either to disordered c-Myc or Max and thus prevented their cofolding and heterodimerization. Yin, *et al*,[8] tested this hypothesis by determining the ability of seven initial hit compounds to disrupt a variety of homomeric and heteromeric HLH, HLHZip or leucine zipper transcription factor complexes, including Myc/Max and the Max homodimer. Interestingly, the seven compounds selectively disrupted Myc/Max heterodimers, suggesting that they interacted specifically with Myc. Further studies based on measurements of intrinsic fluorescence anisotropy for two of the seven original compounds, 10058-F4 and 10074-G5, demonstrated specific interaction with the bHLHZip domain of monomeric c-Myc but not p21 Max (which does not form homodimers).[125] The binding stoichiometry for Myc bHLHZip was 1:1[6,125] and K_d values were 5.3 ± 0.7 μM and 2.8 ± 0.7 μM for 10058-F4 and 10074-G5, respectively.[6]

Small Molecules That Bind Myc Display Weak Structure-Activity Relationships

Analysis of ~70 synthesized derivatives of compound 10058-F4 revealed weak structure-activity relationships, with most compounds displaying binding to Myc at concentrations up to 50 μM.[125,126] This unusual behavior may reflect the influence of dynamics of the disordered Myc polypeptide chain on interactions with small molecules; intrinsic flexibility within the IDP binding site may allow structural rearrangement to accommodate small molecule ligands with different functional groups and structural features.

Small Molecules Bind to Multiple Sites within Intrinsically Disordered Myc

Initially, it was unknown whether Myc inhibitors belonging to different structural classes bound to the same or different sites within the Myc bHLHZip polypeptide chain.[6,7] Two distinct binding sites, for the structurally unrelated, fluorescent compounds 10058-F4 and 10074-G5 were identified using a fluorescence polarization assay and a series of truncated or mutated variants of the Myc bHLHZip domain.[6] The two binding sites (Sites I and II) were further delineated through binding assays with synthetic peptides; short peptides displayed binding affinities similar to those observed with full-length Myc bHLHZip domain, demonstrating that these sites were of limited length and independent.[6] Further, competition binding assays showed that several nonfluorescent inhibitors also bound to either Site I or II.[7] A third binding site (Site III) for the nonfluorescent compound 10074-A4, which could not displace either of the fluorescent inhibitors from the Myc bHLHZip domain, was identified using a circular dichroism (CD)-based binding assay and the Myc bHLHZip variants.[7] The three binding sites identified within Myc bHLHZip are summarized in Figure 6A.

Small Molecule Binding Sites within Myc Display Similar Sequence Features

Disordered proteins are generally depleted in hydrophobic residues and enriched in hydrophilic and charged residues, with some dependence of disorder on the relative numbers of hydrophobic *versus* charged residues.[127] Sequence analysis algorithms have been developed to accurately predict the location of disordered regions within polypeptide sequences.[70] Analysis of the sequence of the Myc bHLHZip domain using the disorder prediction algorithm PONDR provided insights into the features of Sites I-III relative to other regions that do not interact with small molecules (Fig. 6B). Each of the small molecule binding regions exhibits a small cluster of hydrophobic residues, which cause inflections in the plot of disorder probability *versus* residue number. This observation suggests that sites in IDPs that exhibit relatively high hydrophobicity, which correlates with reduced disorder probability, have the potential to bind small molecules. Interestingly, these sites may coincide with regions that mediate functionally important protein-protein

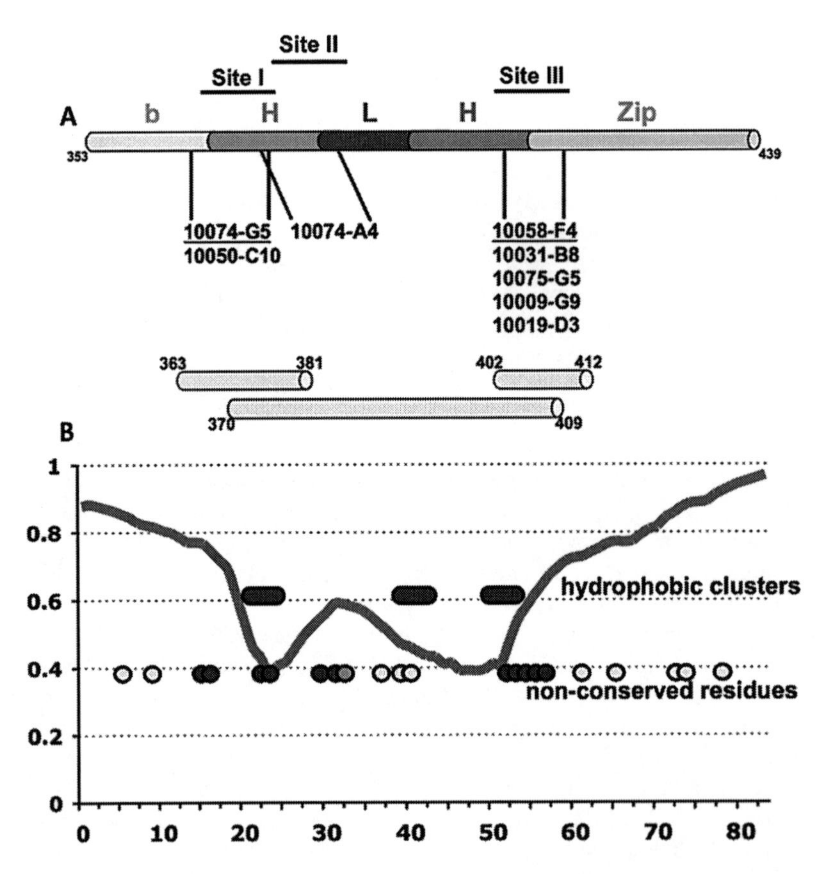

Figure 6. Location and sequence characteristics of inhibitor binding sites on Myc. A) Summary of inhibitor binding sites based on direct binding studies between fluorescent inhibitors and mutated or truncated variants of c-Myc bHLHZip domain and from competition studies between fluorescent and nonfluorescent inhibitors. The protein constructs employed for structural studies are outlined at the bottom of the panel. B) Disorder probability for the c-Myc bHLHZip domain assessed with the PONDR VSL2B algorithm[130]. Regions containing hydrophobic clusters and nonconserved residues are overlaid on the plot. Nonconserved residues found within regions involved in inhibitor binding are highlighted in darker grey. Inhibitor binding sites are located in segments characterized by high hydrophobic content, reduced disorder probability and presence of nonconserved residues. A color version of this image is available at www.landesbioscience.com/curie.

interactions.[128] Additionally, the sequences of Sites I-III are poorly conserved in n-Myc and MAX, providing a basis for binding specificity (Fig. 6B).

Structural Features of Small Molecule Binding Sites within Myc

CD was used to monitor the effects of Myc-binding compounds on the secondary structure of Myc-derived polypeptides. While the effects of binding could not be detected using the full-length c-Myc bHLHZip domain, compounds 10058-F4 and 10074-G5 did cause significant changes in CD spectra of shorter peptides corresponding to Sites I and II, respectively.[6] It was further confirmed that small-molecule binding may occur simultaneously on all sites within full-length c-Myc bHLHZip.[7] NMR spectroscopy was also used to study Myc peptide/small molecule interactions.[6,7] Limited analysis of ^{1}H and natural abundance ^{13}C spectral parameters indicated that small molecule binding restricted peptide dynamics to a very limited extent. For example, $^{1}H\alpha$ secondary chemical shift values, which are a sensitive measure of secondary structure, of residues within Sites I-III in the context of synthetic peptides or full-length c-Myc bHLHZip were slightly but reproducibly perturbed in the presence of the respective, specific small molecule binding partner (Fig. 7A). In addition to chemical shift perturbations, a limited number of intermolecular, small molecule/Myc peptide ^{1}H-^{1}H NOEs were observed for several of the complexes. Computational methods were used to generate structural models of these complexes (Fig. 7B-G). It should be noted that these models are static views of complexes that are believed to be highly dynamic due to fluctuations amongst myriad conformations. Conformational fluctuations cause averaging of NMR parameters; thus, these structural models are likely to reflect average conformations of highly dynamic small molecule/peptide complexes. Despite these limitations, these models are useful for deriving initial insights into how small molecules interact with specific binding sites within the intrinsically disordered Myc polypeptide. In each of the models illustrated (Fig. 7E-G), side chains of hydrophobic residues cluster around hydrophobic portions of the small molecules. These apolar interactions are complemented by electrostatic and H-bond interactions. Thus, these results suggest that small molecules can bind sites in disordered proteins that are slightly enriched in hydrophobic residues and that the mechanism of binding involves, at least in part, hydrophobic collapse which is associated with partial polypeptide folding-upon-binding. The accompanying intermolecular electrostatic and H-bond interactions may mediate the specificity of the interactions. Importantly, these studies suggest that the polypeptide/small molecule complexes remain highly dynamic even as these interactions occur. This phenomenon is reminiscent of interactions between the IDP, Sic1 and the folded globular protein, CDC4. In this system, Sic1 remains highly dynamic when interacting with CDC4.[129,130]

CONCLUSION

It is now well recognized that IDPs are highly abundant and that they play critical functional roles in biological systems, with many mediating signaling and regulation. Bioinformatics studies have dramatically increased awareness of IDPs and their properties. However, structural, biophysical and biochemical studies of IDPs have progressed at a slower pace, creating gaps in our knowledge of the relationship between the physical properties of these proteins and their wide ranging biological functions. Our studies of

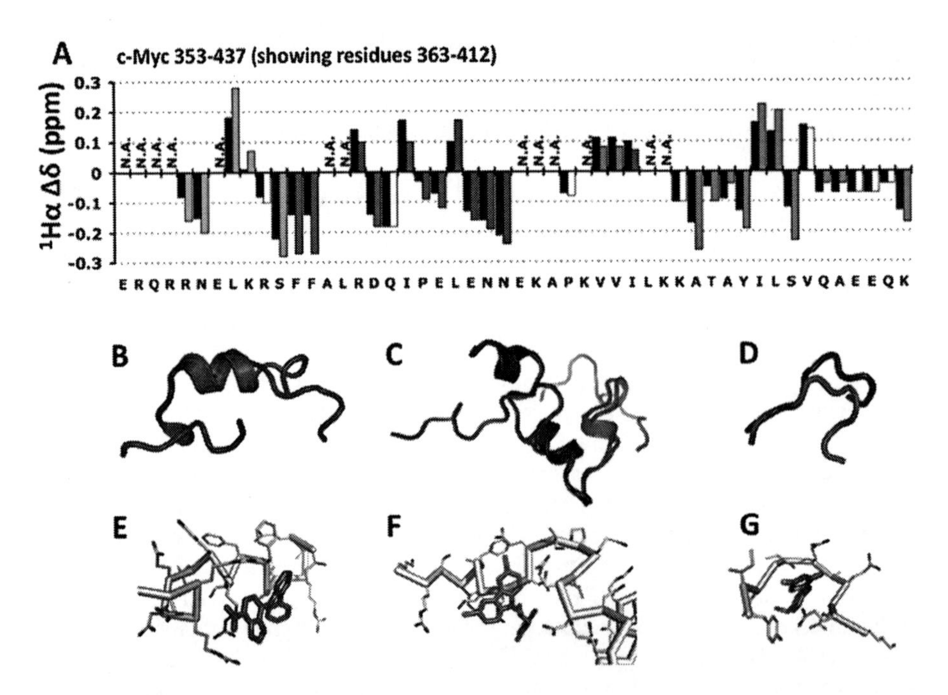

Figure 7. Simultaneous binding of multiple inhibitors to the full-length c-Myc bHLHZip domain. A) ^{1}Hα chemical shift index values for the partially assigned resonances of c-Myc353-437 (only the 363-412 segment is displayed). Values for the free peptide are represented by black bars while colored bars correspond to residues that display changes in chemical shift (>0.02 ppm) upon binding of each inhibitor according to the following scheme: green, 10074-G5; blue, 10074-A4; red, 10058-F4. Values for residues that remain unchanged upon complex formation are indicated as white bars. Secondary chemical shifts for the bound state were deduced from protein spectra acquired in the presence of all three inhibitors; the color-coding for this plot is deduced from the observed changes in chemical shift upon sequential addition of these compounds. Residues that displayed changes in chemical shift upon addition of both 10074-G5 and 10074-A4 (Phe374/375, Asp379 and Ile381) are indicated in shaded blue-green. B-D) Modeled average conformations of binding site peptides in the absence (dark grey) and presence (light grey) of their respective ligand (B: c-Myc363-381, 10074-G5; C: c-Myc370-409, 10074-A4; D: c-Myc402-412, 11058-F4). E-G). Models of each individual inhibitor-peptide complex (in the same order as above). (Adapted from Follis et al. Chem Biol 2008; 15:1149-1155,[6] ©2008 with permission from Elsevier; and Hammoudeh et al. J Am Chem Soc 2009; 131:7390-7401,[7] with permission of the American Chemical Society). A color version of this image is available at www.landesbioscience.com/curie.

the cell cycle regulators, p21 and p27, have revealed many interesting, often unexpected, results that have established important concepts that are likely to apply to many other uncharacterized IDPs. These concepts include the existence of both highly disordered and partially folded modules within IDPs prior to interacting with their biological targets; the participation of these modules in highly specific, sequential binding events; the role of posttranslational modification (PTM) of residues within these modules in regulating IDP function; and the dynamic linkage of multiple PTMs within individual IDPs into signaling conduits. While IDPs play critical biological roles in cells, their functions are often altered in association with human disease. The expression of IDPs is known to be tightly regulated in eukaryotes[23] and deregulation leading to elevated levels of IDPs is often associated with cancer in humans.[24] A notable example is the proto-oncoprotein Myc, which is overexpressed in a wide range of human cancers. Studies by Evan and coworkers[122] showed that inhibition of Myc function in mice either prevented formation of or cause regression of tumors in a lung tumor model. The work of Follis, Hammoudeh, Metallo and coworkers has demonstrated that it is possible to inhibit Myc function in cells with small molecules. The success of these studies provides incentive to pursue more potent small molecule Myc inhibitors in the future and to fully explore their clinical potential as antiproliferative agents. In conclusion, studies of IDPs have and will continue to provide novel insights into the molecular mechanism which govern protein functions and, in addition, hold great potential as druggable therapeutic targets to combat diseases such as cancer and neurodegenerative diseases.

ACKNOWLEDGEMENTS

Due to space limitations, we could not discuss or cite many important reports related to IDPs; we apologize to authors whose work was overlooked. The authors acknowledge members of the Kriwacki laboratory for stimulating discussions during the preparation of this manuscript. We gratefully acknowledge the American Lebanese Syrian Associated Charities (ALSAC) and NCI (2R01CA082491 to RWK; 5P30CA021765 to St. Jude Children's Research Hospital) for financial support.

REFERENCES

1. Receveur-Brechot V, Bourhis JM, Uversky VN et al. Assessing protein disorder and induced folding. Proteins 2006; 62:24-45.
2. Mittag T, Forman-Kay JD. Atomic-level characterization of disordered protein ensembles. Curr Opin Struct Biol 2007; 17:3-14.
3. Dyson HJ, Wright PE. Intrinsically unstructured proteins and their functions. Nat Rev Mol Cell Biol 2005; 6:197-208.
4. Obradovic Z, Peng K, Vucetic S et al. Exploiting heterogeneous sequence properties improves prediction of protein disorder. Proteins 2005; 61 Suppl 7:176-182.
5. Nair SK, Burley SK. X-ray structures of Myc-Max and Mad-Max recognizing DNA. Molecular bases of regulation by proto-oncogenic transcription factors. Cell 2003; 112:193-205.
6. Follis AV, Hammoudeh DI, Wang H et al. Structural rationale for the coupled binding and unfolding of the c-Myc oncoprotein by small molecules. Chem Biol 2008; 15:1149-1155.
7. Hammoudeh DI, Follis AV, Prochownik EV et al. Multiple independent binding sites for small molecule inhibitors on the c-Myc oncoprotein. J Am Chem Soc 2009; 131:7390-7401.
8. Yin X, Giap C, Lazo JS et al. Low molecular weight inhibitors of Myc-Max interaction and function. Oncogene 2003; 22:6151-6159.
9. Clark SS, McLaughlin J, Crist WM et al. Unique forms of the abl tyrosine kinase distinguish Ph1-positive CML from Ph1-positive ALL. Science 1987; 235:85-88.
10. Chu I, Sun J, Arnaout A et al. P27 Phosphorylation by src regulates inhibition of cyclin E-Cdk2. Cell 2007; 128:281-294.
11. Kiessling A, Sperl B, Hollis A et al. Selective inhibition of c-Myc/Max dimerization and DNA binding by small molecules. Chem Biol 2006; 13:745-751.
12. Dunker AK, Brown CJ, Lawson JD et al. Intrinsic disorder and protein function. Biochemistry 2002; 41:6573-6582.
13. Tompa P. Intrinsically unstructured proteins. Trends Biochem Sci 2002; 27:527-533.
14. Uversky VN. Natively unfolded proteins: a point where biology waits for physics. Protein Sci 2002; 11:739-756.
15. Oldfield CJ, Cheng Y, Cortese MS et al. Comparing and combining predictors of mostly disordered proteins. Biochemistry 2005; 44:1989-2000.

16. A list of currently available disorder predictors is available on the DisProt web site at http://www.disprot.org/predictors.php.
17. Dunker AK, Obradovic Z, Romero P et al. Intrinsic protein disorder in complete genomes. Genome Inform Ser Workshop Genome Inform 2000; 11:161-171.
18. Ward JJ, Sodhi JS, McGuffin LJ et al. Prediction and functional analysis of native disorder in proteins from the three kingdoms of life. J Mol Biol 2004; 337:635-645.
19. Xie H, Vucetic S, Iakoucheva LM et al. Functional anthology of intrinsic disorder. 1. Biological processes and functions of proteins with long disordered regions. J Proteome Res 2007; 6:1882-1898.
20. Vucetic S, Xie H, Iakoucheva LM et al. Functional anthology of intrinsic disorder. 2. Cellular components, domains, technical terms, developmental processes and coding sequence diversities correlated with long disordered regions. J Proteome Res 2007; 6:1899-1916.
21. Liu J, Perumal NB, Oldfield CJ et al. Intrinsic disorder in transcription factors. Biochemistry 2006; 45:6873-6888.
22. Iakoucheva LM, Brown CJ, Lawson JD et al. Intrinsic disorder in cell-signaling and cancer-associated proteins. J Mol Biol 2002; 323:573-584.
23. Gsponer J, Futschik ME, Teichmann SA et al. Tight regulation of unstructured proteins: from transcript synthesis to protein degradation. Science 2008; 322:1365-1368.
24. Vavouri T, Semple JI, Garcia-Verdugo R et al. Intrinsic protein disorder and interaction promiscuity are widely associated with dosage sensitivity. Cell 2009; 138:198-208.
25. Ayed A, Mulder FA, Yi GS et al. Latent and active p53 are identical in conformation. Nat Struct Biol 2001; 8:756-760.
26. Hoh JH. Functional protein domains from the thermally driven motion of polypeptide chains: a proposal. Proteins 1998; 32:223-228.
27. Tskhovrebova L, Trinick J. Titin: properties and family relationships. Nat Rev Mol Cell Biol 2003; 4:679-689.
28. Alber F, Dokudovskaya S, Veenhoff LM et al. The molecular architecture of the nuclear pore complex. Nature 2007; 450:695-701.
29. Spolar RS, Record MT Jr. Coupling of local folding to site-specific binding of proteins to DNA. Science 1994; 263:777-784.
30. Demarest SJ, Martinez-Yamout M, Chung J et al. Mutual synergistic folding in recruitment of CBP/p300 by p160 nuclear receptor coactivators. Nature 2002; 415:549-553.
31. Lacy ER, Filippov I, Lewis WS et al. P27 binds cyclin-CDK complexes through a sequential mechanism involving binding-induced protein folding. Nat Struct Mol Biol 2004; 11:358-364.
32. Tompa P, Fuxreiter M. Fuzzy complexes: polymorphism and structural disorder in protein-protein interactions. Trends Biochem Sci 2008; 33:2-8.
33. Frederick KK, Marlow MS, Valentine KG et al. Conformational entropy in molecular recognition by proteins. Nature 2007; 448:325-329.
34. Leung DW, Rosen MK. The nucleotide switch in Cdc42 modulates coupling between the GTPase-binding and allosteric equilibria of Wiskott-Aldrich syndrome protein. Proc Natl Acad Sci USA 2005; 102:5685-5690.
35. Wright PE, Dyson HJ. Intrinsically unstructured proteins: re-assessing the protein structure-function paradigm. J Mol Biol 1999; 293:321-331.
36. Kriwacki RW, Hengst L, Tennant L et al. Structural studies of p21(waf1/cip1/sdi1) in the free and Cdk2-bound state: Conformational disorder mediates binding diversity. Proc Natl Acad Sci USA 1996; 93:11504-11509.
37. Besson A, Dowdy SF, Roberts JM. CDK inhibitors: cell cycle regulators and beyond. Dev Cell 2008; 14:159-169.
38. Joerger AC, Fersht AR. Structural Biology of the Tumor Suppressor p53. Annu Rev Biochem 2008; 77:557-582.
39. Dunker AK, Cortese MS, Romero P et al. Flexible nets. The roles of intrinsic disorder in protein interaction networks. Febs J 2005; 272:5129-5148.
40. Kim PM, Sboner A, Xia Y et al. The role of disorder in interaction networks: a structural analysis. Mol Syst Biol 2008; 4:179-185.
41. Schnell S, Fortunato S, Roy S. Is the intrinsic disorder of proteins the cause of the scale-free architecture of protein-protein interaction networks? Proteomics 2007; 7:961-964.
42. Bode AM, Dong Z. Post-translational modification of p53 in tumorigenesis. Nat Rev Cancer 2004; 4:793-805.
43. Grimmler M, Wang Y, Mund T et al. Cdk-inhibitory activity and stability of p27(Kip1) are directly regulated by oncogenic tyrosine kinases. Cell 2007; 128:269-280.
44. Tompa P, Prilusky J, Silman I et al. Structural disorder serves as a weak signal for intracellular protein degradation. Proteins 2007; 71:903-909.
45. Tsvetkov P, Asher G, Paz A et al. Operational definition of intrinsically unstructured protein sequences based on susceptibility to the 20S proteasome. Proteins 2007; 70:1357-1366.
46. Lavin MF, Gueven N. The complexity of p53 stabilization and activation. Cell Death Differ 2006; 13:941-950.
47. Tsvetkov P, Reuven N, Shaul Y. The nanny model for IDPs. Nat Chem Biol 2009; 5:778-781.

48. Shoemaker BA, Portman JJ, Wolynes PG. Speeding molecular recognition by using the folding funnel: the fly-casting mechanism. Proc Natl Acad Sci USA 2000; 97:8868-8873.
49. Sugase K, Dyson HJ, Wright PE. Mechanism of coupled folding and binding of an intrinsically disordered protein. Nature 2007; 447:1021-1025.
50. Huang Y, Liu Z. Kinetic advantage of intrinsically disordered proteins in coupled folding-binding process: a critical assessment of the "fly-casting" mechanism. J Mol Biol 2009; 393:1143-1159.
51. Hilser VJ, Thompson EB. Intrinsic disorder as a mechanism to optimize allosteric coupling in proteins. Proc Natl Acad Sci USA 2007; 104:8311-8315.
52. Morgan DO. Principles of CDK regulation. Nature 1995; 374:131-134.
53. Sherr CJ, Roberts JM. Cdk inhibitors: positive and negative regulators of G1-phase progression. Genes Dev 1999; 13:1501-1512.
54. Sherr CJ, Roberts JM. Living with or without cyclins and cyclin-dependent kinases. Genes Dev 2004; 18:2699-2711.
55. Poon RY, Hunter T. Expression of a novel form of p21Cip1/Waf1 in UV-irradiated and transformed cells. Oncogene 1998; 16:1333-1343.
56. Reynisdottir I, Massague J. The subcellular locations of p15(Ink4b) and p27(Kip1) coordinate their inhibitory interactions with cdk4 and cdk2. Genes Dev 1997; 11:492-503.
57. Waga S, Hannon GJ, Beach D et al. The p21 inhibitor of cyclin-dependent kinases controls DNA replication by interaction with PCNA. Nature 1994; 369:574-578.
58. Watanabe H, Pan ZQ, Schreiber-Agus N et al. Suppression of cell transformation by the cyclin-dependent kinase inhibitor p57KIP2 requires binding to proliferating cell nuclear antigen. Proc Natl Acad Sci USA 1998; 95:1392-1397.
59. Montagnoli A, Fiore F, Eytan E et al. Ubiquitination of p27 is regulated by Cdk-dependent phosphorylation and trimeric complex formation. Genes Dev 1999; 13:1181-1189.
60. Nguyen H, Gitig DM, Koff A. Cell-free degradation of p27(kip1), a G1 cyclin-dependent kinase inhibitor, is dependent on CDK2 activity and the proteasome. Mol Cell Biol 1999; 19:1190-1201.
61. Kamura T, Hara T, Kotoshiba S et al. Degradation of p57Kip2 mediated by SCFSkp2-dependent ubiquitylation. Proc Natl Acad Sci USA 2003; 100:10231-10236.
62. Matsuoka S, Edwards MC, Bai C et al. P57KIP2, a structurally distinct member of the p21CIP1 Cdk inhibitor family, is a candidate tumor suppressor gene. Genes Dev 1995; 9:650-652.
63. Russo AA, Jeffrey PD, Patten AK et al. Crystal structure of the p27Kip1 cyclin-dependent-kinase inhibitor bound to the cyclin A-Cdk2 complex. Nature 1996; 382:325-331.
64. Harper JW, Adams PD. Cyclin-dependent kinases. Chem Rev 2001; 101:2511-2526.
65. Lacy ER, Wang Y, Post J et al. Molecular Basis for the Specificity of p27 Toward Cyclin-dependent Kinases that Regulate Cell Division. J Mol Biol 2005; 349:764-773.
66. Adkins JN, Lumb KJ. Intrinsic structural disorder and sequence features of the cell cycle inhibitor p57Kip2. Proteins 2002; 46:1-7.
67. Bienkiewicz EA, Adkins JN, Lumb KJ. Functional consequences of preorganized helical structure in the intrinsically disordered cell-cycle inhibitor p27(Kip1). Biochemistry 2002; 41:752-759.
68. Prilusky J, Felder CE, Zeev-Ben-Mordehai T et al. FoldIndex: a simple tool to predict whether a given protein sequence is intrinsically unfolded. Bioinformatics 2005; 21:3435-3438.
69. Dosztanyi Z, Csizmok V, Tompa P et al. IUPred: web server for the prediction of intrinsically unstructured regions of proteins based on estimated energy content. Bioinformatics 2005; 21:3433-3434.
70. Romero P, Obradovic Z, Li X et al. Sequence complexity of disordered protein. Proteins 2001; 42:38-48.
71. Harper JW, Elledge S, Keyomarsi K et al. Inhibition of cyclin-dependent kinases by p21. Mol Biol Cell 1995; 6:387-400.
72. Polyak K, Lee MH, Erdjument-Bromage H et al. Cloning of p27Kip1, a cyclin-dependent kinase inhibitor and a potential mediator of extracellular antimitogenic signals. Cell 1994; 78:59-66.
73. Hengst L, Dulic V, Slingerland JM et al. A cell cycle-regulated inhibitor of cyclin-dependent kinases. Proc Natl Acad Sci USA 1994; 91:5291-5295.
74. Sivakolundu SG, Bashford D, Kriwacki RW. Disordered p27(Kip1) Exhibits Intrinsic Structure Resembling the Cdk2/Cyclin A-bound Conformation. J Mol Biol 2005; 353:1118-1128.
75. Galea CA, Nourse A, Wang Y et al. Role of intrinsic flexibility in signal transduction mediated by the cell cycle regulator, p27(Kip1). J Mol Biol 2007; 376:827-838.
76. Blain SW, Massague J. Breast cancer banishes p27 from nucleus. Nat Med 2002; 8:1076-1078.
77. Chu IM, Hengst L, Slingerland JM. The Cdk inhibitor p27 in human cancer: prognostic potential and relevance to anticancer therapy. Nat Rev Cancer 2008; 8:253-267.
78. Prakash S, Tian L, Ratliff KS et al. An unstructured initiation site is required for efficient proteasome-mediated degradation. Nat Struct Mol Biol 2004; 11:830-837.
79. Prochownik EV. c-Myc: linking transformation and genomic instability. Curr Mol Med 2008; 8:446-458.
80. Meyer N, Penn LZ. Reflecting on 25 years with MYC. Nat Rev Cancer 2008; 8:976-990.

81. Eilers M, Eisenman RN. Myc's broad reach. Genes Dev 2008; 22:2755-2766.
82. Bahram F, von der Lehr N, Cetinakaya C et al c-Myc hot spot mutations in lymphomas result in inefficient ubiquitination and decreased proteasome-mediated turnover. Blood 2000; 95:2104-2110.
83. Neel BG, Hayward WS, Robiinson HL et al. Avian leukosis virus-induced tumors have common proviral intregration sites and synthesize discrete new RNAs: oncogenesis by promoter. Cell 1981; 23:323-334.
84. Steffen D. Proviruses are adjacent to c-myc in some murine leukemia virus-induced lymphomas. Proc Natl Acad Sci USA 1984; 81:2097-2101.
85. Crews S, Barth R, Hood L et al. Mouse c-myc oncogene is located on chromosome 15 and traslocated to chromosome 12 in plasmacytomas. Science 1982; 218:1319-1321.
86. Boxer LM, Dang CV. Translocations involving c-myc and c-myc function. Oncogene 2001; 20:5595-5610.
87. Collins S, Groudine M. Amplification of endogenous myc-related DNA sequences in a human myeloid leukaemia cell line. Nature 1982; 298:679-681.
88. Alitalo K, Shwab M, Lin CC et al. Homogeneously staining chromosomal regions contain amplified copies of an abundantly expressed cellular oncogene (c-myc) in malignant neuroendocrine cells from a human colon carcinoma. Proc Natl Acad Sci USA 1983; 80:1707-1711.
89. Leder A, Pattengale PK, Kuo A et al. Consequences of widespread deregulation of the c-myc gene in transgenic mice: multiple neoplasms and normal development. Cell 1986; 45:485-495.
90. Weng AP, Millholland JM, Yashiro-Ohtani Y et al. C-Myc is an important direct target of Notch1 in T-cell acute lymphoblastic leukemia/lymphoma. Genes Dev 2006; 20:2096-2109.
91. Hann SR. Role of posttranslational modifications in regulating c-Myc proteolysis, transcriptional activity and biological function. Sem Cancer Biol 2006; 16:288-302.
92. Vervoorts J, Luscher-Firzlaff JM, Luscher B. The ins and outs of MYC regulation by posttranslational mechanisms. J Biol Chem 2006; 281:34725-34729.
93. Dang CV. c-Myc target genes involved in cell growth, apoptosis and metabolism. Mol Cell Biol 1999; 19:1-11.
94. Dang CV, O'Donnell KA, Zeller KI et al. The c-Myc target gene network. Sem Cancer Biol 2006; 16:253-264.
95. Obaya AJ, Mateyak MK, Sedivy JM. Mysterious liaisons: the relationship between c-Myc and the cell cycle. Oncogene 1999; 18:2934-2941.
96. Luscher B, Eisenman RN. New light on Myc and Myb. Part I. Myc. Genes Dev 1990; 4:2025-2035.
97. Marcu KB, Bossone SA, Patel AJ. Myc function and regulation. Annu Rev Biochem 1992; 61:809-860.
98. Grandori C, Gomez-Roman N, Felton-Edkins ZA et al c-Myc binds to human ribosomal DNA and stimulates transcription of rRNA genes by RNA polymerase I. Nat Cell Biol 2005; 7:311-318.
99. Stone J, de Lange T, Ramsay G et al. Definition of regions in human c-myc that are involved in transformation and nuclear localization. Mol Cell Biol 1987; 7:1697-1709.
100. Kato GJ, Barrett J, Villa-Garcia M et al. An amino-terminal c-myc domain required for neoplastic transformation activates transcription. Mol Cell Biol 1990; 10:5914-5920.
101. Blackwell TK, Kretzner L, Blackwood EM et al. Sequence-specific DNA binding by the c-Myc protein. Science 1990; 250:1149-1151.
102. McMahon SB, Wood MA, Cole MD. The essential cofactor TRRAP recruits the histone acetyltransferase hGCN5 to c-Myc. Mol Cell Biol 2000; 20:556-562.
103. Grace Cheng SW, Davles KP, Yung E et al. C-MYC interacts with INI1/hSNF5 and requires the SWI/SNF complex for transactivation function. Nat Genet 1999; 22:102-105.
104. Eberhardy SR, Farnham PJ. C-Myc mediates activation of the cad promoter via a postRNA polymerase II recruitment mechanism. J Biol Chem 2001; 276:48562-48571.
105. Eberhardy SR, Farnham PJ. Myc recruits P-TEFb to mediate the final step in the transcriptional activation of the cad promoter. J Biol Chem 2002; 277:40156-40162.
106. Blackwood EM, Eisenman RN. Max: a helix-loop-helix zipper protein that forms a sequence-specific DNA-binding complex with Myc. Science 1991; 251:1211-1217.
107. Blackwood EM, Kretzner L, Eisenman RN. Myc and Max function as a nucleoprotein complex. Curr Opin Genet Dev 1992; 2:227-235.
108. Amati B, Brooks MW, Levy N et al. Oncogenic activity of the c-Myc protein requires dimerization with Max. Cell 1993; 72:233-245.
109. Lavigne P, Crump MP, Gagne SM et al. Insights into the mechanism of heterodimerization from the 1H-NMR solution structure of the c-Myc-Max heterodimeric leucine zipper. J Mol Biol 1998; 281:165-181.
110. Mao DY, Watson JD, Yan PS et al. Analysis of Myc bound loci identified by CpG island arrays shows that Max is essential for Myc-dependent repression. Curr Biol 2003; 13:882-886.
111. Wang YH, Liu S, Zhang G et al. Knockdown of c-Myc expression by RNAi inhibits MCF-7 breast tumor cells growth in vitro and in vivo. Breast Cancer Res 2005; 7:R220-R228.
112. Balaji KC, Koul H, Mitra S et al. Antiproliferative effects of c-myc antisense oligonucleotide in prostate cancer cells: a novel therapy in prostate cancer. Urology 1997; 50:1007-1015.

113. Siddiqui-Jain A, Grand CL, Bearss DJ et al. Direct evidence for a G-quadruplex in a promoter region and its targeting with a small molecule to repress c-MYC transcription. Proc Natl Acad Sci USA 2002; 99:11593-11598.
114. Mo H, Henriksson M. Identification of small molecules that induce apoptosis in a Myc-dependent manner and inhibit Myc-driven transformation. Proc Natl Acad Sci USA 2006; 103:6344-6349.
115. Claasen G, Brin E, Crogan-Grundy C et al. Selective activation of apoptosis by a novel set of 4-aryl-3-(3-aryl-1-oxo-2- propenyl)-2(1H)-quinolinones through a Myc-dependent pathway. Cancer Lett 2008; 274:243-249.
116. Giorello L, Clerico L, Pescarolo MP et al. Inhibition of cancer cell growth and c-Myc transcriptional activity by a c-Myc helix 1-type peptide fused to an internalization sequence. Cancer Res 1998; 58:3654-3659.
117. Berg T, Cohen SB, Desharnais J et al. Small-molecule antagonists of Myc/Max dimerization inhibit Myc-induced transformation of chicken embryo fibroblasts. Proc Nat Acad Sci USA 2002; 99:3830-3835.
118. Xu Y, Shi J, Yamamoto N et al. A credit-card library approach for disrupting protein-protein interactions. Bioorg Med Chem 2006; 14:2660-2673.
119. Kiessling A, Wiesinger R, Sperl B et al. Selective inhibition of c-Myc/Max dimerization by a pyrazolo[1,5-a] pyrimidine. ChemMedChem 2007; 2:627-630.
120. Prochownik EV. c-Myc as a therapeutic target in cancer. Expert Rev Anticanc 2004; 4:289-302.
121. Ponzielli R, Katz S, Barsyte-Lovejoy D et al. Cancer therapeutics: targeting the dark side of Myc. Eur J Cancer 2005; 41:2485-2501.
122. Soucek L, Whitfield J, Martins CP et al. Modelling Myc inhibition as a cancer therapy. Nature 2008; 455:679-683.
123. Soucek L, Helmer-Citterich M, Sacco A et al. Design and properties of a Myc derivative that efficiently homodimerizes. Oncogene 1998; 17:2463-2472.
124. Iakoucheva LM, Brown CJ, Lawson JD et al. Intrinsic disorder in cell-signaling and cancer-associated proteins. J Mol Biol 2002; 323:573-584.
125. Wang H, Hammoudeh DI, Follis AV et al. Improved low molecular weight Myc-Max inhibitors. Mol Cancer Ther 2007; 6:2399-2408.
126. Mustata G, Follis AV, Hammoudeh DI et al. Discovery of novel myc-max heterodimer disruptors with a 3-dimensional pharmacophore model. J Med Chem 2009; 52:1247-1250.
127. Uversky VN, Gillespie JR, Fink AL. Why are "natively unfolded" proteins unstructured under physiologic conditions? Proteins 2000; 41:415-427.
128. Uversky VN, Oldfield CJ, Dunker AK. Showing your ID: intrinsic disorder as an ID for recognition, regulation and cell signaling. J Mol Recognit 2005; 18:343-384.
129. Mittag T, Orlicky S, Choy WY et al. Dynamic equilibrium engagement of a polyvalent ligand with a single-site receptor. Proc Natl Acad Sci USA 2008; 105:17772-17777.
130. Mittag T, Kay LE, Forman-Kay JD. Protein dynamics and conformational disorder in molecular recognition. J Mol Recognit 2010; 23:105-116.

CHAPTER 4

INTERPLAY BETWEEN PROTEIN ORDER, DISORDER AND OLIGOMERICITY IN RECEPTOR SIGNALING

Alexander B. Sigalov
SignaBlok, Inc., Shrewsbury, Massachusetts, USA
Email: sigalov@signablok.com

Abstract: Receptor-mediated signaling plays an important role in health and disease. Recent reports have revealed that many proteins that do not adopt globular structures under native conditions, thus termed intrinsically disordered, are involved in cell signaling. Intriguingly, physiologically relevant oligomerization of intrinsically disordered proteins (IDPs) has been recently observed and shown to exhibit unique biophysical characteristics, including the lack of significant changes in chemical shift and peak intensity upon binding. On the other hand, ligand-induced or -tuned receptor oligomerization is known to be a general feature of various cell surface receptors and to play a crucial role in signal transduction. In this work, I summarize several distinct features of protein disorder that are especially important as related to signal transduction. I also hypothesize that interactions of IDPs with their protein or lipid partners represent a general biphasic process with the electrostatic-driven "no disorder-to-order" fast interaction which, depending on the interacting partner, may or may not be accompanied by the hydrophobic-driven slow formation of a secondary structure. Further, I suggest signaling-related functional connections between protein order, disorder and oligomericity and hypothesize that receptor oligomerization induced or tuned upon ligand binding outside the cell is translated across the membrane into protein oligomerization inside the cell, thus providing a general platform, the Signaling Chain HOmoOLigomerization (SCHOOL) platform, for receptor-mediated signaling. This structures our current multidisciplinary knowledge and views of the mechanisms governing the coupling of recognition to signal transduction and cell response. Importantly, this approach not only reveals previously unrecognized striking similarities in the basic mechanistic principles of function of numerous functionally diverse and unrelated surface membrane receptors, but also suggests the similarity between therapeutic targets, thus opening new horizons for both fundamental and clinically relevant studies.

Fuzziness: Structural Disorder in Protein Complexes, edited by Monika Fuxreiter and Peter Tompa.

INTRODUCTION

Cell surface receptors are integral membrane proteins and, as such, consist of three basic domains: Extracellular (EC) ligand-binding domains, transmembrane (TM) domains and cytoplasmic (CYTO) signaling (or effector) domains. Upon recognition and binding of a specific ligand, cell surface receptors transmit this information into the interior of the cell, activating intracellular signaling pathways and resulting in a cellular response such as proliferation, differentiation, apoptosis, degranulation, the secretion of preformed and newly formed mediators, phagocytosis of particles, endocytosis, cytotoxicity against target cells, etc. The importance of receptors in health and disease[1,2] makes the molecular understanding of signal transduction critical in influencing and controlling this process, thus modulating the cell response.

Ligand-induced receptor oligomerization is frequently employed as a key factor in receptor triggering.[2-4] For many receptors, oligomerization is mediated by homointeractions between folded and well-ordered domains, representing a signaling-related functional link between protein order and oligomericity. On the other hand, intrinsic disorder serves as the native and functional state for many signaling proteins[5] with phosphorylation, one of the critical and obligatory events in cell signaling, occurring predominantly within intrinsically disordered regions (IDRs).[6] In addition, long IDRs preferentially reside on the CYTO side of many human TM proteins.[7,8] In this context, the recently reported surprising ability of intrinsically disordered CYTO domains of immune receptor signaling subunits to form specific dimers[9-11] represents a functional link between protein intrinsic disorder and oligomericity. This phenomenon resolves a long-standing puzzle of receptor-mediated signaling and has important fundamental and clinical applications.

Here, I summarize our knowledge on the recently reported distinct features of signaling-related intrinsically disordered proteins (IDPs), including the lack of folding upon binding to protein and lipid partners. I also hypothesize that receptor oligomerization induced or tuned upon ligand binding outside the cell is translated across the membrane into protein oligomerization inside the cell, thus providing a general platform, the Signaling Chain HOmoOLigomerization (SCHOOL) platform, for receptor-mediated signaling. I also demonstrate how our improved understanding of the recently suggested functional link between protein intrinsic disorder and oligomericity as a key and missing element of transmembrane signal transduction provides novel insight into the molecular mechanisms of cell signaling and has important applications in biology and medicine.

INTRINSICALLY DISORDERED PROTEINS

By definition, IDPs are proteins that lack a well-defined ordered structure under physiological conditions in vitro.[12] To predict whether a given protein or protein region assumes a defined fold or is intrinsically disordered, several computational methods have been developed.[12-18] Experimentally, protein disorder can be detected by far-UV circular dichroism (CD) and nuclear magnetic resonance (NMR) spectroscopy. CD spectroscopy allows the estimation of the secondary structure content of a protein in solution. However, while for an ordered protein the CD signal gives information about each molecule in the sample, because nearly all the molecules are in the same structural state, it is different for an IDP that consists of a broad ensemble of molecules each having a different conformation.[19] In this context, NMR is unparalleled in its ability to provide detailed structural and dynamic information on IDPs and has emerged as a particularly important tool for studies of IDP folding and interactions.[20,21] NMR chemical shifts and line widths are extremely sensitive to subtle changes in protein conformational ensembles and are indispensable for detecting protein disorder (poor proton chemical shift dispersion and broad lines are indicative of disorder) and determining propensities of secondary structure formation on a residue-by-residue basis in unfolded and partly folded proteins.

Despite the fact that the existence of IDPs and IDRs has been recognized for many years, their functional role in crucial areas such as transcriptional regulation, translation and cellular signal transduction has only recently been reported due to progress in biochemical methodology.[22-28] IDPs and IDRs are involved in various signaling and regulatory pathways, via specific protein-protein, protein-nucleic acid and protein-ligand interactions.[22,24-26,28-32] Many of those post-translational modifications that rely on the low affinity, high specificity binding interactions (for example, phosphorylation) are associated primarily with IDPs and IDRs.[6,25,27]

The major functional benefits of protein intrinsic disorder include increased binding specificity at the expense of thermodynamic stability, increased speed of interaction and the ability to recognize and bind multiple distinct partners without sacrificing specificity.[22,23,27,28,31,32]

Binding with Folding

The generally accepted view is that upon binding to their interacting partners and targets, IDPs undergo transitions to more ordered states or fold into stable secondary or tertiary structures—that is, they undergo coupled binding and folding processes (Fig. 1A).[23,33-36]

Protein Partner

Currently, the most characterized examples of folding being driven by binding are protein complexes formed by IDPs with their folded (ordered) protein partners (Fig. 1A). This subject has been addressed in detail in many recent reviews and other articles.[19,21-23,28,30,32,34] A classic example is binding of the kinase-inducible transcriptional-activation domain (KID) of cyclic AMP response element-binding protein (CREB) to CREB-binding protein (CBP). Upon binding to CBP, the intrinsically disordered KID polypeptide[37,38] folds with the formation a pair of orthogonal helices.[39] Coupled binding and folding can involve just a few residues[39-43] or an entire protein domain.[44]

Specific complex formation between IDPs is quite unusual, but not unprecedented.[45] Homodimerization of IDPs was first reported in 2004 for a novel class of signaling-related IDPs[9] and later confirmed for other IDPs[46-49] extending the phenomenon to different classes of IDPs and suggesting physiological relevance. It should be noted that in most of these studies, dimerization is accompanied by a mutual or "synergistic"[45] folding of two IDP molecules at the interaction interface (Fig. 1A). Thus, interactions between the constituents of such homodimers represent specific interactions between folded regions involved in complex formation.

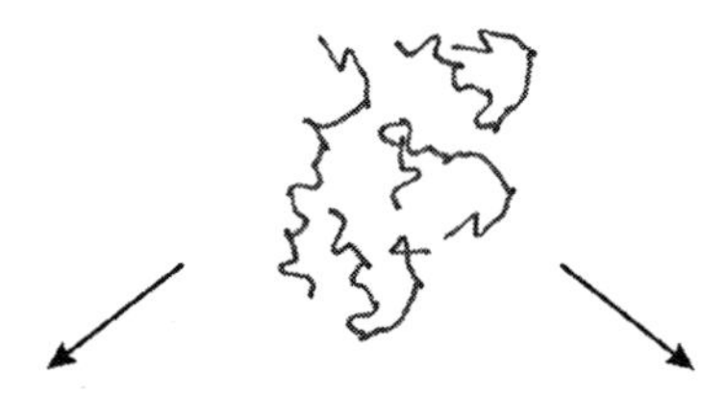

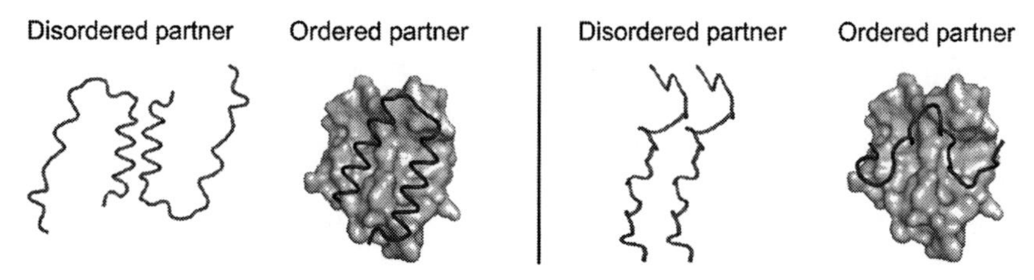

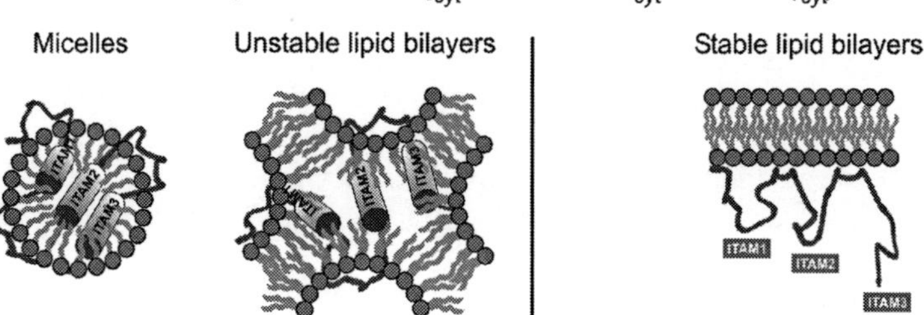

Figure 1. Binding of an intrinsically disordered protein (IDP) to its target. Upon binding, an IDP may (A) or may not (B) fold on its interacting partner. A) Binding of an IDP to another IDP or to a folded and well-ordered protein is coupled to secondary structure induction (left upper panel). In this widely observed scenario, interactions at the binding interface that mediate specific complex formation represent the interactions between folded domains (induction of helices is shown for illustrative purposes). Helical structure induction upon binding of an IDP to micelles or unstable lipid bilayers (left bottom panel) has been also widely reported.[35,36,51,52] B) Binding of an IDP to another IDP or to a folded and well-ordered protein is not accompanied by a disorder-to-order transition (right upper panel). In the recently discovered surprising scenario of IDP homooligomerization,[9,11] interactions at the binding interface that mediate a specific oligomer formation represent the interactions between disordered domains making these interactions unusual and intriguing. The lack of secondary structure induction upon binding of an IDP to a folded and well-ordered protein has been also reported.[58] Binding of an IDP to stable lipid bilayers without a disorder-to-order transition (right bottom panel) has been also demonstrated.[51] Images were created using PyMol (www.pymol.org) from PDB file 1AVV for the HIV-1 Nef core domain (shown as an example of a folded interacting partner) and using arbitrary idealized structural elements to represent the ensemble of unfolded conformations of an IDP. Polar head groups and hydrophobic tales of detergent and lipid molecules are denoted by gray filled circles and lines, respectively. Adapted from Sigalov AB. Mol Biosyst 2010; 6(3):451-61;[50] ©2010, with permission of The Royal Society of Chemistry; and from Sigalov AB et al. Biochem Biophys Res Commun 2009; 389:388-393;[51] ©2009, with permission from Elsevier.

Lipid Partner

Prevalence of IDRs in the CYTO domains of many human TM proteins in general,[7,8] and in particular, in the CYTO domains of signaling-related proteins (Table 1),[9,10,50] raises the question if these regions exert membrane-binding activity and if affirmed, whether this activity has a physiological role. Recent studies of the intrinsically disordered CYTO domains of the ζ and CD3ε signaling subunits (ζ_{cyt} and CD3ε_{cyt}, respectively) of the T-cell receptor (TCR) have demonstrated that these proteins bind to acidic dimyristoylphosphatidylglycerol (DMPG) vesicles and undergo a helical folding transition upon binding.[35,36] ζ_{cyt} and CD3ε_{cyt} contain an immunoreceptor tyrosine-based activation motif (ITAM), tyrosines of which are phosphorylated upon receptor triggering and the authors[35,36] hypothesized that helical folding of ITAMs upon membrane binding represents a conformational switch to control TCR activation.

On the other hand, it has been shown that binding of ζ_{cyt} and CD3ε_{cyt} as well as ITAM-containing CYTO domain of FcεRI receptor γ subunit, FcεRγ_{cyt}, to acidic phospholipid vesicles is mediated by clusters of positively charged amino acids but not the ITAM residues.[10,35,51] In contrast, helical folding of ITAMs observed in the presence of DMPG vesicles[35,36,52] can be also promoted by physiologically irrelevant helical inducers such as trifluoroethanol and detergents.[35,36,52,53] This supports a hypothesis that the association of these IDPs with negatively charged membranes is a biphasic process with a fast rate for an electrostatic-driven protein-liposome interaction and a slow rate for the hydrophobic-driven formation of an amphipathic helix.[10,51] Similar biphasic process has been reported recently for binding of the HIV-1 Nef protein to model membranes.[54]

Lipid bilayers are self-assembled structures, the mechanical properties of which are derived from noncovalent forces such as the hydrophobic effect, steric forces and electrostatic interactions. In this context, the electrostatic force is of special interest because biological membranes are rich in anionic lipids[55] and are therefore charged in aqueous solution. Importantly, electrostatic interactions play the critical role in membrane stability.[56] Thus, considering that net charges of ζ_{cyt}, CD3ε_{cyt} and FcεRγ_{cyt} are +5, +11 and +3 (Table 1), respectively, binding of these proteins to acidic phospholipids can potentially destabilize and disrupt lipid bilayers.

In lipid binding studies of ζ_{cyt} and CD3ε_{cyt},[35,36] the authors used DMPG vesicles to mimic the cell membrane. However, the lipid bilayers of these vesicles are not sufficiently stable and fuse upon binding ζ_{cyt}, as recently confirmed by dynamic light scattering (DLS) and electron microscopy (EM) experiments,[51] suggesting that the observed hydrophobic-driven helical folding of the ITAMs of ζ_{cyt}, CD3ε_{cyt} and FcεRγ_{cyt} likely occurs in the membrane stalk intermediates of fusion (Fig. 1A) and is similar to that promoted by physiologically irrelevant helical inducers such as trifluoroethanol and detergents (Fig. 1A).[35,36,51-53] This not only questions the utility of micelles, detergents and unstable lipid vesicles (i.e., DMPG vesicles) as an appropriate model of the cell membrane but also highlights the importance of ensuring the integrity of model membranes upon protein binding in protein-lipid experiments in general and in particular, in studies of IDP-lipid interactions.

Binding without Folding

IDPs are often referred to as "remaining predominantly disordered" or "largely unfolded" upon dimerization or interaction with other proteins or lipids,[35,37-43,46-49] meaning that the protein regions flanking the interaction interface but not the interface

Table 1. Summary of disorder[a] and secondary structure[b] predictions for cytoplasmic domains of MIRR signaling subunits

Parameter	Protein								
	ζ	CD3ε	CD3δ	CD3γ	Igα	Igβ	FcεRIγ	DAP10	DAP12
net charge[c]	+5	+11	0	0	−9	−10	+3	+3	+3
$\lvert\langle R\rangle\rvert$[d]	0.044	0.193	0.000	0.000	0.143	0.196	0.068	0.125	0.058
$\langle H\rangle_b$[e]	0.429	0.483	0.413	0.413	0.465	0.484	0.438	0.458	0.434
$\langle H\rangle$[f]	0.349	0.315	0.374	0.317	0.389	0.386	0.380	0.357	0.395
$\langle H\rangle_b - \langle H\rangle$[g]	0.080	0.168	0.039	0.096	0.075	0.098	0.058	0.101	0.039
Results of secondary structure prediction[h,i]									
Alpha helix	40.87	14.04	19.57	2.17	22.22	0	25.00	0	19.23
Extended strand	5.22	0	0	0	0	21.57	18.18	20.83	3.85
Random coil	53.91	85.96	80.43	97.83	77.78	78.43	56.82	79.17	76.92

[a] Using the algorithm of Uversky et al.[12]
[b] Using the hierarchical neural network algorithm.[18]
[c] Using the Swiss Institute of Bioinformatics (SIB) server ExPASy.[142]
[d] The mean net charge, defined by Uversky et al[12] as the absolute value of the difference between the numbers of positively and negatively charged residues at pH 7.0, divided by the total residue number.
[e] The boundary $\langle H\rangle$ value, calculated using the Uversky equation $\langle H\rangle_b = (\lvert\langle R\rangle\rvert + 1.151)/2.785$.
[f] The mean hydrophobicity, defined using the Kyte/Doolittle scale,[143] as the sum of all residue hydrophobicities, divided by the total number of residues and rescaled to a range of 0-1.
[g] The positive difference between $\langle H\rangle_b$ and $\langle H\rangle$ indicates that a protein is unfolded.
[h] The values are indicated in %.
[i] No significant fraction was predicted for 3_{10} helix, Pi helix, beta bridge, beta turn, bend region, ambiguous and other states.

itself remain disordered. Recently, it has been suggested to term this mode of interaction "the flanking fuzziness" in contrast to "the random fuzziness" when the IDP remains entirely disordered in the bound state.[57] In this context, "the flanking fuzziness" is a part of the "coupled binding and folding" paradigm. Recent studies of a novel class of IDPs demonstrated that binding of IDPs is not necessarily accompanied by a disorder-to-order transition even within the interaction interface,[9-11,51,58] thus going beyond the classical paradigm. The latter proteins are directly involved in receptor-mediated signaling, which makes these findings particularly interesting and important.

Protein Partner

Little structural information is available for complex formation of IDPs with disordered or ordered partners that is not accompanied by a disorder-to-order transition both outside and within the interaction interface (Fig. 1B). First example of this unusual phenomenon was reported in 2004[9] when using a variety of biophysical and biochemical techniques, the ITAM-containing CYTO domains of immune receptor signaling subunits namely, TCRζ_{cyt}, CD3ε_{cyt}, CD3δ_{cyt} and CD3γ_{cyt}, B-cell antigen receptor Igα_{cyt} and Igβ_{cyt} and FcεRIγ_{cyt}, all were shown to form specific homodimers without a disorder-to-order transition upon dimerization, thus revealing for the first time the existence of specific interactions between disordered protein molecules. Interestingly, for ζ_{cyt}, the oligomerization behavior is best described by a two-step monomer-dimer-tetramer fast dynamic equilibrium with dissociation constants in the order of approximately 10 μM (monomer-dimer) and approximately 1 mM (dimer-tetramer).[9] In contrast to the other ITAM-containing proteins, Igα_{cyt} forms stable dimers and tetramers even below 10 μM.[9] Phosphorylation of the ζ_{cyt} and FcεRIγ_{cyt} ITAM Tyr residues neither significantly alters their homooligomerization behavior nor is accompanied by folding.[9] As shown by CD and NMR spectroscopy for random coil ζ_{cyt}, this IDP does not undergo a transition between disordered and ordered states upon dimerization and remains unfolded both outside and within the interaction interface(s) in the ζ_{cyt} dimer.[9-11] Since its discovery in 2004,[9] the unusual biophysical phenomenon of IDP homooligomerization has become of more and more interest to biophysicists and biochemists,[32,59] and one can expect that further multidisciplinary studies will shed light on the possible structural basis of these interesting IDP features.

Later, NMR studies of a direct, physiologically relevant interaction between ζ_{cyt} and the well-structured core domain of the simian immunodeficiency virus (SIV) Nef protein revealed that random coil ζ_{cyt} bound to Nef with micromolar affinity remains unfolded at the interaction interfaces in the 1:1 ζ_{cyt}-Nef complex,[58] thus extending the "binding without folding" mode observed in IDP-IDP complexes to interactions of IDPs with folded partner proteins (Fig. 1B).

Intriguingly, NMR studies of ζ_{cyt} dimer[9,11] and ζ_{cyt}-Nef complex[58] revealed a new, previously unknown NMR phenomenon—the lack of significant changes in chemical shift and peak intensity upon a specific protein complex formation.[9,11,58] No chemical shift changes and significant changes in peak intensities are observed in the ^{1}H-^{15}N heteronuclear single quantum coherence (HSQC) spectra of ^{15}N-labeled ζ_{cyt} upon dimerization (Fig. 2A)[11] or binding to the Nef protein (Fig. 2B).[58] Importantly, while the ζ_{cyt} dimerization interface is not yet known, the amino acid residues of ζ_{cyt} involved in the ζ_{cyt}-Nef interaction are well-established,[60] but also do not exhibit chemical shift changes upon binding (Fig. 2B; cross-peak positions of SNID residues are marked).[58] ^{1}H-^{15}N HSQC spectra represent a fingerprint of the protein backbone and are widely

Binding of intrinsically disordered TCRζ_{cyt}

No significant changes in chemical shift and peak intensity upon binding

A. TCRζ_{cyt} dimerization

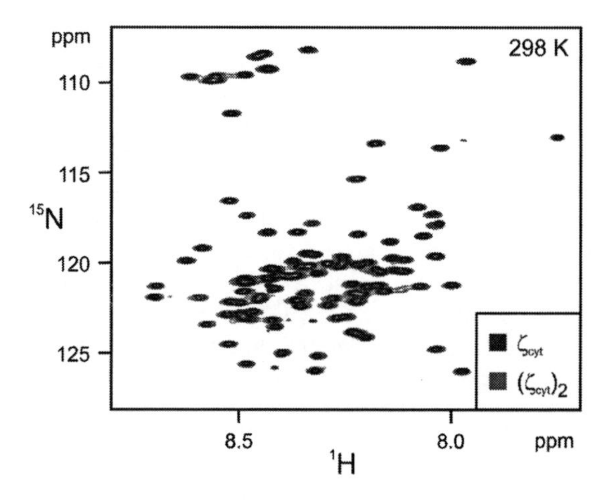

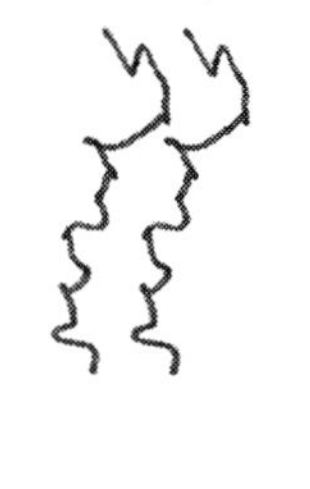

B. TCRζ_{cyt} binding to SIV Nef

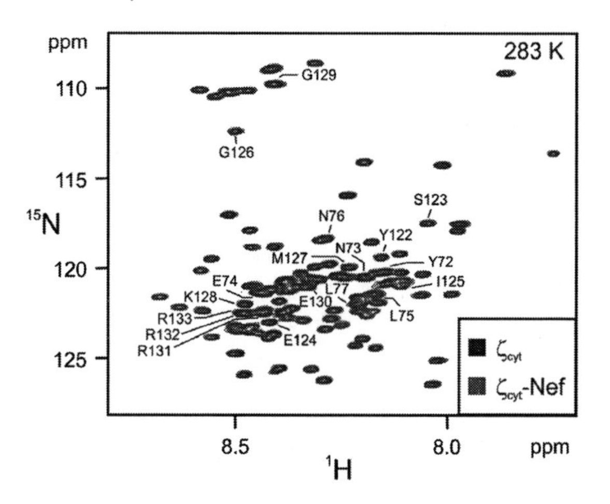

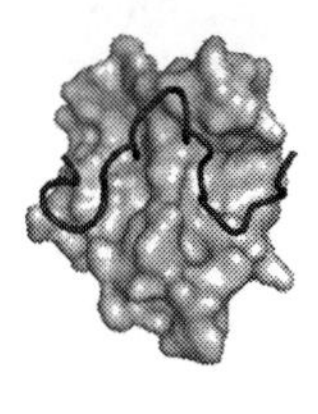

Figure 2. Intrinsically disordered cytoplasmic domain of T-cell receptor ζ chain (ζ_{cyt}) does not fold upon binding to its unfolded (A) or folded (B) protein target. The ^{1}H-^{15}N heteronuclear single quantum correlation (HSQC) spectra of ^{15}N-labeled ζ_{cyt} at 298 K (A) or 283 K (B) in the absence (blue) or presence (red) of its interacting partner, another ζ_{cyt} molecule (A) or the well-folded and ordered SIV Nef protein (B). In both scenarios, a new, previously unobserved NMR phenomenon, the lack of significant changes in chemical shift and intensity upon a specific protein complex formation, has been reported.[9,11,58] Cross-peak positions of the SIV Nef interaction domains (SNIDs) residues are marked to highlight the lack of chemical shift changes for these residues upon binding to Nef. Adapted with permission from Sigalov AB et al. Biochimie 2007; 89:419-421;[11] ©2007 Elsevier; and Sigalov AB et al. Biochemistry 2008; 47:12942-12944;[58] ©2008 American Chemical Society.

used as a quick, informative probe of changes in backbone conformation, particularly as it relates to structural studies of IDPs and their complexes.[20,21] Thus, the unique and unprecedented NMR phenomenon observed[9-11,58] likely highlights an unusual nature of specific interactions of IDPs upon binding without folding on their targets and opens a new line of research in the field of IDPs.

Lipid Partner

In 2000,[36] it has been shown that α-helical folding transition of random coil ζ_{cyt} upon binding to acidic phospholipids prevents ITAM phosphorylation. The authors concluded that this folding transition can represent a conformational switch to regulate TCR triggering,[36] Later, the other group of investigators extended these findings to intrinsically disordered CD3ε_{cyt} and showed that binding of this protein to acidic phospholipids is accompanied by folding of ITAM, leading to inaccessibility of the ITAM tyrosines for phosphorylation in vitro.[35] This led the authors[35,61] to the conclusion that the conformational model of TCR activation previously suggested for ζ_{cyt}[36] can be extended to CD3ε_{cyt}. On the contrary, in other studies, it has been shown that binding of ζ_{cyt}, CD3ε_{cyt} and FcεRγ_{cyt} to acidic phospholipids is not accompanied by a disorder-to-order structural transition.[10] In contrast to micelles and DMPG vesicles,[35,36,52] intrinsically disordered ζ_{cyt}, CD3ε_{cyt} and γ_{cyt} have been shown to bind to acidic 1-palmitoyl-2-oleoyl-phosphatidylglycerol (POPG) vesicles without folding.[10]

A molecular explanation for this paradox was reported in 2009[51] when two different membrane binding modes for ζ_{cyt}, CD3ε_{cyt} and FcεRγ_{cyt} were shown, depending on the bilayer stability: (mode I) coupled binding and folding (Fig. 1A) and (mode II) binding without folding (Fig. 1B). It has been suggested[51] that in both modes, initially, clusters of basic amino acids in the regions outside ITAMs bind to polar heads of acidic phospholipids while the ITAM residues do not contribute to binding at this stage. Then, in micelles (mode I), hydrophobic interactions between ITAMs and detergent tails promote folding of ITAMs, thus making ITAM tyrosines inaccessible for kinases as it has been shown for ζ_{cyt}/lysomyristoylphosphatidylglycerol (LMPG) micelles system.[36,52] In vesicles, depending on the bilayer stability, initial protein binding to the membrane may (mode I) or may not (mode II) induce vesicle fusion and rupture and promote formation of ITAM helixes stabilized by hydrophobic interactions with lipid tails in ruptured bilayers. As shown by DLS and EM, in the POPG vesicles used, protein binding does not disturb the lipid bilayer structure and does not cause vesicle fusion, thus explaining the lack of a disorder-to-order transition.[10,51] Interestingly, phosphorylation of two and six ITAM tyrosines in FcεRIγ_{cyt} and ζ_{cyt}, respectively, reduces the corresponding net charges from +3 to-0.5 for FcεRIγ_{cyt} and from +5 to-5.5 for ζ_{cyt} but does not abrogate binding to POPG vesicles,[10] further confirming a hypothesis that not only does the overall net charge but also the presence of clustered basic amino acid residues prove to be important for lipid-binding activity of these IDPs.[10]

Thus, these findings[10,51] clearly demonstrate that binding to IDPs can induce membrane fusion and rupture in the lipid bilayer vesicles unstable under the experimental conditions used. Importantly, as shown,[51] the destructive effect of protein binding on bilayer lipid membrane is not dependent on vesicle size (small versus large unilamellar vesicles; SUV or LUV, respectively) or technique of SUV preparation (sonication vs. extruding). The membrane rupture is known to result in monolayer fusion of the membranes, i.e., in the formation of a bridge connecting the monolayers, which is usually named the stalk or hemifusion intermediate.[62] Interestingly, tight coupling between the loop-to-helix structural transition and stalk formation as a result of deformation of the target and viral membranes has been reported for influenza hemagglutinin.[63,64] Thus, protein binding-induced membrane perturbation and disruption can represent a molecular basis for the observed formation of the ITAM helixes in the presence of DMPG vesicles.[35,36] It should be also noted that in mode II, ITAMs do not participate in binding to lipid bilayers[51] and the ITAM tyrosines are therefore likely to be easily accessible for phosphorylation (Fig. 1B).

Thus, the use of not only micelles but also lipid vesicles can result in opposite conclusions regarding membrane-binding activity of IDPs and its physiological relevance. This highlights the importance of the choice of an appropriate membrane model in studies of protein-lipid interactions, particularly as it relates to IDPs. This also challenges the field of NMR studies where many lipid bilayer models cannot be used because of the particle size.

Summary

As widely discussed in the literature,[23,28,33-36,57,59,65] IDPs often function through molecular recognition and binding to their folded or unfolded protein partners which is accompanied by induced folding of the IDP interaction (recognition) interface, the coupled folding and binding (or disorder-to-order transition) scenario (Fig. 1A). Another scenario, the "conformational selection" model, in which IDPs bind and fully fold through conformational selection following a two-state model, has been also proposed.[66-68] In the consensus synergistic model,[66] conformational selection has been proposed to play the most important role in the specific encounter, while coupled folding and binding has been suggested to be essential for the formation of the fully-bound complexes.

Recently discovered functionally relevant binding of IDPs to unfolded and folded proteins without folding on the interacting partner[9-11,58] revealed a novel insight into IDP interactions, thus demonstrating the existence of two different modes of IDP binding to their protein partners: with and without folding (the "disorder-to-order" and "no disorder-to-order" scenarios). In the context of signal transduction, the "no disorder-to-order" IDP interactions have been suggested as a novel therapeutic target for a variety of diseases,[2,50,69-73] thus making studies of the basis of the interactions not only of fundamental scientific importance but also of great clinical value. These interactions of low affinity in the micromolar range[9,58] could be of a electrostatic nature, which has been proposed to be exploited in IDPs by generating a "polyelectrostatic effect".[59,74]

The existence of two different modes of IDP binding to lipid bilayer membranes[51] raises an important question: Which mode of action is of physiological relevance? Considering that α-helical folding of ITAMs of signaling-related IDPs is not observed in the presence of those vesicles that are stable upon protein binding, this folding transition is unlikely to play a significant role in transmembrane signaling and cell activation. However, it does not necessarily mean that mode II (binding without folding) is also physiologically irrelevant. For example, within the SCHOOL model,[69,72] homointeractions between CYTO domains of the ITAM-containing receptor signaling subunits are suggested to be necessary and sufficient to trigger the receptor. Thus, membrane binding of ζ_{cyt}, $CD3\varepsilon_{cyt}$ and $Fc\varepsilon R\gamma_{cyt}$ might prevent homooligomerization of these CYTO domains[9] in receptor clusters on the surface of resting cells and during random encounters of receptors diffusing in the cell membrane. Considering the reported prevalence of IDRs in the CYTO domains of many other human TM proteins,[7,8] one can expect that the distinctive features of the ITAM-containing IDRs observed in membrane binding studies[10,51] can be found for other CYTO IDRs, as well. Further studies will have to test this hypothesis of potential physiological significance.

Summarizing, I suggest a novel mechanistic hypothesis that describes interactions of IDPs with their protein or lipid partners as a general biphasic process with a fast rate for the electrostatic-driven "no disorder-to-order" Phase I interaction and a slow rate for the hydrophobic-driven ("disorder-to-order") Phase II formation of a secondary structure (e.g., an amphipathic helix). Within this hypothesis, a Phase II may (folding upon binding) or may not (binding without folding) follow a Phase I depending on the interacting partner.

RECEPTOR SIGNALING

Structural Classification of Receptors

Based on location of binding and signaling (effector) domains, functionally diverse and unrelated cell surface receptors can be structurally classified into two main families: Those in which binding and signaling domains are located on the same protein chain, the so-called single-chain receptors (SRs, Fig. 3) and those in which binding and signaling domains are intriguingly located on separate subunits, the so-called multichain receptors (Fig. 4).[2,50,69,72,75,76] Because many multichain activating receptors are immune receptors, they are all commonly referred to as multichain immune recognition receptors (MIRRs).[2,70,75]

Assuming that the similar structural architecture of the receptors dictates similar mechanisms of receptor triggering and subsequent transmembrane signaling, one can suggest that the targets revealed by these mechanisms are similar in seemingly unrelated diseases. This builds the structural basis for the development of novel pharmacological approaches as well as the transfer of clinical knowledge, experience and therapeutic strategies between various disorders.

Protein Intrinsic Disorder and Receptors

Signaling subunits of MIRRs contain in their cytoplasmic domains the ITAM or the YxxM motif, found in the DAP10 subunit (Fig. 4). Ligand binding results in phosphorylation of the ITAM/YxxM tyrosines, triggering the intracellular signaling cascade. Extracellular structure of signaling subunits varies from the short sequences found in ζ, γ, DAP10 and DAP12 to the Ig-like folds present in CD3ε, CD3δ, CD3γ, Igα and Igβ (Fig. 4). In contrast, as revealed by computational methods, CD and NMR spectroscopy, most of their cytoplasmic domains, namely, ζ_{cyt}, γ_{cyt}, CD3ε_{cyt}, CD3δ_{cyt}, CD3γ_{cyt}, Igα_{cyt} and Igβ_{cyt} represent a novel class of IDPs (Table 1).[9-11] Interestingly, for DAP10$_{cyt}$ and DAP12$_{cyt}$, secondary structure prediction using the hierarchical neural network algorithm[18] exhibits high (about 80%) percentage of random coil conformation (Table 1).[50] Disorder prediction using the algorithm of Uversky et al[12] reveals the boundary <H> values of 0.101 and 0.039 for DAP10$_{cyt}$ and DAP12$_{cyt}$, respectively.[50] These values are characteristic for IDPs and close to those calculated for other ITAM-containing sequences (Table 1).[10] Thus, the cytoplasmic domains of signaling subunits of many different receptors expressed on various cells are surprisingly all intrinsically disordered (Fig. 4). This fits with recent findings that IDRs are prevalent in the cytoplasmic domains of human transmembrane proteins[7] and that protein phosphorylation predominantly occurs within IDRs,[6] further suggesting an important role of intrinsic disorder in receptor-mediated signaling.

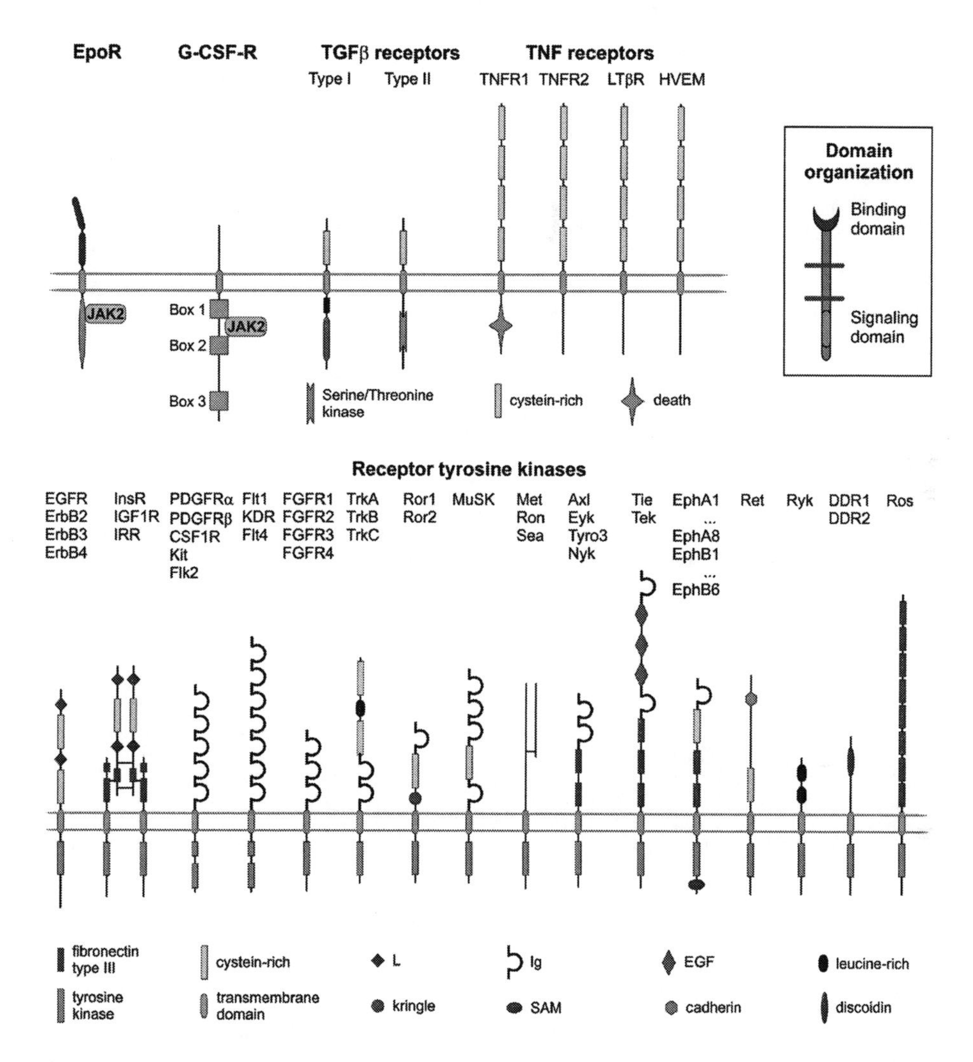

Figure 3. Single-chain receptors (SRs). The extracellular portion of the receptors is on top and the cytoplasmic portion is on bottom. The lengths of the receptors as shown are only approximately to scale. The inset shows SR domain organization. Abbreviations: EpoR, erythropoietin receptor; G-CSF-R, granulocyte colony-stimulating factor receptor; TGFβ, transforming growth factor-beta; TNF, tumor necrosis factor; JAK, Janus kinase; EGFR, epidermal growth factor receptor; InsR, insulin receptor; IGF1R, insulin-like growth factor I receptor; IRR, insulin receptor-related receptor; PDGFR, platelet-derived growth factor receptor; CSF1R, colony-stimulating-factor 1 receptor; FGFR, fibroblast growth factor receptor; MuSK; muscle-specific receptor tyrosine kinase; Eph, ephrin; DDR, discoidin domain receptor; Flt1, KDR and Flt4, vascular endothelial growth factor (VEGF) receptors. Adapted with permission from Sigalov AB. Self/Nonself 2010; 1(1):4-39.

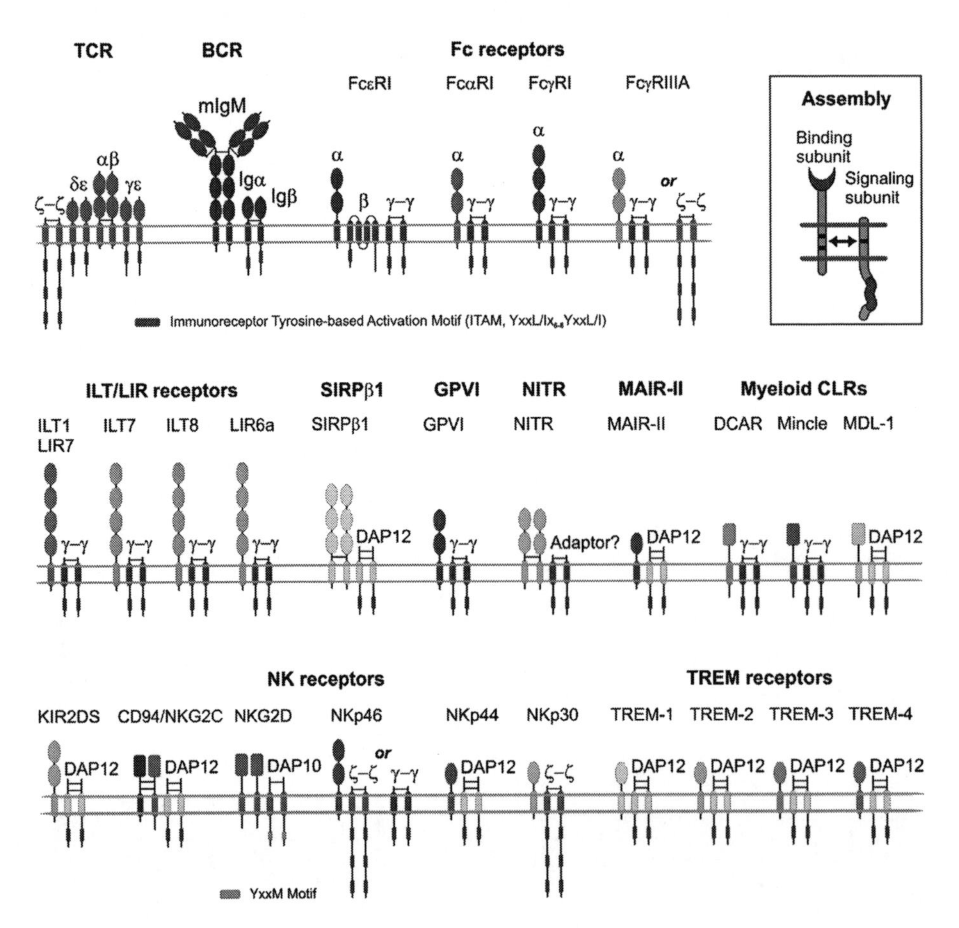

Figure 4. Multichain immune recognition receptors. Schematic presentation of the MIRRs expressed on many different immune cells including T and B-cells, natural killer cells, mast cells, macrophages, basophils, neutrophils, eosinophils, dendritic cells and platelets. The inset shows MIRR assembly. Cytoplasmic domains of the MIRR signaling subunits represent a novel class of intrinsically disordered proteins and are shown to be dimeric. Curved line depicts protein disorder. Abbreviations: ITAM, immunoreceptor tyrosine-based activation motif; BCR, B-cell receptor; DAP-10 and DAP-12, DNAX adapter proteins of 10 and 12 kD, respectively; DCAR, dendritic cell immunoactivating receptor; GPVI, glycoprotein VI; ILT, Ig-like transcript; KIR, killer cell Ig-like receptor; LIR, leukocyte Ig-like receptor; MAIR-II, myeloid-associated Ig-like receptor; MDL-1, myeloid DAP12-associating lectin 1; NITR, novel immune-type receptor; NK, natural killer cells; SIRP, signal regulatory protein, TCR, T-cell receptor; TREM receptors, triggering receptors expressed on myeloid cells. Adapted with permission from Sigalov AB. Self/Nonself 2010; 1(1):4-39.

Receptor and Protein Oligomericity

Binding of multivalent but not monovalent ligand and subsequent receptor clustering are required for induction of the signaling cascade.[2,3,69,77,78] This raises the question: What is the molecular mechanism by which clustering of the extracellular binding domains leads to the generation of the activation signal by intracellular signaling domains? For many receptors, ligand-induced clustering is known to result in oligomerization of receptor transmembrane and cytoplasmic domains.[71,72,79-81] The subsequent formation of competent signaling oligomers in cytoplasmic milieu provides the necessary and sufficient event to trigger the receptor. However, for receptors that signal through ITAM/YxxM modules (Fig. 4), this mechanism has been a long-standing open issue until recently, when formation of ITAM-containing cytoplasmic signaling oligomers was suggested to play a crucial role in transmembrane signaling mediated by these receptors.[50,69,70,72,73] Interestingly, the homooligomerization of these IDPs is best described by a two-step monomer-dimer-tetramer fast dynamic equilibrium with monomer-dimer dissociation constants in the micromolar affinity range.[9,11] These findings are in line with the known dependence of the overall binding affinity between proteins on the function of the protein complex. For example, obligate homodimers are strongly associated with nano- or picomolar binding affinity while, in contrast, proteins that associate and dissociate in response to changes in their environment, such as the majority of signal transduction mediators, tend to bind more weakly.

As mentioned, the first evidence of an IDP's propensity for specific homodimerization distinct from nonspecific aggregation behavior seen in many systems[82] has been recently reported[9] and suggested to play an important role in transmembrane signaling.[69,70,72] Later, other IDPs have also been found to form specific homodimers[46-49] and shown to function through dimer formation,[46,49] further demonstrating a direct functional link between protein intrinsic disorder and oligomericity. In the context of receptor-mediated signal transduction, this link represents a key and missing element in our understanding of transmembrane signal transduction. One can suggest that for the vast majority of receptors, receptor oligomericity (clustering upon binding of multivalent ligand) is translated across the membrane into protein oligomericity (formation of competent cytoplasmic signaling oligomers), thus providing a general platform for receptor-mediated signaling (Fig. 5).[50,73]

Single-Chain Receptor Signaling: Functional Link between Protein Order and Oligomericity

Single-chain receptors (SRs) are receptors with binding and signaling domains located on the same protein chain (Fig. 3). Importantly, EC, TM and CYTO regions of these receptors represent folded and well-ordered domains.

Examples of SRs include receptor tyrosine kinases (RTKs) that are TM glycoproteins consisting of a variable EC N-terminal domain, a single membrane spanning domain and a large CYTO portion composed of a juxtamembrane domain, the highly conserved tyrosine kinase domain and a C-terminal regulatory region (Fig. 3).[83] RTKs activate numerous intracellular signaling pathways, leading to a variety of cell responses. These receptors are triggered by the binding of their cognate ligands and transduce the recognition signal to the cytoplasm by phosphorylating CYTO tyrosine residues on the receptors themselves (autophosphorylation) and on downstream signaling proteins. The proteins of the tumor necrosis factor (TNF) receptor superfamily[84] are a group of SRs critically involved in the maintenance of homeostasis of the immune system (Fig. 3). Triggered by their corresponding ligands, these receptors either induce cell death or promote cell survival of immune cells. Transforming growth factor-β (TGF-β) is a potent regulatory cytokine which inhibits the development of immunopathology to self or nonharmful antigens without compromising immune responses to pathogens.[85] The TGF-β superfamily functions via binding to Type I and II TM serine/threonine kinase receptors that belong to the SR family (Fig. 3).

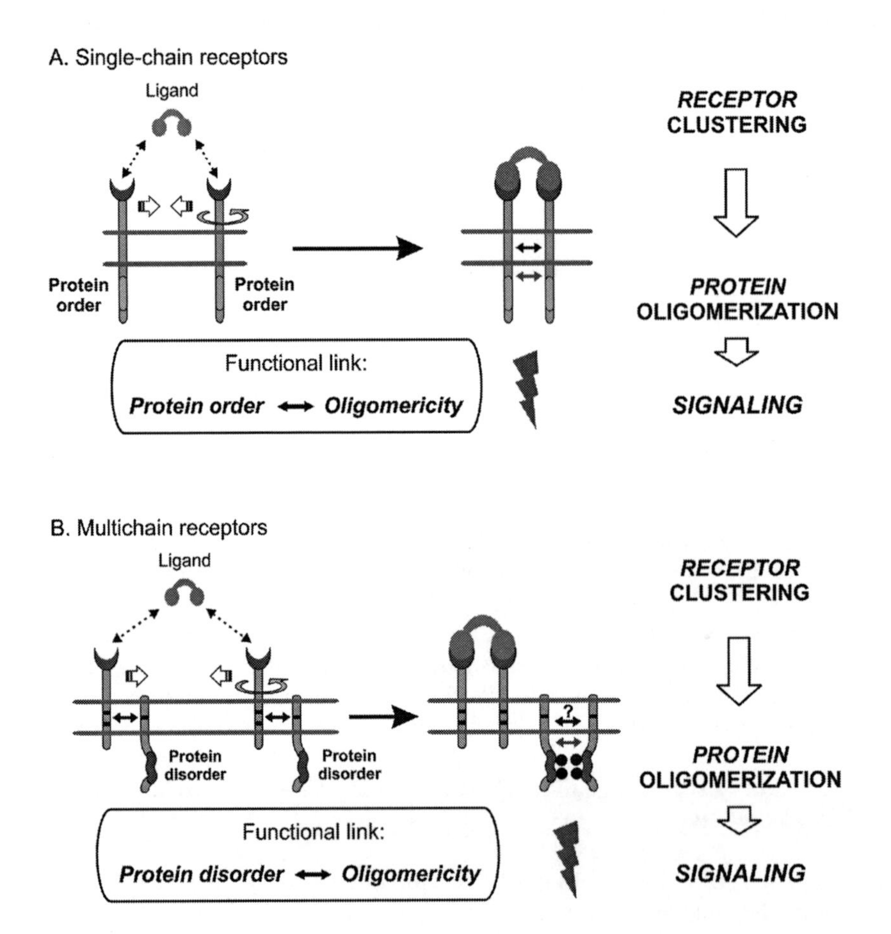

Figure 5. SCHOOL principles of receptor signaling. Single- (A) or multichain (B) receptor oligomerization (clustering) induced upon ligand binding outside the cell is translated across the membrane into protein oligomerization inside the cell with cytoplasmic homointeractions representing the major driving force of receptor triggering. For SRs (A), small solid black and gray arrows indicate specific inter-unit homointeractions between transmembrane and cytoplasmic domains, respectively. For MIRRs (B), small solid black and gray arrows indicate specific inter-unit hetero- and homointeractions between transmembrane and cytoplasmic domains, respectively. Circular arrow indicates ligand-induced receptor re-orientation. Curved line depicts disorder of the cytoplasmic domains of MIRR signaling subunits (B). Phosphate groups are shown as dark circles. Abbreviation: SCHOOL, signaling chain homooligomerization. Adapted with permission from Sigalov AB. Self/Nonself 2010; 1(1):4-39.

According to the SCHOOL platform, signaling chain homooligomerization and formation of competent signaling oligomers in CYTO milieu provides the necessary and sufficient event to trigger receptors and induce cell activation (Fig. 5A).[50,71-73] Within the consensus model of SR signaling, multivalent ligand binding results in receptor re-orientation and dimerization (oligomerization) and subsequent formation of competent signaling oligomers in the cytoplasm and trans-autophosphorylation at defined cytoplasmic tyrosines.[3,50,71,73,77,81,86-89] Some SRs, such as members of the tumor necrosis factor (TNF) receptor superfamily,[88] exist as pre-assembled oligomers on the cell surface. In this scenario, multivalent ligand binding results in re-orientation of individual receptors in the pre-assembled oligomers to adopt an inter-unit geometry permissive for promotion of homointeractions between receptor CYTO domains and further receptor activation.[50,71-73,89]

Thus, in terms of SR signaling, there exists the principal functional link between protein order and oligomericity in CYTO milieu.

MULTICHAIN RECEPTOR SIGNALING: FUNCTIONAL LINK BETWEEN PROTEIN DISORDER AND OLIGOMERICITY

Functionally diverse members of the MIRR family are expressed on many different immune cells, including T and B-cells, natural killer (NK) cells, mast cells, macrophages, basophils, neutrophils, eosinophils, dendritic cells (DCs) and platelets.[2,70,75,90] Figure 4 shows typical examples of MIRRs including the T-cell receptor (TCR) complex, the B-cell receptor (BCR) complex, Fc receptors (e.g., FcεRI, FcαRI, FcγRI and FcγRIII), NK receptors (e.g., NKG2D, CD94/NKG2C, KIR2DS, NKp30, NKp44 and NKp46), immunoglobulin (Ig)-like transcripts and leukocyte Ig-like receptors (ILTs and LIRs, respectively), signal regulatory proteins (SIRPs), dendritic cell immunoactivating receptor (DCAR), myeloid DNAX adapter protein of 12 kD (DAP12)-associating lectin 1 (MDL-1), blood DC antigen 2 protein (BDCA2), novel immune-type receptor (NITR), myeloid-associated Ig-like receptor (MAIR-II), triggering receptors expressed on myeloid cells (TREMs) and the platelet collagen receptor, glycoprotein VI (GPVI). For more information on the structure and function of these and other MIRRs, I refer the reader to recent reviews.[2,91-110]

The MIRR ligand-binding subunits are integral membrane proteins with small CYTO domains that are themselves inert with regard to signaling. Signaling is achieved through the association of the ligand-binding chains with signal-transducing subunits that contain in their CYTO domains one or more copies of the ITAM regions with two appropriately spaced tyrosines (YxxL/Ix_{6-8}YxxL/I; where x denotes nonconserved residues)[111] or the YxxM motif,[112,113] found in the DAP10 CYTO domain[113] (Fig. 4). The association of the MIRR subunits in resting cells is driven mostly by the noncovalent TM interactions between recognition and signaling components (Fig. 4) and plays a key role in receptor assembly, integrity and surface expression.[70,97-99,101,106,109,114-125] Despite extensive studies in the field, the molecular mechanism linking extracellular clustering of MIRR binding subunits to intracellular phosphorylation of ITAM/YxxM tyrosines has been a long-standing mystery.

The intriguing ability of the intrinsically disordered ITAM-containing CYTO domains of MIRR signaling subunits to homooligomerize[9] led to the development of a novel model of MIRR signaling, the Signaling Chain HOmoOLigomerization (SCHOOL) model.[2,50,69-73,76] The model suggests that formation of competent signaling subunit oligomers mediated by homotypic interactions in the cytoplasm, rather than receptor clustering/oligomerization per se, is necessary and sufficient to trigger the receptors and induce the downstream signaling sequence (Fig. 5B). Similar to SRs, some MIRRs such as TCR and major platelet collagen receptor glycoprotein VI (GPVI), can exist as pre-assembled oligomers on the cell surface.[126,127] In these oligomers, multivalent ligand binding results in re-orientation of receptors to adopt an inter-unit geometry permissive for promotion of homointeractions between MIRR signaling subunit CYTO domains and further receptor activation (Fig. 5B).

Thus, in terms of MIRR signaling, there exists the principal functional link between protein disorder and oligomericity in CYTO milieu.

SCHOOL PLATFORM OF RECEPTOR SIGNALING

According to the SCHOOL platform, signaling chain homooligomerization and formation of competent signaling oligomers in CYTO milieu provides the necessary and sufficient event to trigger receptors of both structural families (SRs and MIRRs) and induce cell activation (Fig. 5). Within the platform, receptor oligomerization induced or tuned upon ligand binding outside the cell is translated across the membrane into protein oligomerization in CYTO milieu, thus providing a general platform for receptor-mediated signaling (Fig. 5).

The necessity and sufficiency of formation of competent signaling oligomers mediated by homointeractions between well-structured (SRs) or intrinsically disordered (MIRRs) cytoplasmic signaling domains to trigger receptor function dictates several important mechanistic principles of receptor signaling:

- sufficient interreceptor proximity in receptor dimers/oligomers,
- correct (permissive) relative orientation of the receptors in receptor dimers/oligomers,
- long enough duration of the receptor-ligand interaction that generally correlates with the strength (affinity/avidity) of the ligand, and
- sufficient lifetime of an individual receptor in receptor dimers/oligomers.

These general principles are common for SRs and MIRRs and thus link mechanistically numerous structurally and functionally diverse receptors.

Further, because of the ubiquitous nature of protein-protein interactions and the knowledge that inappropriate protein-protein binding can lead to disease, the specific and controlled inhibition and/or modulation of these interactions provides a promising novel approach for rational drug design. A number of recent reviews have addressed this topic.[128-130] Suggesting important role of TM interactions that mediate ligand-induced SR dimerization (oligomerization) and homointeractions between CYTO domains that result in formation of competent signaling oligomers (Fig. 5A), the SCHOOL model of SR signaling reveals these interactions as important points for intervention to modulate SR signaling.[50,71-73,131,132] Similarly, considering MIRR triggering as the result of the ligand-induced interplay between (1) intrareceptor TM interactions that stabilize and maintain receptor integrity and (2) interreceptor homointeractions between the CYTO domains of MIRR signaling subunits that lead to formation of competent signaling oligomers (Fig. 5B), the SCHOOL models reveals these interactions as important points for intervention to modulate MIRR signaling.[50,70-73,131,132] Importantly, these are common targets for all members of the MIRR family, which means that a general pharmaceutical strategy may be used to treat seemingly disparate disorders such, for example, as T-cell-mediated skin diseases and platelet disorders.[50,70-73,131-133]

APPLICATIONS IN BIOLOGY AND MEDICINE

By revealing specific protein-protein interactions critically involved in receptor-mediated signaling, current SCHOOL models that are based on functional connections between protein order (SRs), disorder (MIRRs) and oligomericity provide molecular explanations for many biological phenomena and processes, represent powerful tools for fundamental and applied research and suggest novel avenues for drug discovery.[2,50,69-73,131-134]

T-Cell Receptor Signaling

Despite TCR being one of the most studied MIRRs, many of the models of TCR signaling suggested to date are descriptive and often fail in trying to explain most of the known immunological data.

Structurally, TCR is a member of the MIRR family (Fig. 4) with its α and β antigen-binding subunits bound by TM interactions with three signaling homo- and heterodimers: $\zeta\zeta$, CD3$\varepsilon\delta$ and CD3$\varepsilon\gamma$.[115] Within the SCHOOL model, distinct TCR signaling is achieved through the ζ and CD3 signaling oligomers,[50,69,70,72,73] and interreceptor TM interactions represent not only a promising therapeutic target but also an important point of viral attack.[70,71,131,135]

The TCR core peptide (CP), a synthetic peptide corresponding to the sequence of the TCRα transmembrane domain, is capable of inhibiting antigen-mediated T-cell activation, whereas T-cell activation via anti-CD3 antibodies is not affected by CP.[136] However, despite extensive studies, the mode of action of this clinically relevant peptide had not been elucidated until 2004 when the SCHOOL model was first introduced.[69]

Recently, inhibition of antigen- but not anti-CD3-stimulated T-cell activation has been reported for the fusion peptide (FP) found in the N terminus of the HIV envelope glycoprotein 41 (gp41).[137] However, the mode of action of this peptide had remained unknown until 2006 when the SCHOOL model was first applied to this area.[70] Within the model, the molecular mechanisms of action for TCR CP and HIV gp41 FP are similar. Briefly, CP and FP compete with TCRα for binding to CD3$\delta\varepsilon$ and $\zeta\zeta$, resulting in functional disconnection of these subunits.[70,131,135,138]

In summary, our current understanding of TCR signaling, together with the lessons learned from the viral pathogenesis,[2,71,73,131,132,135] can be used not only for further fundamental research but also for rational drug design.

Glycoprotein VI Signaling

Activation of circulating platelets by exposed vessel wall collagen is a primary step in the pathogenesis of thrombotic diseases. Despite intensive research efforts in antithrombotic drug discovery, uncontrolled hemorrhage still remains the most common side effect. Intriguingly, the selective inhibition of the GPVI collagen receptor may inhibit thrombosis without affecting hemostasis.[2] However, the mechanism of GPVI signaling has remained unknown until recently,[133,139] therefore hindering the further development of this promising antithrombotic strategy. GPVI belongs to the MIRR family (Fig. 4) and signals through the associated ITAM-containing γ subunit. The application of the

SCHOOL model[71,72,131] resulted in the development of novel mechanistic concept of platelet inhibition and the invention of new platelet inhibitors.[71,133,134,139]

NKG2D Signaling

Despite advances in immune disorder research, there is still a great need for additional targets and agents for effectively reducing inflammatory bowel diseases (IBDs), namely ulcerative colitis (UC) and Crohn's disease (CD) and affect millions of people worldwide. In 2007, a unique subset of CD4+ NKG2D+ T-cells was identified in IBD patients.[140] Later, inhibition of NKG2D, a member of the MIRR family (Fig. 4) that triggers through the associated YxxM motif-containing DAP10, has been proven to be of key importance in successful treatment of UC and CD.[141] Uncovering the molecular mechanisms of NKG2D signaling, the SCHOOL model suggests the NKG2D-DAP10 transmembrane interactions as a promising point of intervention in IBD treatment.[70-72,131,132] Further studies will have to test this concept.

CONCLUSION AND PERSPECTIVES

The crucial role of receptor-mediated signaling in health and disease assumes that our understanding of the underlying molecular mechanisms and methods to modulate the cell response through control of TM signal transduction can contribute significantly towards the improvement of existing therapies and the design of new therapeutic strategies for a diverse set of disorders. For structurally related members of the MIRR family, the functional link between protein intrinsic disorder and oligomericity represents a missing piece of the long-standing puzzle of signaling and reveals striking similarities in the basic mechanistic principles of function of most SRs and MIRRs. In this context, the SCHOOL model of MIRR signaling is similar to the consensus model of SR signaling in regards to both models suggesting that formation of competent signaling oligomers mediated by homointeractions between well-structured (SRs) or intrinsically disordered (MIRRs) CYTO signaling (effector) domains is necessary and sufficient to trigger receptor function. This raises an interesting question: Why for MIRRs, where the recognition and signaling domains are located on separate protein chains, nature selected to use a functional link between protein disorder and oligomericity? One can expect that further multidisciplinary studies will clarify this question of great interest and practical utility.

In conclusion, recent fundamental advances uncovering the molecular mechanisms of receptor-mediated signaling have been accompanied by our improved understanding of unexplained biological phenomena. This opens new horizons in further fundamental and clinical research, research-based education and innovative drug design and discovery.

REFERENCES

1. Rudd CE. Disabled receptor signaling and new primary immunodeficiency disorders. N Engl J Med 2006; 354:1874-1877.
2. Sigalov AB, ed. Multichain Immune Recognition Receptor Signaling: From Spatiotemporal Organization to Human Disease. New York: Springer-Verlag; 2008:357.

3. Klemm JD, Schreiber SL, Crabtree GR. Dimerization as a regulatory mechanism in signal transduction. Annu Rev Immunol 1998; 16:569-592.
4. Metzger H. Transmembrane signaling: the joy of aggregation. J Immunol 1992; 149:1477-1487.
5. Iakoucheva LM, Brown CJ, Lawson JD et al. Intrinsic disorder in cell-signaling and cancer-associated proteins. J Mol Biol 2002; 323:573-584.
6. Iakoucheva LM, Radivojac P, Brown CJ et al. The importance of intrinsic disorder for protein phosphorylation. Nucleic Acids Res 2004; 32:1037-1049.
7. Minezaki Y, Homma K, Nishikawa K. Intrinsically disordered regions of human plasma membrane proteins preferentially occur in the cytoplasmic segment. J Mol Biol 2007; 368:902-913.
8. De Biasio A, Guarnaccia C, Popovic M et al. Prevalence of intrinsic disorder in the intracellular region of human single-pass type I proteins: the case of the notch ligand Delta-4. J Proteome Res 2008; 7:2496-2506.
9. Sigalov A, Aivazian D, Stern L. Homooligomerization of the cytoplasmic domain of the T-cell receptor zeta chain and of other proteins containing the immunoreceptor tyrosine-based activation motif. Biochemistry 2004; 43:2049-2061.
10. Sigalov AB, Aivazian DA, Uversky VN et al. Lipid-binding activity of intrinsically unstructured cytoplasmic domains of multichain immune recognition receptor signaling subunits. Biochemistry 2006; 45:15731-15739.
11. Sigalov AB, Zhuravleva AV, Orekhov VY. Binding of intrinsically disordered proteins is not necessarily accompanied by a structural transition to a folded form. Biochimie 2007; 89:419-421.
12. Uversky VN, Gillespie JR, Fink AL. Why are "natively unfolded" proteins unstructured under physiologic conditions? Proteins 2000; 41:415-427.
13. Le Gall T, Romero PR, Cortese MS et al. Intrinsic disorder in the protein data bank. J Biomol Struct Dyn 2007; 24:325-342.
14. Dunker AK, Brown CJ, Lawson JD et al. Intrinsic disorder and protein function. Biochemistry 2002; 41:6573-6582.
15. Linding R, Jensen LJ, Diella F et al. Protein disorder prediction: implications for structural proteomics. Structure 2003; 11:1453-1459.
16. Ward JJ, McGuffin LJ, Bryson K et al. The DISOPRED server for the prediction of protein disorder. Bioinformatics 2004; 20:2138-2139.
17. Prilusky J, Felder CE, Zeev-Ben-Mordehai T et al. FoldIndex: a simple tool to predict whether a given protein sequence is intrinsically unfolded. Bioinformatics 2005; 21:3435-3438.
18. Oldfield CJ, Cheng Y, Cortese MS et al. Comparing and combining predictors of mostly disordered proteins. Biochemistry 2005; 44:1989-2000.
19. Receveur-Brechot V, Bourhis JM, Uversky VN et al. Assessing protein disorder and induced folding. Proteins 2006; 62:24-45.
20. Dyson HJ, Wright PE. Equilibrium NMR studies of unfolded and partially folded proteins. Nat Struct Biol 1998; 5 Suppl:499-503.
21. Dyson HJ, Wright PE. Unfolded proteins and protein folding studied by NMR. Chem Rev 2004; 104:3607-3622.
22. Dunker AK, Silman I, Uversky VN et al. Function and structure of inherently disordered proteins. Curr Opin Struct Biol 2008; 18:756-764.
23. Dyson HJ, Wright PE. Intrinsically unstructured proteins and their functions. Nat Rev Mol Cell Biol 2005; 6:197-208.
24. Vucetic S, Xie H, Iakoucheva LM et al. Functional anthology of intrinsic disorder. 2. Cellular components, domains, technical terms, developmental processes and coding sequence diversities correlated with long disordered regions. J Proteome Res 2007; 6:1899-1916.
25. Xie H, Vucetic S, Iakoucheva LM et al. Functional anthology of intrinsic disorder. 3. Ligands, post-translational modifications and diseases associated with intrinsically disordered proteins. J Proteome Res 2007; 6:1917-1932.
26. Xie H, Vucetic S, Iakoucheva LM et al. Functional anthology of intrinsic disorder. 1. Biological processes and functions of proteins with long disordered regions. J Proteome Res 2007; 6:1882-1898.
27. Gsponer J, Futschik ME, Teichmann SA et al. Tight regulation of unstructured proteins: from transcript synthesis to protein degradation. Science 2008; 322:1365-1368.
28. Tompa P. The interplay between structure and function in intrinsically unstructured proteins. FEBS Lett 2005; 579:3346-3354.
29. Uversky VN. What does it mean to be natively unfolded? Eur J Biochem 2002; 269:2-12.
30. Uversky VN, Dunker AK. Biochemistry. Controlled chaos. Science 2008; 322:1340-1341.
31. Tompa P. Intrinsically unstructured proteins. Trends Biochem Sci 2002; 27:527-533.
32. Tompa P, Fuxreiter M. Fuzzy complexes: polymorphism and structural disorder in protein-protein interactions. Trends Biochem Sci 2008; 33:2-8.
33. Demchenko AP. Recognition between flexible protein molecules: induced and assisted folding. J Mol Recognit 2001; 14:42-61.

34. Dyson HJ, Wright PE. Coupling of folding and binding for unstructured proteins. Curr Opin Struct Biol 2002; 12:54-60.
35. Xu C, Gagnon E, Call ME et al. Regulation of T-cell receptor activation by dynamic membrane binding of the CD3epsilon cytoplasmic tyrosine-based motif. Cell 2008; 135:702-713.
36. Aivazian DA, Stern LJ. Phosphorylation of T-cell receptor zeta is regulated by a lipid dependent folding transition. Nat Struct Biol 2000; 7:1023-1026.
37. Radhakrishnan I, Perez-Alvarado GC, Dyson HJ et al. Conformational preferences in the Ser133-phosphorylated and nonphosphorylated forms of the kinase inducible transactivation domain of CREB. FEBS Lett 1998; 430:317-322.
38. Richards JP, Bachinger HP, Goodman RH et al. Analysis of the structural properties of cAMP-responsive element-binding protein (CREB) and phosphorylated CREB. J Biol Chem 1996; 271:13716-13723.
39. Radhakrishnan I, Perez-Alvarado GC, Parker D et al. Solution structure of the KIX domain of CBP bound to the transactivation domain of CREB: a model for activator:coactivator interactions. Cell 1997; 91:741-752.
40. Fletcher CM, McGuire AM, Gingras AC et al. 4E binding proteins inhibit the translation factor eIF4E without folded structure. Biochemistry 1998; 37:9-15.
41. Fletcher CM, Wagner G. The interaction of eIF4E with 4E-BP1 is an induced fit to a completely disordered protein. Protein Sci 1998; 7:1639-1642.
42. Tomoo K, Matsushita Y, Fujisaki H et al. Structural basis for mRNA Cap-Binding regulation of eukaryotic initiation factor 4E by 4E-binding protein, studied by spectroscopic, X-ray crystal structural and molecular dynamics simulation methods. Biochim Biophys Acta 2005; 1753:191-208.
43. Bourhis JM, Receveur-Brechot V, Oglesbee M et al. The intrinsically disordered C-terminal domain of the measles virus nucleoprotein interacts with the C-terminal domain of the phosphoprotein via two distinct sites and remains predominantly unfolded. Protein Sci 2005; 14:1975-1992.
44. Zhou P, Lugovskoy AA, McCarty JS et al. Solution structure of DFF40 and DFF45 N-terminal domain complex and mutual chaperone activity of DFF40 and DFF45. Proc Natl Acad Sci USA 2001; 98:6051-6055.
45. Demarest SJ, Martinez-Yamout M, Chung J et al. Mutual synergistic folding in recruitment of CBP/p300 by p160 nuclear receptor coactivators. Nature 2002; 415:549-553.
46. Singh VK, Pacheco I, Uversky VN et al. Intrinsically disordered human C/EBP homologous protein regulates biological activity of colon cancer cells during calcium stress. J Mol Biol 2008; 380:313-326.
47. Danielsson J, Liljedahl L, Barany-Wallje E et al. The intrinsically disordered RNR inhibitor Sml1 is a dynamic dimer. Biochemistry 2008; 47:13428-13437.
48. Lanza DC, Silva JC, Assmann EM et al. Human FEZ1 has characteristics of a natively unfolded protein and dimerizes in solution. Proteins 2009; 74:104-121.
49. Simon SM, Sousa FJ, Mohana-Borges R et al. Regulation of Escherichia coli SOS mutagenesis by dimeric intrinsically disordered umuD gene products. Proc Natl Acad Sci USA 2008; 105:1152-1157.
50. Sigalov AB. Protein intrinsic disorder and oligomericity in cell signaling. Mol Biosyst 2010; 6(3):451-61
51. Sigalov AB, Hendricks GM. Membrane binding mode of intrinsically disordered cytoplasmic domains of T-cell receptor signaling subunits depends on lipid composition. Biochem Biophys Res Commun 2009; 389:388-393.
52. Duchardt E, Sigalov AB, Aivazian D et al. Structure induction of the T-cell receptor zeta-chain upon lipid binding investigated by NMR spectroscopy. Chembiochem 2007; 8:820-827.
53. Laczko I, Hollosi M, Vass E et al. Conformational effect of phosphorylation on T-cell receptor/CD3 zeta-chain sequences. Biochem Biophys Res Commun 1998; 242:474-479.
54. Gerlach H, Laumann V, Martens S et al. HIV-1 Nef membrane association depends on charge, curvature, composition and sequence. Nat Chem Biol 2009; DOI:10.1038/nchembio.268.
55. Langner M, Kubica K. The electrostatics of lipid surfaces. Chem Phys Lipids 1999; 101:3-35.
56. Shoemaker SD, Vanderlick TK. Intramembrane electrostatic interactions destabilize lipid vesicles. Biophys J 2002; 83:2007-2014.
57. Hazy E, Tompa P. Limitations of induced folding in molecular recognition by intrinsically disordered proteins. Chem Phys Chem 2009; 10:1415-1419.
58. Sigalov AB, Kim WM, Saline M et al. The intrinsically disordered cytoplasmic domain of the T-cell receptor zeta chain binds to the Nef Protein of simian immunodeficiency virus without a disorder-to-order transition. Biochemistry 2008; 47:12942-12944.
59. Mittag T, Kay LE, Forman-Kay JD. Protein dynamics and conformational disorder in molecular recognition. J Mol Recognit 2009; DOI: 10.1002/jmr.961.
60. Schaefer TM, Bell I, Fallert BA et al. The T-cell receptor zeta chain contains two homologous domains with which simian immunodeficiency virus Nef interacts and mediates down-modulation. J Virol 2000; 74:3273-3283.
61. Kuhns MS, Davis MM. The safety on the TCR trigger. Cell 2008; 135:594-596.
62. Blumenthal R, Clague MJ, Durell SR et al. Membrane fusion. Chem Rev 2003; 103:53-69.

63. Bullough PA, Hughson FM, Skehel JJ et al. Structure of influenza haemagglutinin at the pH of membrane fusion. Nature 1994; 371:37-43.
64. Carr CM, Kim PS. A spring-loaded mechanism for the conformational change of influenza hemagglutinin. Cell 1993; 73:823-832.
65. Hegyi H, Schad E, Tompa P. Structural disorder promotes assembly of protein complexes. BMC Struct Biol 2007; 7:65.
66. Espinoza-Fonseca LM. Reconciling binding mechanisms of intrinsically disordered proteins. Biochem Biophys Res Commun 2009; 382:479-482.
67. Kumar S, Ma B, Tsai CJ et al. Folding and binding cascades: dynamic landscapes and population shifts. Protein Sci 2000; 9:10-19.
68. Tsai CJ, Ma B, Sham YY et al. Structured disorder and conformational selection. Proteins 2001; 44:418-427.
69. Sigalov AB. Multichain immune recognition receptor signaling: different players, same game? Trends Immunol 2004; 25:583-589.
70. Sigalov AB. Immune cell signaling: a novel mechanistic model reveals new therapeutic targets. Trends Pharmacol Sci 2006; 27:518-524.
71. Sigalov AB. SCHOOL model and new targeting strategies. Adv Exp Med Biol 2008; 640:268-311.
72. Sigalov AB. Signaling chain homooligomerization (SCHOOL) model. Adv Exp Med Biol 2008; 640:121-163.
73. Sigalov AB. The SCHOOL of nature. I. Transmembrane signaling. Self/Nonself 2010; 1:4-39.
74. Borg M, Mittag T, Pawson T et al. Polyelectrostatic interactions of disordered ligands suggest a physical basis for ultrasensitivity. Proc Natl Acad Sci USA 2007; 104:9650-9655.
75. Keegan AD, Paul WE. Multichain immune recognition receptors: similarities in structure and signaling pathways. Immunol Today 1992; 13:63-68.
76. Sigalov A. Multi-chain immune recognition receptors: spatial organization and signal transduction. Semin Immunol 2005; 17:51-64.
77. Cooper JA, Qian H. A mechanism for SRC kinase-dependent signaling by noncatalytic receptors. Biochemistry 2008; 47:5681-5688.
78. Weiss A, Schlessinger J. Switching signals on or off by receptor dimerization. Cell 1998; 94:277-280.
79. Lemmon MA, Schlessinger J. Regulation of signal transduction and signal diversity by receptor oligomerization. Trends Biochem Sci 1994; 19:459-463.
80. Bennasroune A, Fickova M, Gardin A et al. Transmembrane peptides as inhibitors of ErbB receptor signaling. Mol Biol Cell 2004; 15:3464-3474.
81. Siegel RM, Muppidi JR, Sarker M et al. SPOTS: signaling protein oligomeric transduction structures are early mediators of death receptor-induced apoptosis at the plasma membrane. J Cell Biol 2004; 167:735-744.
82. Uversky VN. Amyloidogenesis of natively unfolded proteins. Curr Alzheimer Res 2008; 5:260-287.
83. Hubbard SR, Till JH. Protein tyrosine kinase structure and function. Annu Rev Biochem 2000; 69:373-398.
84. Zhou T, Mountz JD, Kimberly RP. Immunobiology of tumor necrosis factor receptor superfamily. Immunol Res 2002; 26:323-336.
85. Li MO, Wan YY, Sanjabi S et al. Transforming growth factor-beta regulation of immune responses. Annu Rev Immunol 2006; 24:99-146.
86. Jiang G, Hunter T. Receptor signaling: when dimerization is not enough. Curr Biol 1999; 9:R568-R571.
87. Marianayagam NJ, Sunde M, Matthews JM. The power of two: protein dimerization in biology. Trends Biochem Sci 2004; 29:618-625.
88. Chan FK. Three is better than one: preligand receptor assembly in the regulation of TNF receptor signaling. Cytokine 2007; 37:101-107.
89. Dosch DD, Ballmer-Hofer K. Transmembrane domain-mediated orientation of receptor monomers in active VEGFR-2 dimers. FASEB J 2009; DOI: 10.1096/fj.09-132670.
90. Geijtenbeek TB, Gringhuis SI. Signalling through C-type lectin receptors: shaping immune responses. Nat Rev Immunol 2009; 9:465-479.
91. Krogsgaard M, Davis MM. How T-cells 'see' antigen. Nat Immunol 2005; 6:239-245.
92. DeFranco AL. B-cell activation 2000. Immunol Rev 2000; 176:5-9.
93. Dal Porto JM, Gauld SB, Merrell KT et al. B-cell antigen receptor signaling 101. Mol Immunol 2004; 41:599-613.
94. Takai T. Fc receptors and their role in immune regulation and autoimmunity. J Clin Immunol 2005; 25:1-18.
95. Takai T. Fc receptors: their diverse functions in immunity and immune disorders. Springer Semin Immunopathol 2006; 28:303-304.
96. Colonna M, Nakajima H, Navarro F et al. A novel family of Ig-like receptors for HLA class I molecules that modulate function of lymphoid and myeloid cells. J Leukoc Biol 1999; 66:375-381.
97. Borrego F, Kabat J, Kim DK et al. Structure and function of major histocompatibility complex (MHC) class I specific receptors expressed on human natural killer (NK) cells. Mol Immunol 2002; 38:637-660.
98. Moroi M, Jung SM. Platelet glycoprotein VI: its structure and function. Thromb Res 2004; 114:221-233.

99. Barclay AN, Brown MH. The SIRP family of receptors and immune regulation. Nat Rev Immunol 2006; 6:457-464.
100. Kanazawa N, Tashiro K, Miyachi Y. Signaling and immune regulatory role of the dendritic cell immunoreceptor (DCIR) family lectins: DCIR, DCAR, dectin-2 and BDCA-2. Immunobiology 2004; 209:179-190.
101. Biassoni R, Cantoni C, Falco M et al. Human natural killer cell activating receptors. Mol Immunol 2000; 37:1015-1024.
102. Biassoni R, Cantoni C, Marras D et al. Human natural killer cell receptors: insights into their molecular function and structure. J Cell Mol Med 2003; 7:376-387.
103. Aoki N, Kimura S, Xing Z. Role of DAP12 in innate and adaptive immune responses. Curr Pharm Des 2003; 9:7-10.
104. Bakker AB, Baker E, Sutherland GR et al. Myeloid DAP12-associating lectin (MDL)-1 is a cell surface receptor involved in the activation of myeloid cells. Proc Natl Acad Sci USA 1999; 96:9792-9796.
105. van den Berg TK, Yoder JA, Litman GW. On the origins of adaptive immunity: innate immune receptors join the tale. Trends Immunol 2004; 25:11-16.
106. Klesney-Tait J, Turnbull IR, Colonna M. The TREM receptor family and signal integration. Nat Immunol 2006; 7:1266-1273.
107. Takai T. Paired immunoglobulin-like receptors and their MHC class I recognition. Immunology 2005; 115:433-440.
108. Nakahashi C, Tahara-Hanaoka S, Totsuka N et al. Dual assemblies of an activating immune receptor, MAIR-II, with ITAM-bearing adapters DAP12 and FcRgamma chain on peritoneal macrophages. J Immunol 2007; 178:765-770.
109. Fujimoto M, Takatsu H, Ohno H. CMRF-35-like molecule-5 constitutes novel paired receptors, with CMRF-35-like molecule-1, to transduce activation signal upon association with FcRgamma. Int Immunol 2006; 18:1499-1508.
110. Stewart CA, Vivier E, Colonna M. Strategies of natural killer cell recognition and signaling. Curr Top Microbiol Immunol 2006; 298:1-21.
111. Reth M. Antigen receptor tail clue. Nature 1989; 338:383-384.
112. Songyang Z, Shoelson SE, Chaudhuri M et al. SH2 domains recognize specific phosphopeptide sequences. Cell 1993; 72:767-778.
113. Wu J, Cherwinski H, Spies T et al. DAP10 and DAP12 form distinct, but functionally cooperative, receptor complexes in natural killer cells. J Exp Med 2000; 192:1059-1068.
114. Manolios N, Bonifacino JS, Klausner RD. Transmembrane helical interactions and the assembly of the T-cell receptor complex. Science 1990; 249:274-277.
115. Call ME, Pyrdol J, Wiedmann M et al. The organizing principle in the formation of the T-cell receptor-CD3 complex. Cell 2002; 111:967-979.
116. Michnoff CH, Parikh VS, Lelsz DL et al. Mutations within the NH2-terminal transmembrane domain of membrane immunoglobulin (Ig) M alters Ig alpha and Ig beta association and signal transduction. J Biol Chem 1994; 269:24237-24244.
117. Daeron M. Fc receptor biology. Annu Rev Immunol 1997; 15:203-234.
118. Feng J, Garrity D, Call ME et al. Convergence on a distinctive assembly mechanism by unrelated families of activating immune receptors. Immunity 2005; 22:427-438.
119. Feng J, Call ME, Wucherpfennig KW. The assembly of diverse immune receptors is focused on a polar membrane-embedded interaction site. PLoS Biol 2006; 4:e142.
120. Bakema JE, de Haij S, den Hartog-Jager CF et al. Signaling through mutants of the IgA receptor CD89 and consequences for Fc receptor gamma-chain interaction. J Immunol 2006; 176:3603-3610.
121. Varin-Blank N, Metzger H. Surface expression of mutated subunits of the high affinity mast cell receptor for IgE. J Biol Chem 1990; 265:15685-15694.
122. Stevens TL, Blum JH, Foy SP et al. A mutation of the mu transmembrane that disrupts endoplasmic reticulum retention. Effects on association with accessory proteins and signal transduction. J Immunol 1994; 152:4397-4406.
123. Zidovetzki R, Rost B, Pecht I. Role of transmembrane domains in the functions of B- and T-cell receptors. Immunol Lett 1998; 64:97-107.
124. Blum JH, Stevens TL, DeFranco AL. Role of the mu immunoglobulin heavy chain transmembrane and cytoplasmic domains in B-cell antigen receptor expression and signal transduction. J Biol Chem 1993; 268:27236-27245.
125. Ra C, Jouvin MH, Kinet JP. Complete structure of the mouse mast cell receptor for IgE (Fc epsilon RI) and surface expression of chimeric receptors (rat-mouse-human) on transfected cells. J Biol Chem 1989; 264:15323-15327.
126. Schamel WW, Arechaga I, Risueno RM et al. Coexistence of multivalent and monovalent TCRs explains high sensitivity and wide range of response. J Exp Med 2005; 202:493-503.

127. Berlanga O, Bori-Sanz T, James JR et al. Glycoprotein VI oligomerization in cell lines and platelets. J Thromb Haemost 2007; 5:1026-1033.
128. Loregian A, Palu G. Disruption of protein-protein interactions: towards new targets for chemotherapy. J Cell Physiol 2005; 204:750-762.
129. Hershberger SJ, Lee SG, Chmielewski J. Scaffolds for blocking protein-protein interactions. Curr Top Med Chem 2007; 7:928-942.
130. Sillerud LO, Larson RS. Design and structure of peptide and peptidomimetic antagonists of protein-protein interaction. Curr Protein Pept Sci 2005; 6:151-169.
131. Sigalov AB. Transmembrane interactions as immunotherapeutic targets: lessons from viral pathogenesis. Adv Exp Med Biol 2007; 601:335-344.
132. Sigalov AB. New therapeutic strategies targeting transmembrane signal transduction in the immune system. Cell Adh Migr 2010; 4:255-267.
133. Sigalov AB. Novel mechanistic concept of platelet inhibition. Expert Opin Ther Targets 2008; 12:677-692.
134. Sigalov AB. Inhibiting Collagen-induced Platelet Aggregation and Activation with Peptide Variants. US 12/001,258 and PCT PCT/US2007/025389 patent applications filed on 12/11/2007 and 12/12/2007, respectively, claiming a priority to US 60/874,694 provisional patent application filed on 12/13/2006.
135. Sigalov AB. Novel mechanistic insights into viral modulation of immune receptor signaling. PLoS Pathog 2009; 5:e1000404.
136. Wang XM, Djordjevic JT, Kurosaka N et al. T-cell antigen receptor peptides inhibit signal transduction within the membrane bilayer. Clin Immunol 2002; 105:199-207.
137. Quintana FJ, Gerber D, Kent SC et al. HIV-1 fusion peptide targets the TCR and inhibits antigen-specific T-cell activation. J Clin Invest 2005; 115:2149-2158.
138. Sigalov AB. Interaction between HIV gp41 fusion peptide and T-cell receptor: putting the puzzle pieces back together. FASEB J 2007; 21:1633-1634; author reply 1635.
139. Sigalov AB. More on: glycoprotein VI oligomerization: a novel concept of platelet inhibition. J Thromb Haemost 2007; 5:2310-2312.
140. Allez M, Tieng V, Nakazawa A et al. CD4^{+}NKG2D^{+} T-cells in Crohn's disease mediate inflammatory and cytotoxic responses through MICA interactions. Gastroenterology 2007; 132:2346-2358.
141. Ito Y, Kanai T, Totsuka T et al. Blockade of NKG2D signaling prevents the development of murine CD4^{+} T-cell-mediated colitis. Am J Physiol Gastrointest Liver Physiol 2008; 294:G199-207.
142. Appel RD, Bairoch A, Hochstrasser DF. A new generation of information retrieval tools for biologists: the example of the ExPASy WWW server. Trends Biochem Sci 1994; 19:258-260.
143. Kyte J, Doolittle RF. A simple method for displaying the hydropathic character of a protein. J Mol Biol 1982; 157:105-132.

CHAPTER 5

CONSEQUENCES OF FUZZINESS IN THE NFκB/IκBα INTERACTION

Elizabeth A. Komives
Department of Chemistry and Biochemistry, University of California San Diego, San Diego, California, USA
Email: ekomives@ucsd.edu

Abstract: This chapter provides a short review of various biophysical experiments that have been applied to the inhibitor of kappa B, IκBα and its binding partner, nuclear factor kappa B, or NFκB. The picture that emerges from amide hydrogen/deuterium exchange, NMR and binding kinetics experiments is one in which parts of both proteins are "fuzzy" in the free-state and some parts remain "fuzzy" in the NFκB-IκBα complex. The NFκB family of transcription factors responds to inflammatory cytokines with rapid transcriptional activation, in which NFκB enters the nucleus and binds DNA. Just as rapidly as transcription is activated, it is subsequently repressed by newly synthesized IκBα that also enters the nucleus and removes NFκB from the DNA. Because IκBα is an ankyrin repeat protein, it's "fuzziness" can be controlled by mutagenesis to stabilized the folded state. Experimental comparison with such stabilized mutants helps provide evidence that much of the system control depends on the "fuzziness" of IκBα.

INTRODUCTION

The nuclear factor κB (NFκB) pathway transduces extra-cellular signals from various receptors to regulate patterns of gene expression.[1] Although originally discovered in B-cells because it strongly activates the immunoglobulin kappa-chain gene expression,[2] the pathway is ubiquitous and has been implicated in a variety of cellular functions such as cell growth, proliferation, apoptosis and stress responses and is missregulated in numerous diseases.[3,4] The family of NFκB proteins includes p65 (RelA), RelB, c-Rel, p50 and p52 subunits, which form homo- and heterodimers[3] (Fig. 1A). The most prevalent form in most cell types is a p50/p65 heterodimer and the crystal structure of this form bound to a

Fuzziness: Structural Disorder in Protein Complexes, edited by Monika Fuxreiter and Peter Tompa.

canonical κB DNA sequence has been solved[5] (Fig. 1B). The inhibitors of NFκB activity, IκBs, include isoforms IκBα, IκBβ and IκBε, which block the nuclear localization and transcriptional activity of p65 and c-Rel-containing NFκB dimers[6] and others that act in different pathways such as IκBδ.[7] In resting cells, approximately 100,000 NFκB dimers are nearly all bound to IκBs which keep the NFκB in the cytoplasm by sequestering the NFκB nuclear localization signal (NLS).[8,9] The way in which the IκBα binds NFκB was revealed from crystal structures of the NFκB-IκBα.[10,11]

When a cell receives an extracellular signal such as a viral insult or cytokine, extracellular receptors activate the assembly of the IκB kinase (IKK), which in turn phosphorylates the N-terminal signal response domain of NFκB-bound IκBα, leading to subsequent ubiquitination and degradation of the IκBα by the proteasome.[12] NFκB dimers then translocate to the nucleus, bind DNA and regulate transcription of numerous NFκB target genes.[13] NFκB activated genes show widely varying transcription levels, activation kinetics and post-induction repression, but how this single system results in so many different transcription effects is not well understood.[14,15] The gene coding for IκBα is one of the strongly NFκB-activated genes.[16-18] When NFκB transcription is activated,

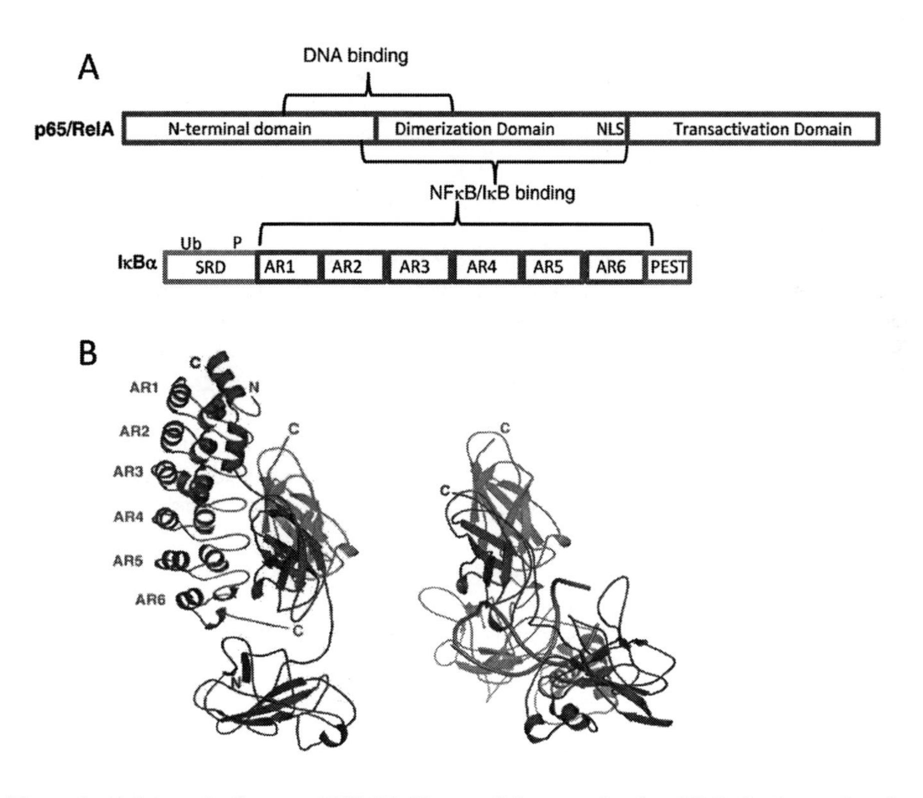

Figure 1. A) Schematic diagram of NFκB(p65) one of the most abundant NFκB family members in the cell and of IκBα, the key member of the inhibitor family. B) LEFT: The crystal structure of IκBα (blue) bound to NFκB(p50, green; p65, red).[11] RIGHT: The crystal structure of NFκB(p50, green; p65, red) bound to κB site DNA (gold).[5] (Figure prepared using PyMOL).[44] A color version of this image is available at www.landesbioscience.com/curie.

the resulting newly synthesized IκBα translocates to the nucleus, binds to NFκB and the complex is exported from the nucleus.[1,19]

Free IκBα is rapidly degraded by a proteasome-dependent but ubiquitin-independent mechanism, with a half-life is less than 10 min. On the other hand, NFκB-bound IκBα is incredibly stable, with an intracellular half-life of many hours consistent with the binding constant of the complex, of 40 pM.[9,20] The IκBα in complex with NFκB is only degraded if it is first phosphorylated, then ubiquitinated and finally degraded by the proteasome in a ubiquitin-dependent fashion resulting in free, active NFκB. A wealth of experimental data now suggests that the various functions of IκBα depend on the partially folded character or "fuzziness" of parts of NFκB and IκBα. These functions include the rapid degradation of the free protein, its tight binding to NFκB and its ability to dissociate NFκB from the DNA. This chapter will briefly summarize the evidence for "fuzziness" and show the functional consequences of "fuzziness" in the NFκB signaling system.

EXPERIMENTAL EVIDENCE OF IκBα "FUZZINESS"

NMR Evidence of IκBα "Fuzziness"

NMR studies of the entire 6-AR ARD of IκBα, residues 67-287, revealed that most of the cross peaks for AR5 and AR6 were missing, which likely indicated conformational exchange processes.[21] In contrast, all of the resonances could be assigned for the 4-AR fragment containing residues 67-206.[21,22] When the chemical shift values for these residues were compared to those of the NFκB-bound IκBα(67-287), they were nearly identical indicating that the structure of this part of IκBα is nearly identical in the free and bound states.[21] In addition, backbone dynamics experiments indicated that this part of the IκBα ARD is rigidly structured.[21,22] Residual dipolar coupling (RDC) measurements were also performed on this fragment. RDCs predicted from the crystal structure of this part of IκBα bound to NFκB did not agree well with the measured values and we surmised that this might be because RDCs report on motions from microseconds to hundreds of milliseconds (Fig. 2A).[22] Indeed, RDCs computed for an ensemble of structures of the IκBα(67-206) generated from accelerated molecular dynamics simulations agreed much better with the experimental data (Fig. 2B). Thus, RDC measurements combined with AMD simulations might more realistically represent the solution ensemble in all its "fuzziness".

Amide Hydrogen/Deuterium Exchange of the IκBα ARD

Full-length IκBα is composed of three regions; an N-terminal signal response region of ~70 amino acids, where phosphorylation and ubiquitination occur, an ankyrin repeat domain (ARD) of ~220 amino acids and a C-terminal PEST sequence that extends from residues 275-317.[10,11] The NFκB binding activity is localized to the ARD and PEST regions, for which high resolution crystal structures were obtained only when in complex with NFκB. Sequence analyses predict intrinsic disorder in both the N-terminal domain and the PEST region of IκBα as well as in a good portion of the ARD (Fig. 3A).[23,24] IκBα has resisted all attempts to crystallize it in the unbound state and its biophysical behavior is consistent with a native state that does not adopt a unique compact fold.[25] It is interesting to note the qualitative agreement between the the predicted disorder (Fig. 3A) and the native state amide exchange (plotted as percent exchanged at 2 min) (Fig. 3B).

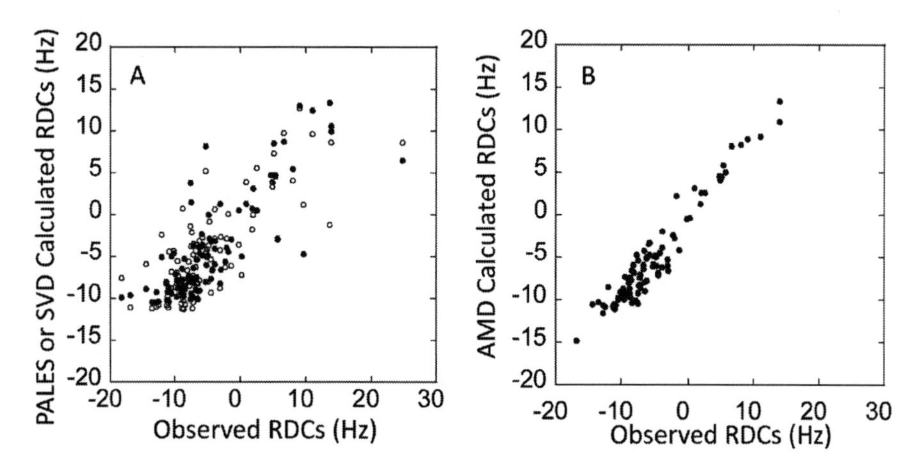

Figure 2. A) Observed vs. theoretical residual dipolar couplings measured by the program PALES for IκBα(67-206) (closed symbols) and SVD (open symbols) using the crystal structure of the IκBα-NFκB complex (PDB accession code 1IK).[10] B) Observed vs. AMD-calculated residual dipolar couplings for IκBα(67-206). The RDCs were measured as previously described.[22]

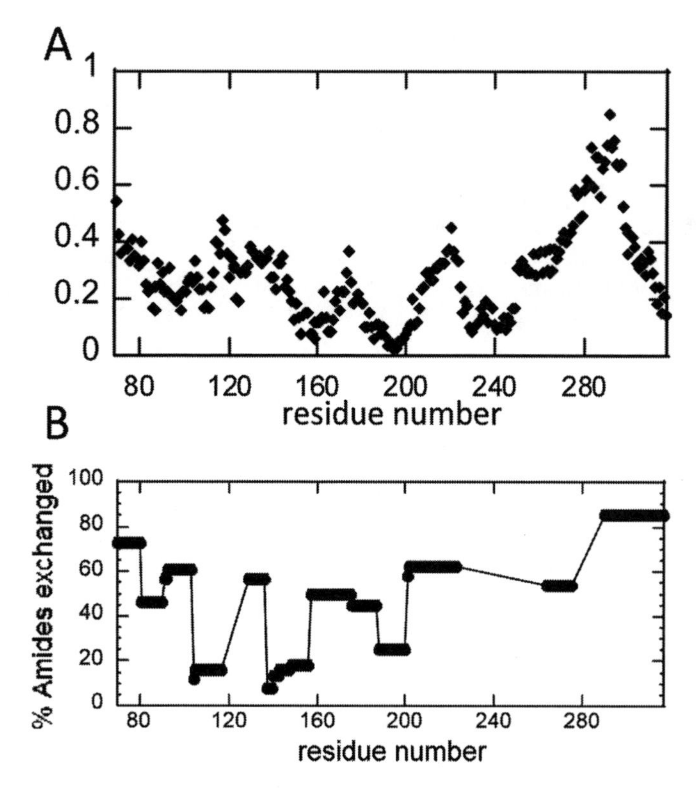

Figure 3. A) IUPRED[23,24] analysis of the intrinsic disorder in the ankyrin repeat domain of IκBα. B) Summary of native state amide H/D exchange results on free IκBα. IκBα(67-317) was allowed to exchange for 2 min and after quenching, the protein was digested with pepsin and the amount of exchange in each peptide was analyzed by MALDI mass spectrometry.[25]

BOTH IκBα AND NFκB FOLD ON BINDING

The NFκB NLS Folds on Binding to IκBα

NFκB family members such as the canonical member of the family, RelA, contain three main domains, the N-terminal domain, the dimerization domain and the transactivation domain. IκBα binds primarily to the dimerization domain whereas DNA binds in between the N-terminal and dimerization domains (Fig. 1A). The crystal structure of the NFκB-IκBα complex shows that IκBα binds to NFκB in a head-to-tail fashion and the small sequence of NFκB containing the nuclear localization signal (NLS) forms two short helices that lay over the top of AR1 of IκBα (Fig. 1B).[10,11] The NLS connects the dimerization domain to the transactivation domain in the full length NFκB(p65) protein. In between the two helices is the KRKR sequence, which constitutes the minimal NLS.[11] Theoretical studies of the binding of the NLS polypeptide (residues 291-325) of NFκB(p65) to IκBα also suggested that this segment of NFκB folds on binding to IκBα.[26] NMR heteronuclear single quantum coherence spectra of the NLS (residues 289-321 of NFκB(p65) clearly show that in the free-state the chemical shifts of the NLS backbone NHs are not well-dispersed and are mostly at random coil chemical shift values. In contrast, when bound to IκBα, the chemical shifts are well-dispersed and at values expected for helical structure (Fig. 4) (Cervantes et al).[27] It has been experimentally observed that this segment binds with a 1 μM K_D to IκBα[9] and with a large $\Delta C_{P,obs}$ for IκBα binding to this NLS segment (−1.30 ± 0.03 kcal mol^{-1} K^{-1}) that could not be accounted for by burial of polar and nonpolar surface area calculations derived from the crystal structures.[28-30] Thus, the thermodynamic signatures of the binding interaction cannot be accounted for by merely docking the individual static structures and larger structural re-arrangements must be implicated, as is often observed for protein-DNA interactions.[31] Chemical shift values in the random coil region are a good indicator of lack of persistent structure in the free NLS, which must, therefore, be more "fuzzy" in the free-state than in the bound state where persistent helical structure is observed. Thus, the "head" of IκBα (ARs 1-3) appears to be folded based on H/D exchange experiments and the "tail" of NFκB (the NLS polypeptide) folds upon binding to it.

IκBα Folds on Binding to NFκB

Native state amide H/D exchange experiments revealed that the fifth and sixth ARs exchange all of their amides within 2 minutes whereas the β-hairpins of AR2 and AR3 were remarkably resistant to exchange. (Fig. 5).[32] The decrease in the number of exchanging amides could not be accounted for just by interface protection suggesting that IκBα undergoes a folding transition upon binding. Amide exchange is an interesting probe of "fuzziness". Although the rate of amide exchange depends on many factors that often cannot be teased apart, it can reliably report on relative differences. Thus, it is possible to compare the β-turns of each AR relative to one another as we did for the free protein and it is also possible to compare the β-turn of a particular AR in the free vs. NFκB-bound state. To separate the decrease in amide exchange due to decreased solvent accessibility at the protein-protein interface from the decreased solvent accessibility due to folding of IκBα upon binding, we compared the solvent accessible surface areas calculated from the crystal structure of the NFκB-IκBα complex to the results from amide exchange. These comparisons revealed that whereas the difference in exchange between

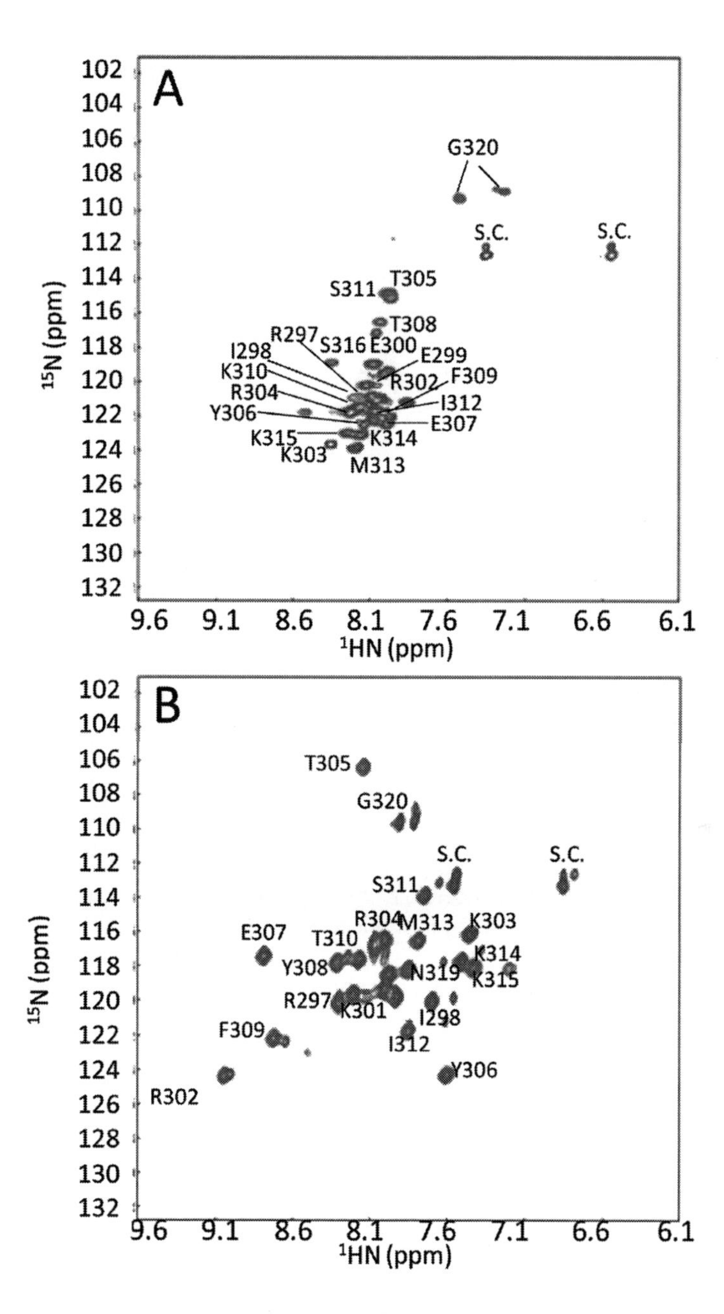

Figure 4. The HSQC spectra of A) free and B) IκBα-bound p65(293-321). The secondary chemical shifts of bound p65(289-321) show the characteristic positive values characteristic of helical regions, conforming to the helical areas seen in these same residues in the crystal structure of the IκBα-NFκB complex (PDB accession code 1NFI).[11] The chemical shifts for free p65(289-321) are indicative of an unfolded peptide.

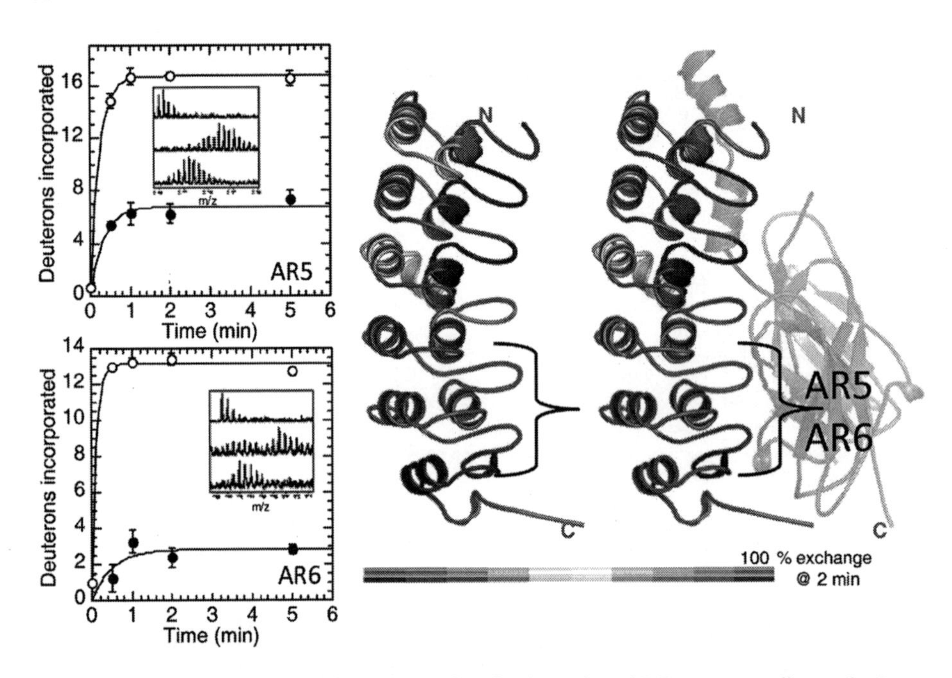

Figure 5. A) Native state amide H/D exchange data for the region of IκBα corresponding to the β-turn of AR5 (TOP) and AR6 (bottom) in the free-state (open circles) and in the NFκB-bound state (closed circles). B) Structural summary of the amide H/D exchange data (red is highly exchanging and blue is slowly exchanging) for IκBα in the free state (LEFT) and in the NFκB-bound state (RIGHT) showing that exchange is similar for most of the IκBα molecule in each state, except for the β-turns of AR5 and AR6 that are exchanging much less in the bound state. A color version of this image is available at www.landesbioscience.com/curie.

bound and free was some 10 amides, the solvent accessible surface area was expected to change only by 2-3 amides if the two folded proteins were brought together into the bound complex (Fig. 5).[33] The large difference in amide exchange upon binding should therefore be attributed to folding of the "fuzzy" parts of the free protein upon binding.

REMAINING FUZZINESS IN THE NFκB-IκBα COMPLEX

At the other end of the IκBα ARD, deletion of the PEST sequence (residues 276-287) reduces the NFκB binding by some 5 kcal/mol.[34] Taken together, the binding affinity losses due to deletion at the ends of the interface are more than enough to account for the entire binding energy of complex formation. The PEST region does not become completely ordered upon binding to NFκB according to high resolution NMR spectroscopy data.[35] The native state of the NFκB-IκBα complex thus retains regions with high dynamic character.

NMR experiments on the NFκB-IκBα complex revealed another interesting feature. Although amide exchange was very low in AR2 and AR3, NMR relaxation experiments on IκBα(67-206) indicated that the backbone of AR3 was more dynamic than was observed for the other ARs.[22] NMR experiments further revealed that the dynamics observed in

AR3 of free IκBα become even more excentuated upon binding to NFκB. In the complex, many of the cross peaks for AR3 are not observed indicating strong conformational exchange to a range of chemical shift values making the peaks so broad as to become unobservable under the conditions of the experiment.[21] These results strongly support the idea that while some parts of IκBα (AR5 and AR6) become less "fuzzy" in the complex, other parts become more "fuzzy".

FUNCTIONAL CONSEQUENCES OF FUZZINESS IN THE NFκB-IκBα COMPLEX

Tight Binding to Multiple Partners

NFκB is a family of homo and heterodimeric molecules made of at least five different proteins. While the most prevalent form in many cell types is NFκB(p50/p65), under certain conditions, the homodimeric form NFκB(p65/p65) also becomes prevalent.[36] The structure of this form of NFκB bound to IκBβ has been solved and it is remarkably similar to the structure of NFκB(p50/p65) bound to IκBα.[37] This is despite the fact that the sequences are not that similar. Kinetic and thermodynamic measurements of binding affinity show that IκBα binds to NFκB(p50/p65) with nearly the same affinity as to NFκB(p65/p65).[9]

"Fuzziness" Determines the Degradation Rate of IκBα

Free IκBα, which is marginally stable,[25] has a very short intracellular half-life of less than 10 minutes.[20,38] This rapid degradation rate depends in part on the presence of the C-terminal PEST sequence.[39-41] The degradation of the free protein appears to be independent of ubiquitinylation, since all of the Lys residues in IκBα can be mutated without changing the degradation rate of the free protein.[41] In addition, although free IκBα can be phosphorylated and ubiquitinylated, its degradation rate is not different in IKK–/– cells indicating that ubiquitin-independent degradation is the primary route for free IκBα.[20]

To further probe how "fuzziness" was related to free IκBα degradation rate, we prepared a mutant in which two residues in AR6 were mutated to more commonly-found residues at those positions in other AR sequences. Taking advantage of the single tryptophan residue in IκBα at position 258, we showed that in the wild-type protein, W258 did not show a co-operative folding transition whereas in the mutant it did (Fig. 6A,B). This is a strong indication that AR6 is not part of the co-operatively folding ARD in wild-type IκBα, but the mutations stabilize AR6 so that now the fluorescence signal from W258 follows the major co-operative transition (Fig. 6B). Importantly, the Y254L, T257A mutant IκBα is degraded more slowly than wild-type IκBα both in vitro by the 20S proteasome and in vivo suggesting that in addition to the PEST sequence, the "fuzzy" AR6 of IκBα is important for rapid ubiquitin-independent degradation (Fig. 6C).[42] In contrast to the marginally-stable free IκBα, the IκBα-NFκB complex has a very long intracellular half-life of the complex, which is completely stable in the absence of IκB kinase (IKK) phosphorylation and subsequent ubiquitinylation (>12 hrs). Thus, IκBα "fuzziness", controlled by binding to NFκB, switches its degradation mechanisms.[41]

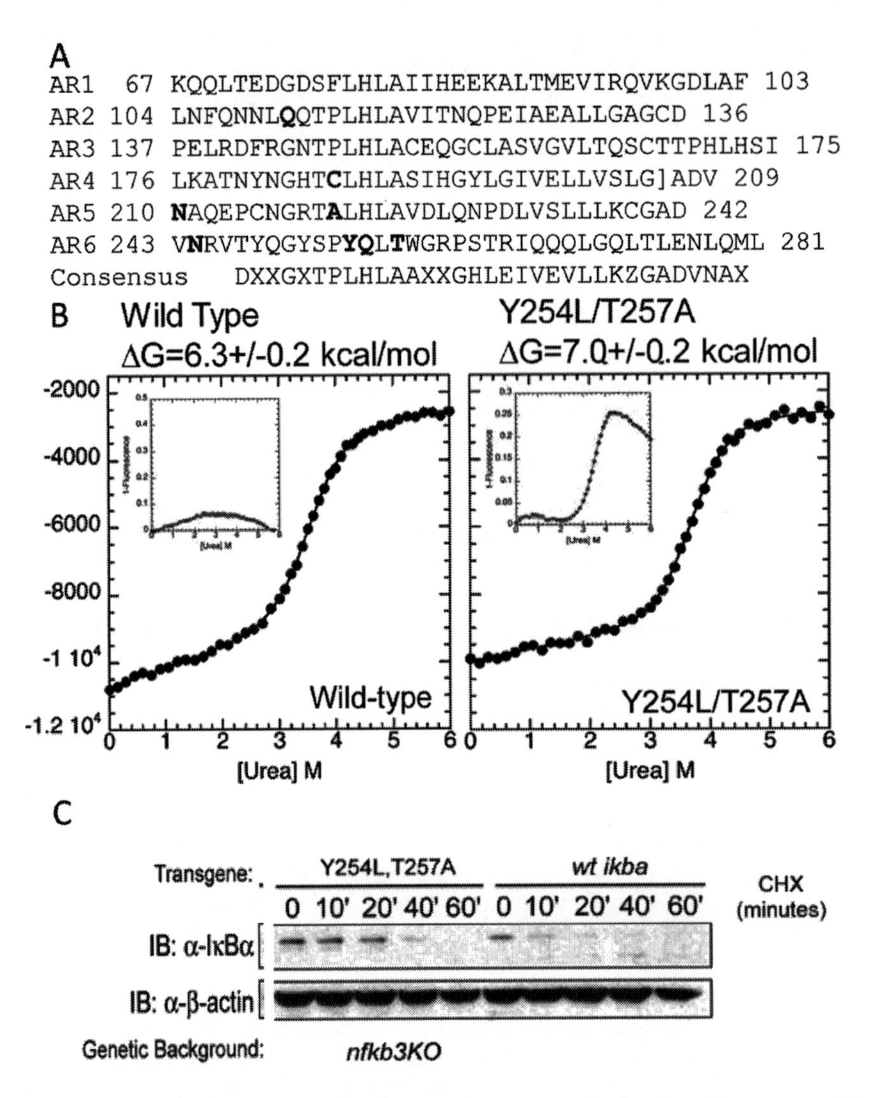

Figure 6. A) Sequence of IκBα compared to the consensus sequence showing sites of consensus mutations (in bold). B) Equilibrium unfolding experiments with wild type IκBα and (B) Y254L,T257A mutant IκBα. The insets show the change in fluorescence of W258, a naturally-occurring Trp258 in AR6. In the wild type protein, this residue does not change fluorescence appreciably with denaturant, however in the stabilized mutant, its fluorescence changes in a manner similar to the CD signal indicating it follows the major co-operative folding transition of the protein. C) Degradation of IκB isoforms in cells after stimulation of NFκB signaling were measured by quantitative western blot.

"Fuzziness" is Important for Rapid Signal Repression by IκBα

A key feature of the NFκB negative feedback is the rapidity with which the transcriptional activation is subsequently repressed (Fig. 7A).[1] Rapid post-induction repression is partly explained by the fact that the gene for IκBα is strongly induced by NFκB, so activation of NFκB immediately produces newly synthesized IκBα. However,

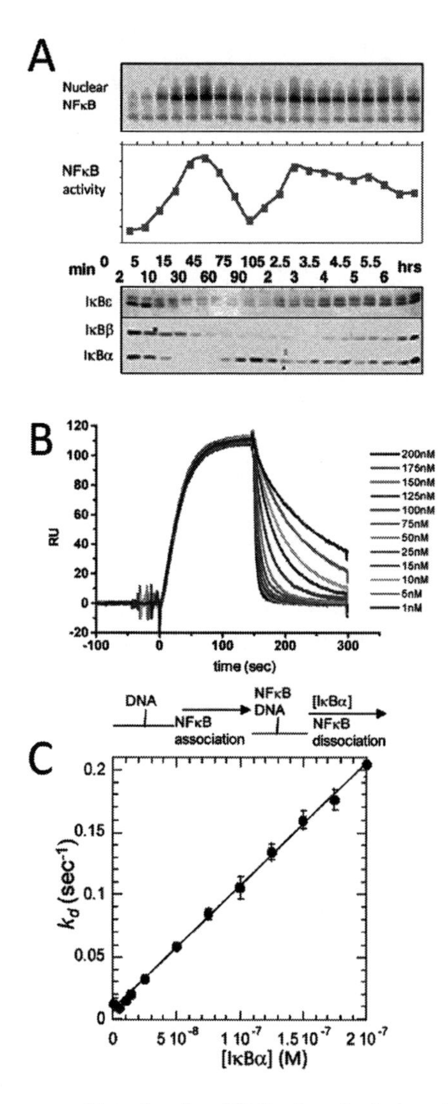

Figure 7. A) Quantitative western blot of nuclear NFκB after stimulation of cells with tumor necrosis factor α. The nuclear NFκB increases after stimulation due to release from inhibition by IκBα and then is rapidly recycled out of the nucleus by newly synthesized IκBα. B) Real-time SPR binding experiment in which κB-site DNA was bound to the streptavidin chip at t = 0, then NFκB(p50$_{(19\text{-}363)}$/p65$_{(1\text{-}325)}$) was allowed to associate with the DNA and finally varying concentrations of IκBα were injected through the second sample loop and the dissociation rate constant (k_d) was measured.[43] A schematic of the binding events is shown below the graph. C) Dissociation rate constants for active dissociation are plotted as a function of IκBα variant concentration.

the new IκBα must still escape proteasome degradation, enter the nucleus and compete for binding to NFκB with the very large number of κB sites in the DNA. We recently discovered an intriguing kinetic phenomenon in which IκBα is able to markedly increase the rate of dissociation of NFκB from the DNA.[43] The phenomenon was initially discovered by flowing nanomolar concentrations of IκBα over the NFκB-DNA complex in a co-injection step in an SPR experiment (Fig. 7B). IκBα is remarkably efficient at increasing the dissociation rate (k_d) of NFκB from the DNA; the apparent second order rate constant for the IκBα-mediated dissociation is 10^6 M^{-1} s^{-1} (Fig. 7C). Several mutant forms of IκBα were also tested for their ability to mediate dissociation of NFκB from the DNA. The mutations had a variety of effects on NFκB binding, some bound with the same affinity and some showed decreased affinity, up to 100-fold. However, all of the thermodynamically stabilized mutants, even the ones that bound with the same affinity, were less able to mediate dissociation of NFκB from the DNA.[43] Thus, an important function of the "fuzzy" AR6 in IκBα may be to facilitate dissociation of NFκB from the DNA to rapidly repress post-induction transcriptional activation.

CONCLUSION

The NFκB signaling regulates many genes and therefore is highly controlled. Biophysical experiments, including amide H/D exchange and NMR reveal that parts of both proteins are "fuzzy". From these experiments, we see that "fuzziness" comes in many flavors and only some parts of each protein are "fuzzy". In addition, the "fuzziness" of IκBα is reduced in some regions and increased in other regions upon binding to NFκB. More importantly, we have used biochemical experiments to show that "fuzziness" in IκBα provides kinetic control of dynamic regulatory processes, including its degradation through Ub-dependent and independent pathways. An important function of the "fuzzy" region of IκBα is to rapidly remove NFκB from transcription sites.

REFERENCES

1. Hoffmann A, Levchenko A, Scott ML et al. The IkappaB-NF-kappaB signaling module: temporal control and selective gene activation. Science 2002; 298(5596):1241-1245.
2. Hoffmann A, Baltimore D. Circuitry of nuclear factor kappaB signaling. Immunol Rev 2006; 210:171-186.
3. Ghosh S, May MJ, Kopp EB. NF-kappa B and Rel proteins: evolutionarily conserved mediators of immune responses. Annu Rev Immunol 1998; 16:225-260.
4. Kumar A, Takada Y, Boriek AM et al. Nuclear factor-kappaB: its role in health and disease. J Mol Med 2004; 82(7):434-448.
5. Chen FE, Huang DB, Chen YQ et al. Crystal structure of p50/p65 heterodimer of transcription factor NF-kappaB bound to DNA. Nature 1998; 391(6665):410-413.
6. Verma IM, Stevenson JK, Schwarz EM et al. Rel/NF-kappa B/I kappa B family: intimate tales of association and dissociation. Genes Dev 1995; 9(22):2723-2735.
7. Basak S, Kim H, Kearns JD et al. A fourth IkappaB protein within the NF-kappaB signaling module. Cell 2007; 128(2):369-381.
8. Baeuerle PA. IkB-NF-kB structures: at the interface of inflammation control. Cell 1998; 95:729-731.
9. Bergqvist S, Croy CH, Kjaergaard M et al. Thermodynamics reveal that helix four in the NLS of NF-kappaB p65 anchors IkappaBalpha, forming a very stable complex. J Mol Biol 2006; 360(2):421-434.
10. Huxford T, Huang DB, Malek S et al. The crystal structure of the IkappaBalpha/NF-kappaB complex reveals mechanisms of NF-kappaB inactivation. Cell 1998; 95(6):759-770.
11. Jacobs MD, Harrison SC. Structure of an IkappaBalpha/NF-kappaB complex. Cell 1998; 95(6):749-758.
12. Traenckner EB, Baeuerle PA. Appearance of apparently ubiquitin-conjugated I kappa B-alpha during its phosphorylation-induced degradation in intact cells. J Cell Sci Suppl 1995; 19:79-84.
13. Pahl HL. Activators and target genes of Rel/NF-kappaB transcription factors. Oncogene 1999; 18(49):6853-6866.
14. Hoffmann A, Leung TH, Baltimore D. Genetic analysis of NF-kappaB/Rel transcription factors defines functional specificities. EMBO J 2003; 22(20):5530-5539.
15. Werner SL, Barken D, Hoffmann A. Stimulus specificity of gene expression programs determined by temporal control of IKK activity. Science 2005; 309(5742):1857-1861.

16. Brown K, Park S, Kanno T et al. Mutual regulation of the transcriptional activator NF-kappa B and its inhibitor, I kappa B-alpha. Proc Natl Acad Sci USA 1993; 90(6):2532-2536.
17. Scott ML, Fujita T, Liou HC et al. The p65 subunit of NF-kappa B regulates I kappa B by two distinct mechanisms. Genes Dev 1993; 7(7A):1266-1276.
18. Sun SC, Ganchi PA, Ballard DW et al. NF-kappa B controls expression of inhibitor I kappa B alpha: evidence for an inducible autoregulatory pathway. Science 1993; 259(5103):1912-1915.
19. Arenzana-Seisdedos F, Turpin P, Rodriguez M et al. Nuclear localization of IkBa promotes active transport of NF-kB from the nucleus to the cytoplasm. J Cell Sci 1997; 110:369-378.
20. O'Dea EL, Barken D, Peralta RQ et al. A homeostatic model of IkappaB metabolism to control constitutive NF-kappaB activity. Mol Syst Biol 2007; 3:111.
21. Sue SC, Cervantes C, Komives EA, Dyson HJ. Transfer of flexibility between ankyrin repeats in IkappaBalpha upon formation of the NF-kappaB complex. J Mol Biol 2008; 380(5):917-931.
22. Cervantes CF, Markwick PRL, Sue SC et al. Functional dynamics of the folded ankyrin repeats of IkappaB alpha revealed by nuclear magnetic resonance. Biochemistry 2009; 48:8023-8031.
23. Dosztányi Z, Csizmók V, Tompa P et al. IUPred: web server for the prediction of intrinsically unstructured regions of proteins based on estimated energy content. Bioinformatics 2005; 21:3433-3434.
24. Dosztányi Z, Csizmók V, Tompa P, I. S. The pairwise energy content estimated from amino acid composition discriminates between folded and intrinsically unstructured proteins. J Mol Biol 2005; 347:827-839.
25. Croy CH, Bergqvist S, Huxford T et al. Biophysical characterization of the free IkappaBalpha ankyrin repeat domain in solution. Protein Sci 2004; 13(7):1767-1777.
26. Latzer J, Papoian GA, Prentiss MC et al. Induced fit, folding and recognition of the NF-kappaB-nuclear localization signals by IkappaBalpha and IkappaBbeta. J Mol Biol 2007; 367:262-274.
27. Cervantes CF, Bergqvist S, Kjaergaard M et al. The RelA Nuclear Localization Signal Folds upon Binding to IκBα. J Mol Biol 2011; 405(3):754-64.
28. Ha JH, Spolar RS, Record MT. Role of the hydrophobic effect in stability of site-specific protein-DNA complexes. J Mol Biol 1989; 209(4):801-816.
29. Livingstone JR, Spolar RS, Record MT. Contribution to the thermodynamics of protein folding from the reduction in water-accessible nonpolar surface-area. Biochemistry 1991; 30(17):4237-4244.
30. Spolar RS, Livingstone JR, Record MT. Use of liquid-hydrocarbon and amide transfer data to estimate contributions to thermodynamic functions of protein folding from the removal of nonpolar and polar surface from water. Biochemistry 1992; 31(16):3947-3955.
31. Spolar RS, Record JMT. Coupling of local folding to site-specific binding of proteins to DNA. Science 1994; 263:777-784.
32. Truhlar SM, Torpey JW, Komives EA. Regions of IkappaBalpha that are critical for its inhibition of NF-kappaB. DNA interaction fold upon binding to NF-kappaB. Proc Natl Acad Sci USA 2006; 103(50):18951-18956.
33. Truhlar SM, Croy CH, Torpey JW et al. Solvent accessibility of protein surfaces by amide H/2H exchange MALDI-TOF mass spectrometry. J Am Soc Mass Spectrom 2006; 17(11):1490-1497.
34. Bergqvist S, Ghosh G, Komives EA. The IkBa/NF-kB complex has two hot-spots, one at either end of the interface. Prot Sci 2008; 17:2051-2058.
35. Sue SC, Dyson HJ. Interaction of the IkappaBalpha C-terminal PEST sequence with NF-kappaB: insights into the inhibition of NF-kappaB DNA binding by IkappaBalpha. J Mol Biol 2009; 388(4):824-838.
36. Shih VF, Kearns JD, Basak S et al. Kinetic control of negative feedback regulators of NF-kappaB/RelA determines their pathogen- and cytokine-receptor signaling specificity. Proc Natl Acad Sci USA 2009; 106(24):9619-9624.
37. Malek S, Huang DB, Huxford T et al. X-ray crystal structure of an IkappaBbeta x NF-kappaB p65 homodimer complex. J Biol Chem 2003; 278(25):23094-23100.
38. Rogers S, Wells R, Rechsteiner M. Amino acid sequences common to rapidly degraded proteins: the PEST hypothesis. Science 1986; 234(4774):364-368.
39. Rice NR, Ernst MK. In vivo control of NF-kappa-B activation by I-kappa-B-alpha. EMBO J 1993; 12:4685-4695.
40. Pando MP, Verma IM. Signal-dependent and-independent degradation of free and NF-kappa B bound IkappaBalpha. J Biol Chem 2000; 275:21278-21286.
41. Mathes E, O'Dea EL, Hoffmann A et al. NF-kappaB dictates the degradation pathway of IkappaBalpha. EMBO J 2008; 27(9):1357-1367.
42. Truhlar SME, Mathes E, Cervantes CF et al. Pre-folding IkappaBalpha alters control of NF-kappaB signaling. J Mol Biol 2008; 380:67-82.
43. Bergqvist S, Alverdi V, Mengel B et al. Kinetic enhancement of NF-kappaB•DNA dissociation by IkappaBalpha. Proc Natl Acad Sci USA 2009; 106(46):19328-19333.
44. The PyMOL Molecular Graphics System [computer program]. Version. San Carlos, CA, USA: DeLano Scientific; 2002.
45. Garner E, Romero P, Dunker AK et al. Predicting binding regions within disordered proteins. Genome Informatics 1999; 10:41-50.

CHAPTER 6

ROLES FOR INTRINSIC DISORDER AND FUZZINESS IN GENERATING CONTEXT-SPECIFIC FUNCTION IN ULTRABITHORAX, A HOX TRANSCRIPTION FACTOR

Sarah E. Bondos* and Hao-Ching Hsiao
Department of Molecular and Cellular Medicine, Texas A&M Health Science Center, College Station, Texas, USA
**Corresponding Author: Sarah E. Bondos—Email: sebondos@tamhsc.edu*

Abstract: Surprisingly few transcription factors drive animal development relative to the number and diversity of final tissues and body structures. Therefore, most transcription factors must function in more than one tissue. In a famous example, members of the Hox transcription factor family are expressed in contiguous stripes along the anterior/posterior axis during animal development. Individual Hox transcription factors specify all tissues within their expression domain and thus must respond to cellular cues to instigate the correct tissue-specific gene regulatory cascade. We describe how, in the *Drosophila* Hox protein Ultrabithorax, intrinsically disordered regions implement, regulate and co-ordinate multiple functions, potentially enabling context-specific gene regulation. The large N-terminal disordered domain encodes most of the transcription activation domain and directly impacts DNA binding affinity by the Ubx homeodomain. Similarly, the C-terminal disordered domain alters DNA binding affinity and specificity, interaction with a Hox binding protein and strongly influences both transcription activation and repression. Phosphorylation of the N-terminal disordered domain and alternative splicing of the C-terminal disordered domain could allow the cell to both regulate and co-ordinate DNA binding, protein interactions and transcription regulation. For regulatory mechanisms relying on disorder to continue to be available when Ubx is bound to other proteins or DNA, fuzziness would need to be preserved in these macromolecular complexes. The intrinsically disordered domains in Hox proteins are predicted to be on the very dynamic end of the disorder spectrum, potentially allowing disorder to persist when Ubx is bound to proteins or DNA to regulate the function of these "fuzzy" complexes. Because both intrinsically disordered regions within Ubx have multiple roles, each region may implement several different regulatory mechanisms identified in fuzzy complexes. These intrinsic disorder-based regulatory mechanisms are likely to be

Fuzziness: Structural Disorder in Protein Complexes, edited by Monika Fuxreiter and Peter Tompa.

critical for allowing Ubx to sense tissue identity and respond by implementing a context-specific gene regulatory cascade.

INTRODUCTION

Surprisingly few transcription factors promote animal development relative to the complex array of tissues and organs that must be generated. Consequently, individual transcription factors must function in multiple tissues, implementing the specific gene regulatory cascade that is uniquely appropriate for each context. The ability of these transcription factors to reliably sense and correctly respond to tissue identity is critical, since incorrect regulation would create catastrophic defects in the developing organism. The need to implement many specific and reliable functions imposes several functional challenges on the transcription factors that drive animal development: (i) they must incorporate mechanisms to sense multiple cellular signals, (ii) they must functionally integrate these signals into a unique spatiotemporal specific response and (iii) they must use these integrated signals to co-ordinate the activities of disparate modular functional domains (e.g., DNA binding, transcription activation and transcription repression domains).

INTRINSIC DISORDER AND CONTEXT-SPECIFIC GENE REGULATION

All of these logistical problems can potentially be elegantly solved using intrinsically disordered protein sequences. A key attribute of intrinsic disorder is that it provides a mechanism to expand the functionality of a protein, allowing variations on a central function, or, in extreme cases, entirely different functions to be mediated by the same protein sequence.[1-4] For instance, intrinsically disordered regions can be used to bind multiple proteins, yielding complexes with unique functionalities.[1,4] Furthermore, intrinsic disorder facilitates amino acid sequence alterations, such as alternative mRNA splicing or post-translational modifications, to further expand functionality.[3,5-7]

By combining sequences for alternative splicing, phosphorylation, conformational changes, protein interaction, or ubiquitination/degredation, intrinsically disordered regions also provide a mechanism by which multiple signals could be integrated to elicit a tissue-specific response.[8-11] For instance, phosphorylation could facilitate or enable protein interactions, placing two input requirements (phosphorylation and protein binding) on the output—unique function of the protein complex. The reverse case, in which protein binding precedes phosphorylation, may also occur, but requires the disordered region retain flexibility (fuzziness) in the protein complex.[8]

Finally, transcription factors that regulate multiple genes must bind different factors or assemble different multiprotein-DNA complexes at each enhancer/promoter. The flexibility of their disordered domains may allow these transcription factors to generate multiple macromolecular complexes with different geometries. Taken together, it is perhaps not surprising that more than 90% of all transcription factors contain intrinsically disordered regions.[12,13] Given the variety of potential regulatory mechanisms, the categorization of fuzzy complexes proposed by Tompa and Fuxrieter[14] is extremely valuable in discussing, comparing and categorizing how intrinsically disordered regions of transcription factors enable context-specific transcription regulation.

Potential Roles for Fuzzy Complexes in Transcription Regulation

When a regulatory mechanism relies on intrinsic disorder, any processe that removes a disordered region or triggers stable folding of the intrinsically disordered segment inactivates the regulatory mechanism. For a transcription factor that relies on multiple disorder-based regulatory mechanisms, the adaptability and the potential of the protein to integrate multiple cellular signals would also be lost. Consequently, a significant level of disorder is likely to be retained even when the protein engages in protein-DNA or protein-protein complexes. In a recent paper, Tompa and Fuxreiter classified different types of macromolecular complexes containing disorder—generally termed "fuzzy complexes"—and discussed the role of disorder in these complexes.[14] Given the potential of intrinsically disordered regions to impart context-specific function, one would expect many examples of fuzziness to occur in transcription factor complexes, particularly those involved in animal development. Since transcription factors have a modular domain structure and are often composed of both structured and disordered domains, one would also expect such context-specific functions to require extensive interplay between structured and disordered regions of the protein. Many such regulatory functions enabled by intrinsically disordered regions have been observed in the Hox protein Ubx, a transcription factor that specifies and maintains tissue identity during animal development. In this chapter, we will describe the roles of intrinsic disorder in modulating the formation of macromolecular complexes by the *Drosophila* Hox protein Ultrabithorax to ultimately create a tissue-specific response.

ULTRABITHORAX, A HOX TRANSCRIPTION FACTOR, AS A MODEL SYSTEM

Role of Hox Proteins in Animal Development

In all bilaterally symmetric animals, Hox proteins operate during development to generate unique tissues, organs and appendages from serially repeated structures. Individual members of the Hox transcription factor family are expressed in contiguous, non-overlapping regions along the anterior-posterior axis, where they specify the fate of structures in every tissue layer. These functions are remarkably conserved in vertebrates and invertebrates.[15,16] The basal role of Hox proteins in determining tissue fate allows misexpression of a Hox protein to transform one structure or region of the body into another,[17,19] revealing dramatic phenotypes that underscore the requirement for reliable Hox function in vivo.

A given Hox protein specifies the fate of multiple body structures. Consequently, Hox proteins must sense and respond to their environment to regulate different subsets of their target genes in order to generate unique fates for each tissue, organ, or appendage in which the Hox protein is expressed.[10,20-22] For instance, the *Drosophila melanogaster* Hox protein Ubx specifies the posterior-most legs, the halteres (balancing organs used during flight) and the posterior aorta, as well as portions of the midgut, ectoderm, musculature and central and peripheral nervous systems.[17,23-25] To correctly guide development of each structure, each Hox protein must regulate a distinct subset of its own downstream targets in each tissue or region within a tissue in which that Hox protein is expressed.[20,21,26] This phenomenon, termed "context-specific gene regulation", requires different DNA sequences to be bound by the same Hox protein in each cellular context. Furthermore, once bound, the Hox protein must determine whether to activate or repress the downstream gene in a tissue-dependent manner. For instance, Ubx activates the gene *decapentaplegic* in the midgut, but represses this same gene in the developing haltere.[27-29] Therefore, understanding Hox function in animal development requires elucidating how a single Hox protein senses spatio-temporal information to instigate a variety of context-specific transcription cascades. Intrinsically disordered regions in Hox proteins appear crucial for these processes.

Alternative Splicing in Ubx

This chapter will primarily focus on the *Drosophila melanogaster* Hox protein Ultrabithorax (Ubx), for which the most regulatory mechanisms, protein interactions and gene targets have been described.[9,10,20,30-33] Intrinsically disordered regions that flank a structured or functional domain can still influence function.[14] This phenomenon occurs several times within the Ubx protein, but for the C-terminal disordered region there is an added twist: This region is also alternatively spliced, providing the potential for tissue-specific mRNA splicing to contribute to context-specific Hox function.

Different Ubx mRNA splicing isoforms are produced in a stage- and tissue-specific manner by alternative splicing of three microexons (Fig. 1). Isoforms containing the 9-a.a. "b element" are a minor component in all tissues.[34] Inclusion of the two 17 a.a. microexons—mI and mII—is determined by tissue identity, germ layer and developmental stage. Based on these complex expression patterns, alternative splicing may be a source of contextual information for directing different Ubx functions.[34] Indeed, ectopic expression of Ubx isoforms differentially transforms the peripheral nervous system.[35,36] Furthermore, Ubx isoforms differ in their ability to activate the Ubx target gene *decapentaplegic* and generate the correct muscle development patterns.[37] Among Drosophilid species, the amino acid sequence of the optional microexons, as well as isoform-specific expression patterns are remarkably conserved.[38] Furthermore, key amino acids in this same region N-terminal to the homeodomain are conserved in Hox paralogue groups in vertebrates.[39]

Identification of Intrinsically Disordered Regions in Ubx

Ubx is also the only Hox protein in which predicted intrinsically disordered regions have been experimentally validated.[9] Whereas 7% of a typical protein sequence is the flexible amino acid glycine,[40-43] the Ubx sequence is 17% glycine and its activation domain is 27% glycine. This extremely high glycine content, combined with the few proline residues, strongly suggested portions of Ubx are intrinsically disordered. Indeed, computational algorithms predicted much of the Ubx protein is intrinsically disordered (Fig. 1A).[9] The existence and location of the disordered domains were experimentally verified by native state proteolysis.[9] Since proteases can only cleave sites embedded in at least 10 unstructured, solvent-exposed amino-acids,[44] protease cleavage marks unstructured or disordered regions within a protein. Ubx is much more protease-sensitive than ApoMb or unliganded Lac Repressor, both proteins with a well-characterized disordered region. Proteolytic fragments were identified by size and epitope content of the resulting peptide fragments.[9,45,46] In general, cleaved sites were located within regions predicted to be disordered, whereas protected sites occurred within regions expected to have structure. Ubx contains several intrinsically disordered regions, including much of

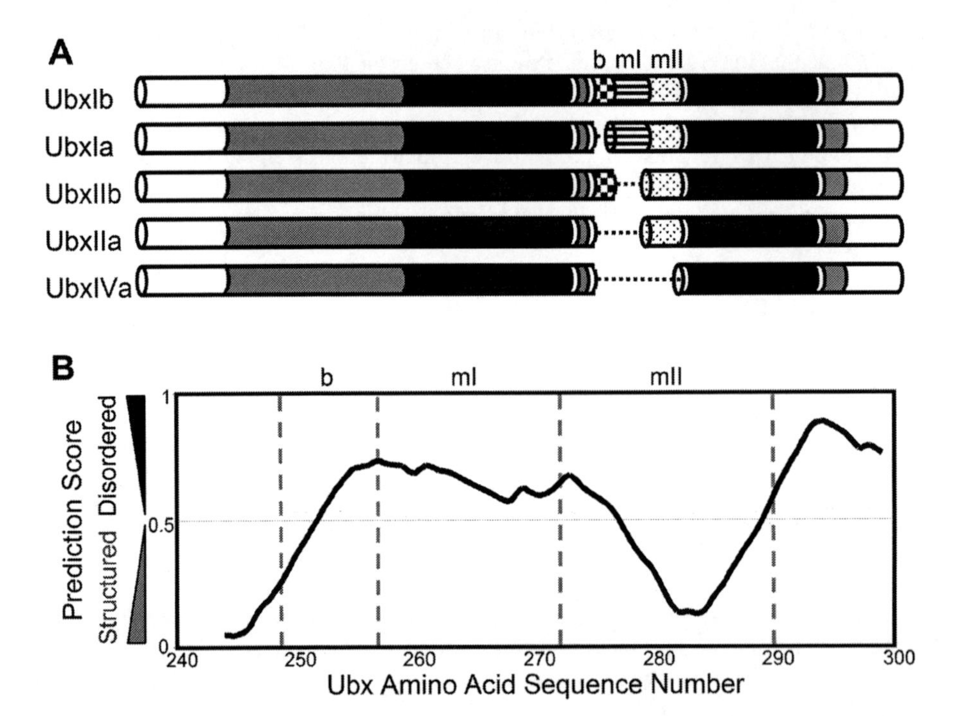

Figure 1. Alternative splicing influences both the sequence and disorder content of Ubx. A) Schematics of alternative splicing isoforms of Ubx produced in vivo. Ubx Ia and Ubx Ib are produced primarily in the embryonic mesoderm, Ubx IIa and Ubx IIb are generated in larval imaginal discs—tissues that will develop into appendages and Ubx IVa is the only isoform produced in the central nervous system in embryos and larva.[34,45] Microexons removed by alternative splicing are depicted by a dashed line. B) The microexon region in general is intrinsically disordered, although IUPred predicted disorder scores vary between individual microexons.[9]

the transcription activation domain and the alternatively spliced microexons. Disordered regions are frequently involved in protein or ligand interactions. These regions can also be post-translationally modified or alternatively spliced.[3,5,47] *Thus, the intrinsically disordered regions in Ubx have potential regulatory functions*. Extensive experimental studies have demonstrated that these regions in Ubx do indeed mediate or regulate protein interactions, transcription activation and DNA binding (see below refs. 9,10,33).

In fact, Ubx has fewer predicted disordered amino acids than any other *Drosophila* Hox protein (Fig. 2A). Furthermore, the magnitude of these scores is generally higher for other *Drosophila* Hox proteins, suggesting a larger degree of disorder similar than Ubx. These disordered regions can be located on either or both sides of the structured homeodomain. Although the position and the degree of predicted disorder varies significantly between Hox homologues, these features are remarkably conserved among Hox orthologues (Fig. 2B).

Human Hox proteins are also predicted to be significantly disordered, with the length and magnitude varying significantly between proteins (Fig. 3). Disorder scores for selected human Hox proteins linked to skeletal defects and carcinogenesis are mapped onto sequence in Figure 3A. The extent of intrinsic disorder score varies significantly between human Hox proteins, ranging from 18% for Hox D13 to 85% for Hox A3. The portion

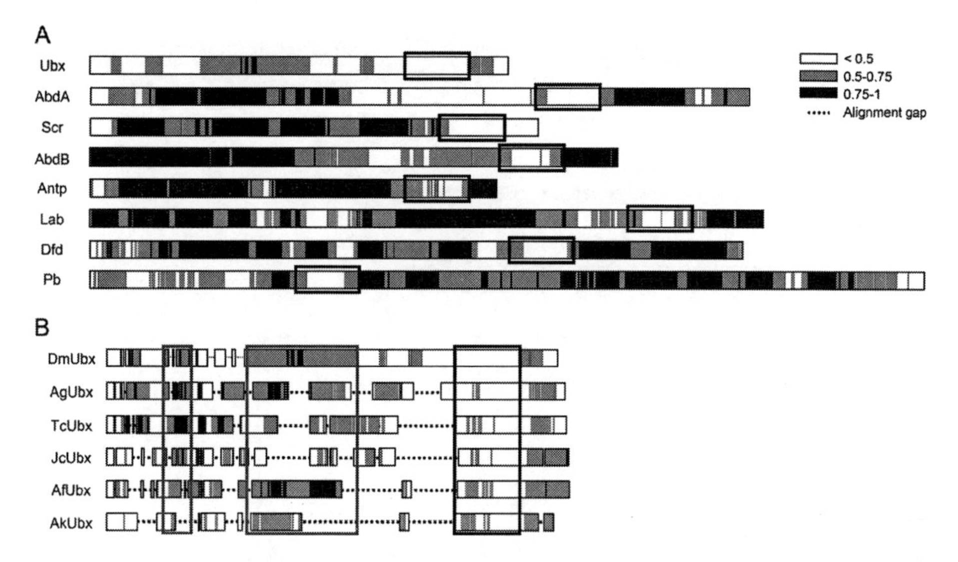

Figure 2. Variations in the content, magnitude and location of IUPred predicted intrinsic disorder in Hox proteins. Each bar represents a protein sequence, in which regions predicted to have a disorder score of 0.5 to 0.75 are shaded grey and regions predicted to be very disordered (score 0.75 to 1.0) are in black. Protein schematics are aligned by the conserved homeodomain, which is indicated by a black box. A) The eight Hox proteins in *Drosophila melanogaster*, arranged by increasing numbers of disordered amino acids, all have extensive intrinsically disordered regions. B) The *Drosophila* Ubx (DmUbx) orthologues from mosquito (AgUbx), red flour beetle (TcUbx), butterfly (JcUbx), brine shrimp (AfUbx) and velvet worm (AkUbx) reflect 540 million years of evolution. Predictions were generated using IUPred. Evolutionarily conservation of disordered regions (grey boxes) suggests the disorder mediated regulation observed in Ubx may be based on ancient regulatory mechanisms.

of the sequence predicted to be disordered is more consistent between Hox paralogues (~21% for Hox A13, ~30% disordered for Hox B13, ~36% for Hox C13 and 18% for Hox D13) (Fig. 3B). Consequently, not only are disorder-based regulatory mechanisms still likely to be active in human Hox proteins, but variation in disorder content may also contribute to differential function within the Hox protein family.

Experimental Approaches to Studying Intrinsic Disorder in Ubx In Vitro

Identifying functions or regulatory processes mediated by intrinsically disordered regions is a challenging task, since the logic and experimental approaches typically used to identify functional or regulatory roles for portions of structured proteins cannot be easily applied to intrinsically disordered regions. First, since intrinsically disordered regions often mediate protein interactions, conditions must be identified that maintain soluble protein monomers. For Ubx, a filter-based aggregation assay was used to rapidly screen for buffers that maintain solubility.[48] Second, since structural information is frequently unavailable for intrinsically disordered regions and is only available for less than one fifth of the Ubx sequence,[49] a guide is lacking to probe function by point mutagenesis. Furthermore, intrinsically disordered regions are generally insensitive to mutation. As a result, most experiments on Ubx exploited a series of truncation mutants to progressively define functional or regulatory regions.[9,30,33] The start sites for N-terminal truncations

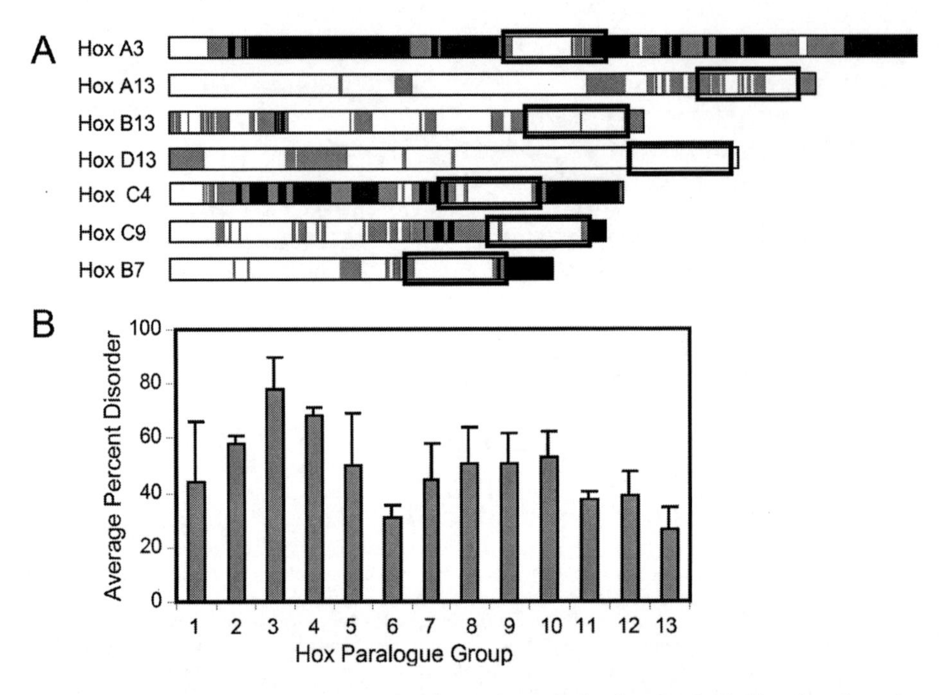

Figure 3. Human Hox proteins are also significantly intrinsically disordered. A) Disorder schematics for human Hox proteins with known roles in developmental malformations or carcinogenesis. Whereas *Drosophila* have one cluster of Hox genes, humans have four clusters, each represented by a different letter, which arose by genome duplication. Hox proteins at similar positions in different clusters (paralogues, with the same number) have related functions during human development. B) For each paralogue group, e.g., Hox A4, Hox B4, Hox C4 and Hox D4, the average predicted percent of intrinsically disordered residues is depicted. Most Hox paralogue groups are approximately 50% disordered, however some groups are significantly more (Hox group 3 and 4) or less (Hox group 6 and 13) disordered. The extent of disorder is generally well conserved within a Hox paralogue group.

and the end points for C-terminal truncations were chosen based on the location of evolutionarily conserved sequences, predicted structured and disordered domains and regions enriched in particular amino acids. The resulting truncation mutants were tested for both solubility and activity prior to use in experiments.[9,30,33] To ensure gross structural re-arrangements would not generate false results, conclusions based on data from truncation mutants were verified in full-length Ubx using small internal deletions and, when possible, point mutants.

ROLES FOR INTRINSIC DISORDER IN IMPLEMENTING OR REGULATING Ubx FUNCTION

A critical role of intrinsically disordered regions in Ubx is to communicate positional information from the cell and to co-ordinate the activity of several functional domains. Nearly every region of Ubx impacts the function of many other regions of Ubx. To describe multiple complex regulatory functions, we will first discuss identification of each functional domain and subsequently describe the regulatory interactions that modulate their activity.

DNA Binding

The Hox protein family was first discovered in *Drosophila*. Using DNA from homeotic mutants, in which entire regions of the body partially or completely develop as different body regions,[50] chromosome walk experiments discovered a series of genes in region of the genome responsible for these transformations in the body plan.[18,51,52] Subsequent DNA hybridization experiments revealed that these genes all shared a very similar DNA sequence, termed the homeobox, in the protein coding region.[53] The portion of the protein encoded by the homeobox was called the homeodomain.[54] Homeodomain-containing Hox proteins are present in all bilaterally symmetric animals.[54]

An individual Hox protein determines the expression of hundreds, if not thousands of genes,[55-57] which, in turn, direct the patterns of growth, differentiation, proliferation and apoptosis specific to that region of the developing organism.[17,20,26,29,58] Hox proteins bind their target DNA sequences via a 60 amino acid homeodomain (Fig. 2), in which the C-terminal helix contacts the major groove of DNA and the disordered N-terminal arm binds the minor groove. Due, in part, to charge-charge interactions between the extremely positively charged homeodomain (net charge of +11 in Ubx) and the phosphate backbone of DNA, the isolated homeodomain of a Hox protein binds DNA with extremely high affinity (~60 pM).[9,10] Most Hox homeodomains prefer to bind the DNA sequence 5′-TAAT-3′,[59-61] although base substitutions are permitted at multiple positions.[20] DNA sequences outside this 4 bp motif only have moderate effects on affinity,[62,63] perhaps because the homeodomain•DNA interface tolerates significant structural heterogeneity.[64,65] Consequently, the DNA binding specificity of Hox homeodomains is notoriously poor and anticipated to be insufficient to distinguish cognate and noncognate target sequences in vivo.[66-68] While chromatin remodeling will limit the total number of exposed binding sites,[69] the breadth of target sequences bound by Hox proteins implies a significant number of noncognate binding sites remain exposed. Thus protein regions outside the homeodomain, most of which are intrinsically disordered, must contribute to DNA binding specificity.

INTRINSICALLY DISORDERED REGIONS MODULATE DNA BINDING AFFINITY BY THE STRUCTURED UBX HOMEODOMAIN

The Ubx homeodomain (UbxHD) has a 2.5-fold higher affinity for the optimal Ubx DNA binding sequence than full-length Ubx. Consequently, amino acid sequences outside the homeodomain, which are primarily disordered, must impact DNA recognition. The location of sequences that impact DNA binding was determined using N- and/or C-terminal truncation mutants. Sequential removal of each section tested whether a region enhances, inhibits, or has no effect upon DNA binding and revealed three regions that alter binding affinity (Fig. 4).[9] The I1 region, located between amino acids 235 and 286 and containing the YPWM motif and adjacent microexon region, weakly inhibits binding, whereas the I2 region (a.a. 174-216, located in the core activation domain) strongly inhibits binding. Binding is restored by the R region (a.a. 1-174) in a manner that is linearly dependent on the length of R (Fig. 4), a feature consistent with this region's intrinsically disordered character. These data reveal several key points:[9]

i. Most of the nonhomeodomain regions of the Ubx protein are capable of impacting DNA binding (83%).
ii. Over half of the regions that regulate DNA binding are intrinsically disordered (51%) (Fig. 4).
iii. Regions that regulate affinity include sequences that are post-translationally modified[70] or alternatively spliced,[34,35] potentially allowing cellular factors to impact DNA affinity.
iv. The Exd interaction motif and the activation domain both directly impact DNA binding affinity, potentially allowing Exd binding and transcription regulation to influence or be influenced by DNA interactions.

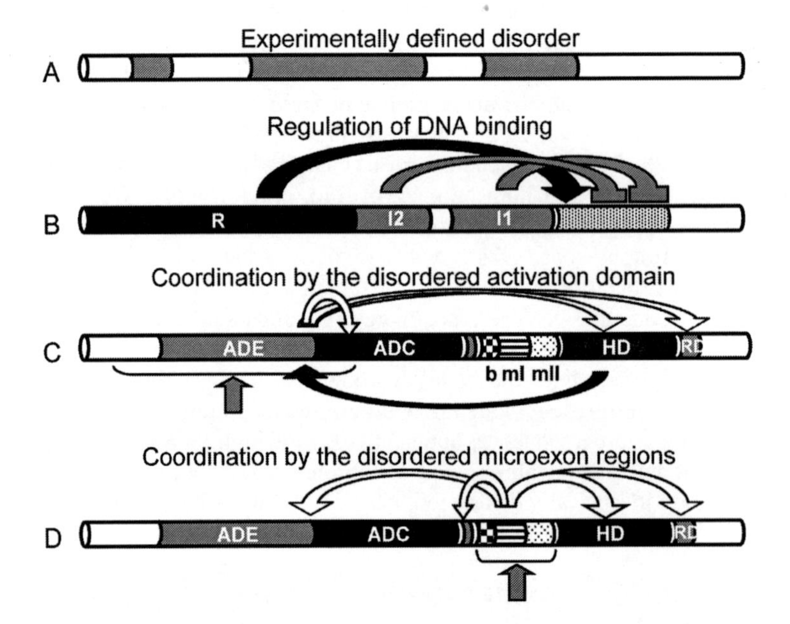

Figure 4. Intrinsically disordered regions contribute to regulatory mechanisms and long-range communication in Ubx. A) A schematic of Ubx shading gray the experimentally defined borders of intrinsically disordered regions.[9] B) The regions of Ubx that impact DNA binding affinity to an optimal DNA binding site.[9] The R region (black shading) restores most of the loss of affinity by the homeodomain (stippled) caused by the I1 and I2 regions (grey shading). Subsections of the I1 region, which includes the YPWM Exd-interaction motif and the alternatively spliced microexons, directly impacts DNA binding specificity and also modulate regulation of binding by other subsections. C,D) Schematics of Ubx depicting communication between functional domains mediated by intrinsic disorder. ADE, Activation domain enhancer; ADC, activation domain core; Y, YPWM motif used for Exd binding; b, mI and mII, alternatively spliced microexons; HD, homeodomain; RD, a partial repression domain. C) cellular factors (grey arrow), by phosphorylating sites located within the N-terminal disordered region (bracketed area), could regulate the function of this region in mediating transcription activation, altering DNA binding specificity, or regulating the balance between transcription activation and repression (open arrows). Reciprocally, DNA sequences bound by the homeodomain influence transcription activation (black arrow). D) Cellular factors which regulate spatio-temporal-dependent alterantive splicing by Ubx (grey arrow) alter the microexon content of the protein, which, in turn, impacts transcription activation, Exd binding, DNA binding and transcription repression (open arrows).

Intrinsically Disordered Regions Directly Contact the Homeodomain to Alter Binding

Intrinsically disordered regions might inhibit DNA binding either by (i) preventing DNA from approaching the homeodomain via dynamic conformational fluctuations through the space surrounding the homeodomain or (ii) specifically interacting with the homeodomain to alter its structure or dynamics.[9] The first mechanism should be independent of DNA sequence and therefore is unlikely to contribute to context-specific DNA site selection.

To determine which mechanism is used to modulate affinity by each deleted region, we exploited an environmentally sensitive ionizable residue in the homeodomain.[9] DNA binding by UbxHD is enhanced 20-fold at pH = 6.0.[71] Point mutagenesis revealed this enhancement is not due to histidine, the only amino acid with a pK_a near 6.0.[9] Therefore, an ionizable residue whose pK_a is shifted by the local environment must cause this improved binding. Since binding by full-length Ubx is independent of pH, nonhomedomain regions of Ubx must interact with the homeodomain to perturb the environment around this amino acid, shifting its pK_a back towards its solution value. Repeating truncation mutant binding studies at pH = 6.0 enabled location of the regions of Ubx that interact with the homeodomain to alter this pK_a and hence impact homeodomain structure or dynamics. Two regions were identified that restore DNA binding affinity—one overlapping the I1 inhibitory domain and one in the middle of the N-terminal half of the protein.[9] These two regions directly modulate DNA binding affinity by altering homeodomain structure or dynamics and thus are excellent candidates for regions that impact DNA binding specificity.

The YPWM Motif and the Intrinsically Disordered Alternatively Spliced Microexons Modulate DNA Binding Specificity by Ubx

Given the DNA binding specificity of the Ubx homeodomain is likely insufficient to recognize the appropriate target genes in vivo,[66-68] could full-length Ubx bind with greater specificity than UbxHD? This possibility was examined by binding full-length Ubx and UbxHD to DNA oligos based on a variety of natural Ubx binding sites utilized in vivo that differ in DNA sequence and the number and spacing of Ubx binding sites, the developmental stage and tissue in which Ubx regulates the gene, the requirement for interacting proteins to bind DNA and the consequence (activation vs. repression) of binding (Table 1). A large variation in the binding affinity for these sequences reflects high sequence specificity, whereas similar affinities for all DNA sequences indicates low specificity.

A comparison of the DNA binding affinities of full-length Ubx and UbxHD for these DNAs revealed striking differences in binding specificity. UbxHD bound with similar high affinities to all DNAs, including the Dll binding site which is bound in vivo in conjunction with the Hox cofactor Exd. In contrast, Ubx exhibited a more than 10-fold difference in affinity for binding to these same sequences.[10] Thus a significant portion of DNA binding specificity information is clearly encoded in amino acid sequences outside the homeodomain.

The search for determinants of DNA binding specificity began in the I1 region because it directly alters DNA binding affinity,[9] it is necessary for *dll* regulation in vivo,[72] and it is located close to the N-terminal arm of the homeodomain, which forms most of the base-specific contacts with DNA.[49] A series of internal deletion mutants within

Table 1. DNA binding sequences utilized in gel shift experiments

Oligo	Top Strand Sequence	Reference	Devel. Stage	Tissue	Activation vs. Repression
Number and spacing of monomer Ubx binding sites varies					
Dpp	5′-GG**TAAT**GG**TAAT**AGTAACGG**TAAT**GA**TAAT**GG**TA** **AT**GG**TAAT**GG-3′	98	Unknown	Unknown	Unknown
UA	5′-GGAA**TAAT**AACAA**TAAT**GCCGCTGA**TAAT**GTGGA **TAAT**AAGG-3′	98	Embyro	Visceral mesoderm	Activate
Sal	5′-TACTGCTTTT**TAAT**AAGTTATGCGAACTGATGGGT TTGGTTAGGCATGCTCGATGTCTTTTA**TAAT**GTGCCCGTC T**TAAT**ATGATTTTG-3′	77	Larva	Haltere (Ectoderm)	Repress
A1	5′-GGTA**TAATATAATAAT**AAAAA**TAATAATAATAAT** **AATAAT**ATATGG-3′	76	Embryo	Visceral mesoderm	Repress
Sequence and spacing between Hox and Exd binding sites varies					
Dll	5′-GCACTATAAAACTGTCCGCGGGAA**TGATTTAAT**T TCCCAAATAT-3′	99,100	Embryo	Ectoderm	Repress with Exd
Dppe4	5′-GTAGCTTC**ATCGATAAA**ACT**TAAT**GGCCACTATAAC-3′	101	Embryo	Mesoderm	Activate with Exd
Repeat 3	5′-GGGG**TGATGGAT**GGGCGCTG-3′	73	N/A	N/A	

this region, combined with point mutants within the YPWM motif, identified several interesting trends (Fig. 4):

i. In the UbxIa splicing isoform, the YPWM motif inhibits DNA binding *only* to composite DNA sequences normally bound by both Ubx and Exd.[10] Thus, this motif may act as a "policeman" to prevent inappropriate Ubx monomer binding and mis-regulation. The analogous motif in a severe truncation mutant of another *Drosophila* Hox protein, Labial, inhibits DNA binding affinity,[73] suggesting this may be a general Hox regulatory mechanism.
ii. The region surrounding the YPWM motif independently alters DNA specificity: Internal deletions in this region cause DNA binding affinity to change in a DNA sequence-dependent manner. One of these deletions removes a portion of the mI microexon.
iii. If the 14 amino acids surrounding the YPWM motif are removed, the YPWM gains the capacity to inhibit binding to all DNA sequences examined. The deleted region included part of the mI microexon and also affects binding affinity and specificity.

Together, these data suggest that alternative splicing modulates Ubx•DNA binding affinity and also alters the requirement for Exd to bind to Ubx and relieve inhibition by the YPWM motif. Thus, tissue-specific alternative splicing and Exd availability may combine to target Ubx to a subset of its DNA binding sites.

Transcription Regulation

Unlike the highly conserved homeodomain, the sequence and activity of transcription activation domains vary significantly between Hox proteins.[33,74] The search for the Ubx activation domain relied on the yeast one-hybrid approach, in which a LexA-Ubx fusion binds to a series of tandem LexA DNA recognition sites, positioning the Ubx activation domain to stimulate transcription of a β-galatosidase reporter gene.[33] Screening a series of Ubx N- and/or C-terminal truncation mutants in yeast revealed the extent to which each variant retained the ability to activate transcription. Because DNA binding is mediated by LexA, the Ubx homeodomain could be deleted without compromising the assay. The results indicated the activation domain is divided into two functional regions: The core domain is the minimal element capable of transcription activation, albeit at ~50% of the strength of full-length protein, whereas the enhancing domain boosts the activity of the core to levels comparable to the full-length protein.[33] Most of the activation domain is intrinsically disordered, a common feature in transcription factors.[12,75] However, a region at the C-terminus of the core activation domain is predicted to form an α-helix and is required for activity in yeast. To determine whether this helix is also required for activation by full-length Ubx in its native context, point mutations were examined using a promoter-reporter assay in *Drosophila* S2 cells using a genomic promoter regulated by Ubx in vivo.[76] The strength of activation correlates with the predicted stability of this α-helix: Alanine mutations stabilize the helix, thus creating hyper-active Ubx, whereas proline mutations disrupt helical structure and abrogate transcription activation without impeding DNA binding or transcription repression.[33] To our knowledge, this is the first reported instance of a structural element regulating the activity of a large (>150 amino acid) disordered region.

By generating chimeras between *Drosophila* Ubx and its orthologues, both the McGinnis and Carroll labs simultaneously located a partial repression domain near the C-terminus of Ubx.[77,78] This region, also predicted to form a helix, is sufficient to mediate transcription repression, although removal of this domain from Ubx does not completely abrogate repression. Evolution of this region correlates with the functional complexity of Ubx.

Mechanisms for the Cellular Context to Regulate Transcription by Ubx

Many regulatory events have the potential to impact transcription regulation by Ubx in a cell context-dependent manner, including phosphorylation, alternative splicing, protein interactions and DNA binding. Ubx is phosphorylated at least five times, with most of these phosphorylation sites lying in the transcription activation domain.[70] Although a functional role has not been ascribed to these phosphorylation events, the overlap of these sites with the activation domain suggested that phosphorylation has the potential to modulate transcription regulation by Ubx. If the factors that phosphorylate Ubx are differentially available in various cell contexts, then phosphorylation could regulate transcription activation by Ubx in a tissue-specific manner. Phosphorylation is one of the few Ubx-related activities not affected by alternative splicing.[70]

Ubx mRNA is alternatively spliced in a tissue-specific manner during development and the resulting Ubx protein isoforms activate and repress transcription to different degrees.[70,76] Ubx Ib, Ubx Ia and Ubx IIa all activate and repress transcription with similar efficiencies, whereas Ubx IIb and Ubx IVa activate transcription to a much greater degree but are less able to repress transcription. Thus the microexon content of Ubx, which is regulated in a stage- and tissue-specific manner during development, can also influence the strength of both transcription activation and repression.

Protein interactions also alter the ability of Ubx to activate or repress transcription. The double stranded RNA binding protein DIP1 binds the C-terminal half of the Ubx sequence.[30] Although this region of Ubx does not contain any portion of the activation domain, DIP1 binding inhibits transcription activation by Ubx. Likewise, interaction with the general Hox cofactor Exd can either enhance or suppress transcription activation by Hox proteins, with the effect likely dependent on the identity of the Hox protein as well as on the tissue in which the interaction takes place.[79,81] Exd binds two small motifs in the C-terminal half of Ubx.[32,82]

Finally, a single Hox splicing isoform can both activate and repress transcription, even in a single tissue. Consequently, once Ubx binds the target DNA sequence, the DNA must subsequently have some mechanism of communicating to Ubx whether to activate or repress the gene. One obvious possibility is that heterologous transcription factors which also bind the enhancer determine this balance, either by binding Ubx and altering its function[30-32,83,84] or by providing additional activation or repression domains to enhance or compete with transcription regulation by Ubx.[79,85]

While such protein interactions undoubtedly can have a major impact on Ubx activity, the DNA sequence also influences the choice between transcription activation and repression. In both *Drosophila* S2 cells and in vitro assays in which known Hox binding proteins are absent, DNA sequence alone is sufficient to determine whether Ubx activates or represses transcription.[33] Activation-deficient mutants repress transcription from an "activating" DNA sequence in cell culture,[33] suggesting activation and repression compete for dominance and the DNA sequence favors one mode of transcription regulation. Therefore, both Ubx domains and target DNA sequences bound by the homeodomain contribute to the ability to activate or repress a gene, a process that requires long-range communication between opposite ends of the protein sequence (Fig. 4).

Protein Interactions

Interactions with heterologous proteins, especially other transcription factors, have long been hypothesized to be a major source of context-specific information for Hox proteins.[21,30-32,84,86,87] During development, the availability of Hox partners in the nucleaus can depend on the developmental stage, tissue identity, or position within the tissue. For instance, nuclear localization of Exd, as well as cell-signaling regulated Hox partners such as Armadillo and the Smad family are regulated during development.[21,31,88-90] Furthermore, if two Hox-interacting proteins bind the same surface of a Hox protein, their relative concentrations and affinities at that position would determine the dominant complex assembled. As described above, both DNA site selection and transcription regulation by Hox proteins can be significantly influenced by such protein interactions.

Alternative Splicing Alters Exd Interaction

A key Hox binding protein in *Drosophila* is Exd, which is homologous to Pbx proteins in vertebrates. The Exd/Pbx family of transcription factors expands Hox function by expanding the variety of DNA sites recognized as well as influencing the decision to activate or repress transcription.[32,79,83,85] The best characterized Hox-interacting protein is Extradenticle. In many Hox proteins, including Ubx, the YPWM/hexapeptide motif is separated from the DNA-binding homeodomain by the alternatively spliced microexons. Given this proximity, it is perhaps not surprising that alternative splicing impacts the ability of Ubx to bind Extradenticle.[91] Isoforms containing the b-element interact markedly less well with Exd. However, inclusion or exclusion of the mI and mII microexons have no impact on Exd binding. Furthermore, regions flanking the YPWM motif alter Exd interaction with Hox proteins in general.[92] Reciprocally, alternative splicing of the Exd/Pbx protein family may alter complex function with Hox proteins. In mammals, the ability of the Pdx-Pbx complex (similar to Hox-Exd in *Drosophila*) to repress transcription is dependent on the identity of the Pbx isoform.[93]

Other Protein Interactions

Due to their hypothesized importance, an increasing number of Ubx-interacting proteins have been identified, including transcription factors regulated by cell signaling cascades which sub-divide Ubx-specified tissues.[30-32,89] Although the regions of Ubx bound by these proteins is unknown, the high number of interacting proteins suggests intrinsically disordered regions may be involved in partner binding. Indeed, disordered regions are frequently found in the most interactive proteins.[94]

INTEGRATING INFORMATION USING INTRINSIC DISORDER IN LIGHT OF THE FUZZY COMPLEX MODEL

Putting all of these regulatory mechanisms together reveals significant cross-domain interactions in Ubx that use intrinsically disordered regions (Fig. 4). In Ubx, intrinsically disordered domains not only implement or regulate a single function, they appear to co-ordinate multiple functions. These regulatory opportunities offer multiple chances for cellular information not only to impact the function of individual domains, but also to alter communication between that domain and distant regions of the protein. For instance, the transcription activation domain in Ubx is phosphorylated within its intrinsically disordered sequence,[70] and consequently, this phosphorylation may enhance or supress the ability of Ubx to activate transcription. However, since this region also directly impacts DNA binding by the homeodomain,[9] phosphorylation could also regulate DNA binding. Likewise, alternative splicing impacts DNA binding affinity and specificity, protein interactions and transcriptional regulation.

These types of interactions have the potential to also enhance a particular mode of regulation to increase the reliability of transcription regulation. For example, interaction with a subset of DNA binding sites favors transcription activation over transcription repression. Since the activation domain can directly impact DNA affinity, stimulation of transcription activation may stabilize Ubx interaction with a favorable DNA sequence or destabilize interactions with poor sequences. In a second example, the YPWM motif specifically inhibits Ubx interaction with Hox-Exd DNA binding sites, but not sites bound by Ubx monomers.[10,73] This inhibition, which prevents mis-regulation of composite sites by Hox monomers, is relieved when Exd binds the YPWM motif.[73]

Grouping these regulatory interactions by similar features is an important step in understanding how they function as well as determining the breadth of mechanisms enabled by the presence of intrinsic disorder. Many intrinsically disordered proteins fold upon complex formation and the crucial role of intrinsic disorder is to modulate the energy of the final complex[8,95,96] or to provide a scaffold for assembly of a multi-protein complex.[97] In contrast, other intrinsically disordered regions remain disordered in the bound state, creating fuzzy complexes.[14] These residual disordered regions also have regulatory potential.[14]

Due to their length and extremely disordered character, Hox proteins likely form fuzzy complexes with DNA and most, if not all, interacting proteins. All of the regulatory interactions described above rely on or are significantly influenced by the intrinsically disordered regions of Ubx. Consequently, preservation of intrinsic disorder in Hox-DNA complexes is likely to be important in maintaining these regulatory interactions as well as to permit additional regulation. Although structural data for full-length Ubx or its protein or DNA complexes is unavailable, the region linking the YPWM motif and the homeodomain lacks density in a crystal structure of the ternary complex formed by the Exd homeodomain, a Hox-Exd DNA binding sequence and the Ubx homeodomain with additional N-terminal residues including the YPWM motif. The absence of structure in the region linking the YPWM motif and the Ubx homeodomain suggests this region, which includes alternatively spliced microexons in full-length Ubx, remains disordered even in protein and DNA complexes. Therefore, disorder is likely to be retained at least in complexes formed by Hox truncation mutants.[49] The high disorder scores and the glycine-rich sequence of the N-terminal intrinsically disordered region in Ubx further suggests this region may also remain flexible when bound to DNA or other proteins.

Analysis and comparison of the mechanisms by which fuzzy complexes function is crucial to define commonalities between systems, establish the variety of mechanisms available and to predict the mode of regulation based on comparison with known fuzzy complexes. In their review, Tompa and Fuxreiter classified known fuzzy complexes into groups depending on the placement and dynamics of the disordered region in the protein complex.[14] These classifications will be especially helpful in comparing regulatory mechanisms among disordered proteins.

Participation of Disorder in Protein Interaction Domains

The large N-terminal intrinsically disordered domain in Ubx forms the majority of the activation domain and thus must interact with components of the general transcription apparatus. Without any sort of structural information for these interactions, we cannot know for certain whether this extremely disordered region manages to fold (or at least immobilize) upon binding or forms a static or dynamic polymorphic complex, although the extremely high level of disorder[9] suggests some type of dynamic complex is likely. The large number of proteins known to bind Ubx suggests the intrinsically disordered regions of Ubx may be directly involved in binding these proteins as well.[30,31]

Ubx-Exd-DNA: A Ternary Clamp That Allows Variation in DNA Binding Sites

In the parlance of Tompa and Fuxreiter, a clamp protein interaction domain consists of two structured, or at least static, regions separated by a region that remains intrinsically disordered in the bound form.[14] This architecture allows the protein to accommodate a variety of target macromolecules by adjusting both the orientation and the distance between two structured regions. Although Ubx-Exd is a protein complex, rather than a single protein, its DNA bound form retains many of the features as well as the function of dynamic clamp complexes. Both Ubx and Exd have structured DNA-binding homeodomains, which in the complex generates a bipartite DNA recognition sequence. Since the region between the Ubx YPWM motif, which mediates interaction with Exd, and the homeodomain is intrinsically disordered, this portion of Ubx resembles the disordered segment linking the structured DNA binding domains in the Ubx-Exd clamp. As is typical for clamp complexes, the Ubx-Exd complex can recognize a variety of DNA binding sites and even accommodate variation in the distance between the Ubx and Exd recognition sequences (Table 1). In an interesting twist on the clamp functional theme, the linking intrinsically disordered region is alternatively spliced, providing an opportunity for the cell to dramatically alter the length of this linker (8 to 51 amino acids) and thus potentially dictate the range of DNA sequences that can be recognized by the Ubx-Exd complex.[91,92]

Long-Range and Short-Range Regulation by Flanking Disordered Regions

An alternate way to view the microexon region is as a disordered region that independently flanks both the DNA binding homeodomain and the Exd binding YPWM motif. In general, disordered regions that flank a structured binding site can also influence binding in the absence of other factors.[14] Perhaps not surprisingly, both DNA and Exd interactions are influenced by this flanking disordered region. Alternative splicing of this Ubx region is known to impact Exd interaction independent of DNA binding.[91,92]

Likewise, alternative splicing in Ubx has been hypothesized to alter DNA binding, independent of Exd interaction.[10]

In Ubx, both large intrinsically disordered regions alter the function of distant regions of the Ubx sequence, although in the folded protein these regions may be more proximate. The disordered alternatively spliced region impacts the function of the large activation domain, in which the closest edge is 6 residues away,[33,34] and the repression domain, which is separated by 76 residues.[34,77] Likewise, the disordered activation domain directly influences the homeodomain to alter DNA binding affinity.[9] It is not yet known what mechanism or mechanisms underlie these long-range interactions, or whether they share any features with regulation by short-range flanking regions.

CONCLUSION

Due to their need to implement a variety of tissue-specific functions in a reliable manner, transcription factors in animal development require a high degree of internal regulatory mechanisms to sense tissue identity, co-ordinate a response among multiple functional domains and reinforce this response to ensure the correct regulatory pathway is consistently activated. As described for Ubx, the high degree of cell•protein and domain•domain communication requires individual regions of a Hox protein to have multiple functional and regulatory roles, a feat facilitated by exploiting intrinsic disorder in both Hox monomers and in Hox complexes with DNA and other proteins.

ACKNOWLEDGEMENTS

We thank members of the Bondos lab at TAMHSC for helpful discussions and Jordan McIntyre for her assistance with this chapter. S.E.B. acknowledges the American Heart Association (Beginning Grant-in-Aid 0865064F) for supporting her lab's research on the role of intrinsic disorder in regulating Hox protein function.

REFERENCES

1. Kriwacki RW, Hengst L, Tennant L et al. Structural studies of p21Waf1/Cip1/Sdi1 in the free and Cdk2-bound state: conformational disorder mediates binding diversity. Proc Natl Acad Sci USA 1996; 93:11504-11509.
2. Tompa P, Szász C, Buday L. Structural disorder throws new light on moonlighting. Trends Biochem Sci 2005; 30:484-489.
3. Romero PR, Zaidi S, Fang YY et al. Alternative splicing in concert with protein intrinsic disorder enables increased functional diversity in multicellular organisms. Proc Natl Acad Sci USA 2006; 103:8390-8395.
4. Vilasi S, Ragone R. Abundance of intrinsic disorder in SV-IV, a multifunctional androgen-dependent protein secreted from rat seminar vesicle. FEBS J 2008; 275:763-774.
5. Iakoucheva LM, Radivojac P, Brown CJ et al. The importance of intrinsic disorder for protein phosphorylation. Nucl Acids Res 2004; 32:1037-1049.
6. Lu X, Hamkalo B, Parseghian MH et al. Chromatin condensing functions of the linker histone C-terminal domain are mediated by specific amino acid composition and intrinsic protein disorder. Biochemistry 2009; 48:164-172.
7. Sandhu KS. Intrinsic disorder explains diverse nuclear roles of chromatin remodeling proteins. J Mol Recognit 2009; 22:1-8.
8. Galea CA, Wang Y, Sivakolundu SG et al. Regulation of cell division by intrinsically unstructured proteins: intrinsic flexibility, modularity and signaling conduits. Biochemistry 2008; 47:7598-7609.

9. Liu Y, Matthews KS, Bondos SE. Multiple intrinsically disordered sequences alter DNA binding by the homeodomain of the Drosophila Hox protein Ultrabithorax. J Biol Chem 2008; 283:20874-20887.
10. Liu Y, Matthews KS, Bondos SE. Internal regulatory interactions determine DNA binding specificity by a Hox transcription factor. J Mol Biol 2009; 390:760-774.
11. Phng LK, Gerhardt H. Angiogenesis: a team effort coordinated by Notch. Dev Cell 2009; 16:196-208.
12. Liu J, Perumal NB, Oldfield CJ et al. Intrinsic disorder in transcription factors. Biochemistry 2006; 45:6873-6888.
13. Ward JJ, Sodhi JS, McGuffin LJ et al. Prediction and functional analysis of native disorder in proteins from the three kingdoms of life. J Mol Biol 2004; 337:635-645.
14. Tompa P, Fuxreiter M. Fuzzy complexes: polymorphism and structural disorder in protein-protein interactions. Trends Biochem Sci 2008; 33:2-8.
15. Halder G, Callaerts P, Gehring WJ. Induction of ectopic eyes by targeted expression of the eyeless gene in Drosophila. Science 1995; 267:1788-1792.
16. Lo PCH, Frasch M. Establishing A-P polarity in the embryonic heart tube: a conserved function of Hox genes in Drosophila and vertebrates. Trends Cardiovasc Med 2003; 13:182-187.
17. Hughes CL, Kaufman TC. Hox genes and the evolution of the arthropod body plan. Evol Dev 2002; 4:459-499.
18. Lewis EB. A gene complex controlling segmentation in Drosophila. Nature 1978; 276:565-570.
19. Gellon G, McGinnis W. Shaping animal body plans in development and evolution by modulation of Hox expression patterns. Bio Essays 1998; 20:116-125.
20. Pearson JC, Lemons D, McGinnis W. Modulating Hox gene functions during animal body patterning. Nature Rev Genet 2005; 6:893-904.
21. Bondos SE, Tan XX. Combinatorial transcription regulation: the interaction of transcription factors and cell signaling molecules with homeodomain proteins in Drosophila development. Crit Rev Euk Gene Express 2001; 11:145-171.
22. Mann RS, Lelli KM, Joshi R. Hox specificity: unique roles for cofactors and collaborators. Curr Topics Dev Biol 2009; 88:63-101.
23. Bienz M. Homeotic genes and positional signaling in the Drosophila viscera. Trends Genet 1994; 10:22-26.
24. Rivilin PK, Gong A, Schneiderman AM et al. The role of Ultrabithorax in the patterning of adult thoracic muscles in Drosophila melanogaster. Dev Genes Evol 2001; 211:55-66.
25. Rogulja-Ortmann A, Renner S, Technau GM. Antagonistic roles for Ultrabithorax and Antennapedia in regulating segment-specific apoptosis of differentiated motoneurons in the Drosophila embryonic central nervous system. Development 2008; 135:3435-3445.
26. Graba Y, Aragnol D, Pradel J. Drosophila Hox complex downstream targets and the function of homeotic genes. BioEssays 1997; 19:379-388.
27. Capovilla M, Brandt M, Botas J. Direct regulation of decapentaplegic by Ultrabithorax and its role in Drosophila midgut morphogenesis. Cell 1994; 76:461-475.
28. de Navas LF, Garulet DL, Sánchez-Herrero E. The Ultrabithorax Hox gene of Drosophila controls haltere size by regulating the Dpp pathway. Development 2006; 133:4495-4506.
29. Weatherbee SD, Halder G, Kim J et al. Ultrabithorax regulates genes at several levels of the wing-patterning hierarchy to shape the development of the Drosophila haltere. Genes Dev 1998; 12:1474-1482.
30. Bondos SE, Catanese DJ Jr, Tan XX et al. Hox transcription factor Ultrabithorax Ib physically and genetically interacts with Disconnected Interacting Protein 1, a double-stranded RNA-binding protein. J Biol Chem 2004; 279:26433-26444.
31. Bondos SE, Tan XX, Matthews KS. Physical and genetic interactions link Hox function with diverse transcription factors and cell signaling proteins. Mol Cell Proteomics 2006; 5:824-834.
32. Chan SK, Jaffe L, Capovilla M et al. The DNA binding specificity of Ultrabithorax is modulated by cooperative interactions with Extradenticle, another homeoprotein. Cell 1994; 78:603-615.
33. Tan XX, Bondos S, Li L et al. Transcription activation by Ultrabithorax Ib requires a predicted α-helical region. Biochemistry 2002; 41:2774-2785.
34. López AJ, Artero RD, Perez-Alonso M. Stage, tissue and cell specific distribution of alternative Ultrabithorax mRNAs and protein isoforms in the Drosophila embryo. Roux's Arch Dev Biol 1996; 205:450-459.
35. Mann RS, Hogness DS. Functional dissection of Ultrabithorax proteins in D. melanogaster. Cell 1990; 60:597-610.
36. Subramaniam V, Bomze HM, López AJ. Functional differences between Ultrabithorax protein isoforms in Drosophila melanogaster: evidence from elimination, substitution and ectopic expression of specific isoforms. Genetics 1994; 136:979-991.
37. Reed HC, Hoare T, Thomsen S. Alternative splicing modulates Ubx protein function in Drosophila melanogaster. Genetics 2010; 184:745-758.
38. Bomze HM, López AJ. Evolutionary conservation of the structure and expression of alternatively spliced Ultrabithorax isoforms from Drosophila. Genetics 1994; 136:965-977.

39. Merabet S, Hudry B, Saadaoui M et al. Classification of sequence signatures: a guide to Hox protein function. BioEssays 2009; 31:500-511.
40. Jones DT, Taylor WR, Thornton JM. The rapid generation of mutation data matrices from protein sequences. Compt Appl Biosci 1992; 8:275-282.
41. White SH. Global statistics of protein sequences: implications for the origin, evolution and prediction of protein structure. Annu Rev Biophys Biomol Struct 1994; 23:407-439.
42. Tourasse NJ, Li WH. Selective constraints, amino acid composition and the rate of protein evolution. Mol Biol Evol 2000; 17:656-664.
43. Brooks DJ, Fresco JR. Increased frequency of cysteine, tyrosine and phenylalanine since the last universal ancestor. Mol Cell Proteomics 2002; 1:125-131.
44. Hubbard S, Benyon RJ. Proteolysis of native proteins as a structural probe. In: Benyon R, Bonds JS, eds. Proteolytic Enzymes, 2nd edition. Oxford: Oxford University Press, 2001:248-249.
45. López AJ, Hogness DS. Immunochemical dissection of the Ultrabithorax homeoprotein family in Drosophila melanogaster. Proc Natl Acad Sci USA 1991; 88:9924-9928.
46. White RAH, Wilcox M. Protein products of the bithorax complex in Drosophila. Cell 1984; 39:163-171.
47. Hegedűs T, Serohijos AWR, Dokholyan NV et al. Computational studies reveal phosphorylation-dependent changes in the unstructured R domain of CFTR. J Mol Biol 2008; 378:1052-1063.
48. Bondos SE, Bicknell A. Detection and prevention of protein aggregation before, during and after purification. Anal Bioch 2003; 316:223-231.
49. Passner JM, Ryoo HD, Shen L et al. Structure of a DNA-bound Ultrabithorax-Extradenticle homeodomain complex. Nature 1999; 397:714-719.
50. Bridges CB, Morgan TH. The third-chromosome group of mutant characters of Drosophila melanogaster. Publs Carnegie Instn 1923; 327:1-251.
51. Garber RL, Kuroiwa A, Gehring WJ. Genomic and cDNA clones of the homeotic locus Antennapedia in Drosophila. EMBO J 1983; 2:2027-2036.
52. Scott MP, Weiner AJ, Polisky BA et al. The molecular organization of the Antennapedia complex of Drosophila. Cell 1983; 35:763-776.
53. McGinnis W, Levine MS, Hafen E et al. A conserved DNA sequence in homeotic genes of the Drosophila antennapedia and bithorax complexes. Nature 1984; 308:428-433.
54. McGinnis W, Garber RL, Wirz J et al. A homologous protein-coding sequence in Drosophila homeotic genes and its conservation in other metazoans. Cell 1984; 37:403-408.
55. Hersh BM, Nelson CE, Stoll SJ et al. The UBX-regulated network in the haltere imaginal disc of D. melanogaster. Dev Biol 2007; 302:717-727.
56. Carr A, Biggin MD. A comparison of in vivo and in vitro DNA-binding specificities suggests a new model for homeoprotein DNA binding in Drosophila embryos. EMBO J 1999; 18:1598-1608.
57. Mastick GS, McKay R, Oligino T et al. Identification of target genes regulated by homeotics proteins in Drosophila melanogaster through genetic selection of Ultrabithorax protein-binding sites in yeast. Genetics 1995; 139:349-363.
58. Zhai Z, Stein MAS, Lohmann I. Expression of the apoptosis gene reaper in homeotic, segmentation and other mutants in Drosophila. Gene Expr Patterns 2009; 9:357-363.
59. Damante G, Pelizzari L, Esposito G et al. A molecular code dictates sequence-specific DNA recognition by homeodomains. EMBO J 1996; 15:4992-5000.
60. Berger MF, Badis G, Gehrke AR. Variation in homeodomain DNA binding revealed by high-resolution analysis of sequence preferences. Cell 2008; 133:1266-1276.
61. Noyes MB, Christensen RG, Wakabayashi A et al. Analysis of homeodomain specificities allows the family-wide prediction of preferred recognition sites. Cell 2008; 133:1277-1289.
62. Ekker SC, Young E, von Kessler DP et al. Optimal DNA sequence recognition by the Ultrabithorax homeodomain of Drosophila. EMBO J 1991; 10:1179-1186.
63. Ekker SC, Jackson DG, von Kessler DP et al. The degree of variation in DNA sequence recognition among four Drosophila homeotic proteins. EMBO J 1994; 13:3551-3560.
64. Frazee RW, Taylor JA, Tulius TD. Interchange of DNA-binding modes in the deformed and Ultrabithorax homeodomain of Drosophila. EMBO J 2002; 10:1179-1186.
65. Gutmanas A, Billeter M. Specific DNA recognition by the Antp homeodomain: MD simulations of specific and nonspecific complexes. Proteins 2004; 57:772-782.
66. Hoey T, Levine M. Divergent homeo box proteins recognize similar DNA seuqneces in Drosophila. Nature 1988; 332:858-861.
67. Kalionis B, O'Farrell PH. A universal target sequence is bound in vitro by diverse homeodomains. Mech Dev 1993; 43:57-70.
68. Gehring WJ, Qian YQ, Billeter M et al. Homeodomain-DNA recognition. Cell 1994; 78:211-223.
69. Clapier CR, Cairns BR. The biology of chromatin remodeling complexes. Ann Rev Biochem 2009; 78:273-304.
70. Gavis ER, Hogness DS. Phosphorylation, expression and function of the Ultrabithorax protein family in Drosophila melanogaster. Development 1991; 112:1077-1093.

71. Li L, von Kessler D, Beachy PA et al. pH-dependent enhancement of DNA binding by the Ultrabithorax homeodomain. Biochemistry 1996; 35:9832-9839.
72. Tour E, Hittinger CT, McGinnis W. Evolutionarily conserved domains required for activation and repression functions of the Drosophila Hox protein Ultrabithorax. Development 2005; 132:5271-5281.
73. Chan SK, Pöpperl H, Krumlauf R et al. An Extradenticle-induced conformational change in a Hox protein overcomes an inhibitory function of the conserved hexapeptide motif. EMBO J 1996; 15:2476-2487.
74. Li X, Murre C, McGinnis W. Activity regulation of a Hox protein and a role for the homeodomain in inhibiting transcriptional activation. EMBO J 1999; 18:198-211.
75. Garza AS, Ahmad N, Kumar R. Role of intrinsically disordered protein regions/domains in transcriptional regulation. Life Sci 2009; 84:189-193.
76. Krasnow MA, Saffman EE, Kornfeld K et al. Transcriptional activation and repression by Ultrabithorax proteins in cultured Drosophila cells. Cell 1989; 57:1031-1043.
77. Galant R, Carroll SB. Evolution of a transcriptional repression domain in an insect Hox protein. Nature 2002; 415:910-913.
78. Ronshaugen M, McGinnis N, McGinnis W. Hox protein mutation and the macroevolution of the insect body plan. Nature 2002; 415:914-917.
79. Pinsonneault J, Florence B, Vaessin H et al. A model for Extradenticle-induced function as a switch that changes Hox proteins from repressors to activators. EMBO J 1997; 16:2032-2042.
80. Saleh M, Rammbaldi I, Yang XJ et al. Cell signaling switches Hox-Pbx complexes from repressors to activators of transcription. Mol Cell Biol 2000; 20:8623-8633.
81. Merabet S, Kambris Z, Capovilla M et al. The hexapeptide and linker regions of the AbdA Hox protein regulate its activating and repressive functions. Dev Cell 2003; 4:761-768.
82. Merabet S, Saadaoui M, Sambrani N et al. A unique Extradenticle recruitment mode in the Drosophila Hox protein Ultrabithorax. Proc Natl Acad Sci USA 2007; 104:16946-16951.
83. Mann RS, Chan SK. Extra specificity from Extradenticle: the partnership between HOX and PBX/EXD homeodomain proteins. Trends Genet 1996; 12:258-262.
84. Mann RS, Affolter M. Hox proteins meet more partners. Curr Opin Genet Dev 1998; 8:423-429.
85. van Dijk MA, Murre C. Extradenticle raises the DNA binding specificity of homeotic selector gene products. Cell 1994; 78:617-624.
86. Heuber SD, Lohmann I. Shaping segments: Hox gene function in the genomic age. BioEssays 2008; 30:965-979.
87. Merabet S, Pradel J, Graba Y. Getting a molecular grasp on Hox contextual activity. Trends Genet 2005; 21:477-480.
88. Xu L. Regulation of Smad activities. Biochem Biophys Acta 2006; 1759:503-513.22.
89. Walsh CM, Carroll SB. Collaboration between Smads and a Hox protein in target gene repression. Development 2007; 134:3585-3592.
90. Verheyen EM, Gottardi CJ. Regulation of Wnt/beta1-Catenin signaling by protein kinases. Dev Dyn 2010; 239:34-44.
91. Johnson FB, Parker E, Krasnow MA. Extradenticle protein is a selective cofactor for the Drosophila homeotics: role of the homeodomain and YPWM amino acid motif in the interaction. Proc Natl Acad Sci USA 1995; 92:739-743.
92. Shanmugan K, Featherstone MS, Saragovi HU. Residues flanking the HOX YPWM motif contribute to cooperative interactions with PBX. J Biol Chem 1997; 272:19081-19087.
93. Asahara H, Dutta S, Kao HY et al. Pbx-Hox heterodimers recruit coactivator-corepressor complexes in an isoform-specific manner. Mol Cell Biol 1999; 19:8219-8225.
94. Haynes C, Oldfield CJ, Ji F et al. Intrinsic disorder is a common feature of hub proteins from four eukaryotic interactomes. PLoS Comp Biol 2006; 8:890-901.
95. Hilser VJ, Thompson EB. Intrinsic disorder as a mechanism to optimize allosteric coupling in proteins. Proc Natl Acad Sci USA 2007; 104:8311-8315.
96. Dunker AK, Cortese MS, Romero P. Flexible nets: the roles of intrinsic disorder in protein interaction networks. FEBS J 2005; 272:5129-5148.
97. Hall J, Karplus PA, Barbar E. Multivalency in the assembly of intrinsically disordered dynein intermediate chain. J Biol Chem 2009; 48:33115-33121.
98. Beachy PA, Varkey J, Young KE et al. Cooperative binding of an Ultrabithorax homeodomain protein to nearby and distant DNA sites. Mol Cell Biol 1993; 13:6941-6956.
99. Gebelein B, Culi J, Ryoo HD et al. Specificity of Distalless repression and limb primordia development by Abdominal Hox proteins. Dev Cell 2002; 3:487-498.
100. Gebelein B, McKay DJ, Mann RS. Direct integration of Hox and segmentation gene inputs during Drosophila development. Nature 2004; 431:653-659.
101. Sun B, Hursh DA, Jackson D et al. Ultrabithorax protein is necessary but not sufficient for full activation of decapentaplegic expression in the visceral mesoderm. EMBO J 1995; 14:520-535.

CHAPTER 7

MOLECULAR RECOGNITION BY THE EWS TRANSCRIPTIONAL ACTIVATION DOMAIN

Kevin A.W. Lee

Department of Biology, Hong Kong University of Science and Technology, Hong Kong, China
Email: bokaw@ust.hk

Abstract: Interactions between Intrinsically Disordered Protein Regions (IDRs) and their targets commonly exhibit localised contacts via target-induced disorder to order transitions. Other more complex IDR target interactions have been termed "fuzzy" because the IDR does not form a well-defined induced structure. In some remarkable cases of fuzziness IDR function is apparently sequence independent and conferred by amino acid composition. Such cases have been referred to as "random fuzziness" but the molecular features involved are poorly characterised. The transcriptional activation domain (EAD) of oncogenic Ewing's Sarcoma Fusion Proteins (EFPs) is an ~280 residue IDR with a biased composition restricted to Ala, Gly, Gln, Pro, Ser, Thr and Tyr. Multiple aromatic side chains (exclusively from Try residues) and the particular EAD composition are crucial for molecular recognition but there appears to be no other major geometrically constrained requirement. Computational analysis of the EAD using PONDR (Molecular Kinetics, Inc. http://www.pondr.com) complements the functional data and shows, accordingly, that propensity for structural order within the EAD is conferred by Tyr residues. To conclude, molecular recognition by the EAD is extraordinarily malleable and involves multiple aromatic contacts facilitated by a flexible peptide backbone and, most likely, a limited number of weaker contributions from amenable side chains. I propose to refer to this mode of fuzzy recognition as "polyaromatic", noting that it shares some fundamental features with the "polyelectrostatic" (phosphorylation-dependent) interaction of the Sic1 Cdk inhibitor and Cdc4. I will also speculate on more detailed models for molecular recognition by the EAD and their relationship to native (non-oncogenic) EAD function.

Fuzziness: Structural Disorder in Protein Complexes, edited by Monika Fuxreiter and Peter Tompa.

INTRODUCTION

Aberrant genomic fusion of members of the TET family[1,2] (TAF15, EWS and TLS) to several different cellular partners, gives rise to the Ewing's family of oncogenic proteins (EWS-Fusion-Proteins, or EFPs)[2-5] and associated tumors (EFTs) (see Fig. 1). EFPs are potent gene-specific activators with the n-terminal ~250 residues of EWS providing a transcriptional activation domain (EWS-Activation-Domain, or EAD) and the fusion partner conferring DNA-binding/promoter specificity and hence tumor phenotype.[3] Aberrant transcriptional activation by EFPs is most likely central to EFT oncogenesis but other effects of EFPs, including gene-specific transcriptional repression[4] or perturbation of pre-mRNA splicing,[4] may also be important. Detailed molecular studies of the EAD may offer additional opportunities for therapeutic targeting of the entire family of currently fatal EFTs.[6]

Progress in determining structure/function relations for the EAD has been challenging for several reasons. First, comparative studies have yielded few pointers concerning how the EAD may work because, despite similarities, the EAD is not obviously related to known Transcriptional Activation Domains (TADs). Likewise native TET family members have not been informative for the EAD because (in contrast to EFPs) TETs only weakly activate transcription.[7,8] Second, the EAD is an Intrinsically Disordered Protein Region

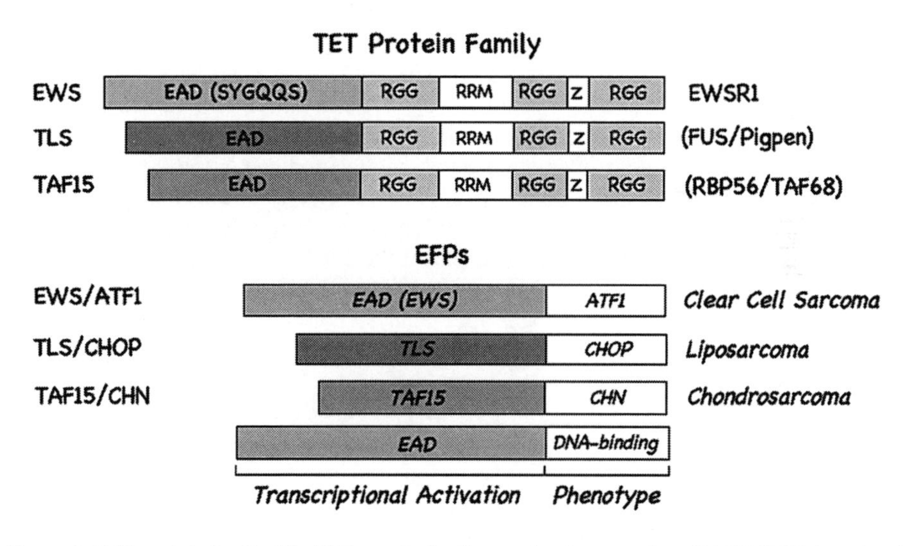

Figure 1. TET protein family. The TET protein family contains three members (TAF15 [TBP-Associated Factor 15], EWS [Ewing's Sarcoma oncoprotein] and TLS [Translocated in Liposarcoma]). TETs are a sub-family of RNA-binding proteins containing an N-terminal region referred to here as the EWS-Activation Domain (EAD, purple boxes) and a C-terminal RNA-binding domain (RBD). The EAD is described in detail Figure 3. The RBD contains two elements [an RNA-Recognition Motif (RRM) and RGG boxes] commonly found in RNA-binding proteins and a C2–C2 zinc finger (Z). The RRM harbors unique features that define the TET sub-family.[14] **EFP protein family.** EWS-Fusion Proteins (EFPs) are oncoproteins that arise due to aberrant chromosomal translocations involving a TET protein and a transcription factor partner. Three representative EFP (and their associated malignancies, right) are shown. All EFPs contain the EAD (at least residues 1-264 in the case of EWS) and a DNA-binding domain contributed by the fusion partner. EFPs are potent EAD-dependent transcriptional activators and the DNA-binding function largely determines tumor type. A color version of this image is available at www.landesbioscience.com/curie.

(IDR) and is not amenable to classical structural analysis. Third, the EAD probably interacts with a complex array of proteins[9] as a network hub[9,10] or scaffold protein[11] although functionally relevant EAD-interacting proteins have yet to be characterised. This latter limitation is largely due to the extended and repetitive EAD sequence and resulting lack of informative EAD mutants, a problem only recently addressed[12] via application of gene synthesis technology. A final barrier to molecular characterization of the EAD is the surprising lack of a cell free assay.[13]

Detailed definition of molecular recognition by IDRs has commonly been achieved only upon identification of IDR/target complexes that allow the induced structure of the bound IDR to be revealed. The models for molecular recognition by the EAD presented herein are based solely on functional and computational data and are necessarily speculative. I will describe the experimental approaches employed, results obtained and some plausible working models for molecular recognition by the EAD.

TET PROTEINS

Three genes (TAF15, EWS and TLS) encode the TET family of proteins[14] that represent a sub-family of RNA-binding proteins (Fig. 1). The N-terminal region (NTR) of TETs functions as a potent transcriptional activation domain in the context of EFPs and in EWS the NTR is referred to as the EWS-Activation Domain (EAD). Molecular recognition by the EAD and the role that this plays in transcription and cancer is the focus of this chapter. The RNA-binding domain of TETs (not present in EFPs) resides in the C-terminal region and the function of the EAD in native TET proteins is poorly understood. TETs are implicated in remarkably diverse cellular functions[2] and EWS forms the hub of a protein interaction network, potentially contacting perhaps ~80 different proteins.[9] Molecular recognition by TETs is clearly of interest to many fields of study in molecular and cellular biology. Hub proteins are often highly disordered[10] and accordingly, computation of disorder using the Predictor of Naturally Disordered Regions[15] (PONDR) shows EWS to be ~80% disordered (Fig. 2).

With respect to molecular recognition by the EAD it is important to stress that the function of native EWS/TET proteins is profoundly different from that of oncogenic EFPs. Firstly the EAD is indirectly affected by the different cellular locations and processes characteristic of TETs versus EFPs.[2] Secondly, although this remains to be verified, it has been suggested that the EAD interacts intramolecularly with the Arg rich EWS RNA-binding domain, thus accounting for differential interaction of TETs and EFPs with distinct partners.[16] Thirdly, transcriptional activation by the EAD is strongly repressed by the TET RNA-binding domain in native EWS.[8,17,18] and thus the transcriptional role of the EAD within TETs remains enigmatic.

EAD PRIMARY STRUCTURE

EAD primary structure is summarised in Figure 3. The fully functional EAD is ~280 residues long and has a repetitive and highly restricted amino acid composition (enriched in Tyr, Gln, Ser/Thr, Ala, Gly and Pro) resulting largely from the presence of ~30 Degenerate Hexapeptide Repeats (DHRs consensus SYGQQS). DHRs and minimal spacers accounts for ~85% of the EAD sequence and two longer spacers (S1 and S2, Fig. 3)

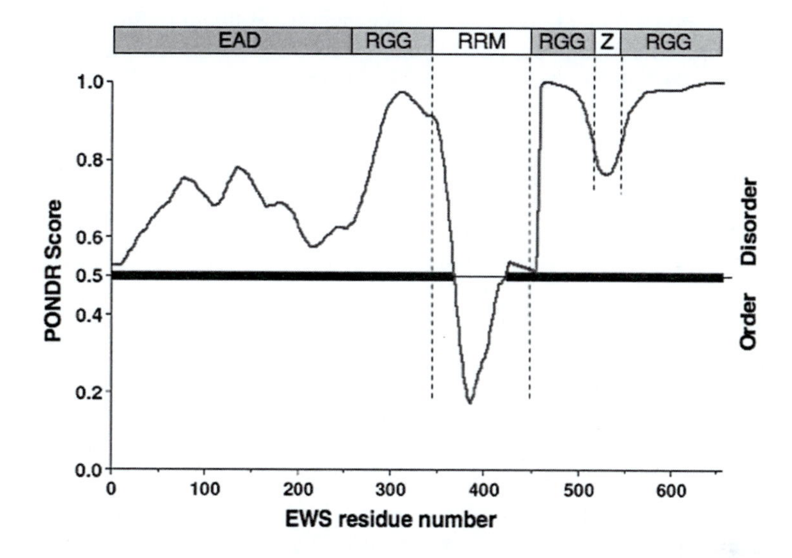

Figure 2. PONDR Analysis of EWS. Computation of disorder was performed using the Predictor of Naturally Disordered Regions[15] (PONDR). Intact EWS is shown to be a highly disordered protein (PONDR scores well above 0.5) including the EAD region. The exception is the RNA Recognition Motif (RRM) that has a well characterised conserved folded structure. A color version of this image is available at www.landesbioscience.com/curie.

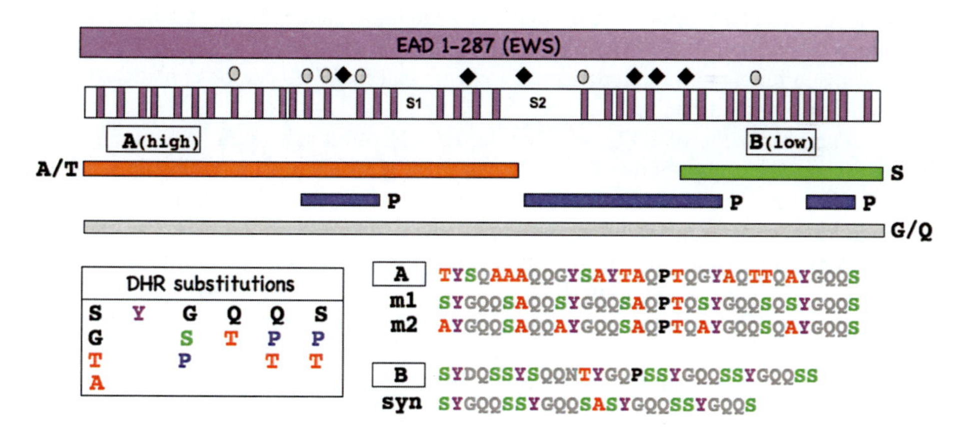

Figure 3. Primary EAD structure. The diagram shows EAD residues 1-287 from EWS. Most Tyr residues (purple boxes) are present within a Degenerate Hexapeptide Repeat (DHR consensus SYGQQS) with absolutely conserved Tyr (position 2) and well conserved Gln (position 4). Multiple conserved SH2-binding sites (grey circles) and SH3-binding sites (PxxP, black triangles) are also indicated. Seven additional nonDHR Tyr residues are dispersed throughout the EAD. DHRs constitute 70% of the EAD sequence, spaced typically by 0-3 residues (except for S1 [12 residues] and S2 [25 residues]). DHR substitutions (table) and differing spacers produce regional variations in amino acid composition. Horizontal colour bars show the location of enriched amino acids (nonTyr) labeled with the one letter code (A, G, P, Q, S, T). The location of two specific EAD peptides (A and B, black boxes) and their relative activity (high/low) are shown. Other sequences are as follows: m1 and m2 correspond to peptide A with all DHRs converted to consensus SYGQQS (m1) or AYGQQS (m2); syn is a synthetic peptide containing only consensus DHR (except for a single A residue) and no spacers.

have no major effect on activity.[12,19] DHR degeneracy results in regional variations in sequence composition within the EAD and this impacts activity (see later).

A number of SH2/SH3 interaction motifs and Tyr phosphorylation sites are present and evolutionarily conserved (between humans and frogs). Compositional bias (lack of order promoting Cys, Val, Leu, Ile, Met and enrichment in disorder promoting Gln, Ser, Pro, Ala and Gly) indicate that the EAD is an Intrinsically Disordered Protein Region (IDR). The severe lack of charged residues within the EAD is however strikingly atypical for an IDR.[20] CD spectrometry,[21] computational[12] and functional analysis[12] all demonstrate that the EAD is an IDR (see Figs. 2 and 7).

TET/EAD EVOLUTION

In relating evolution of EAD structure to function it is crucial to emphasise that EAD activity in EFPs is an out of context function that is not manifest in normal TET proteins.[7] Self evidently the functional properties of EFPs relevant to oncogenesis/transcriptional activation have not undergone evolutionary selection and accordingly, several conserved EAD structural features are dispensable for the EFP activity.[12] The implications of this intriguing observation must be kept in mind in relation to models of molecular recognition by the EAD in the context of normal TET proteins.

TET proteins arose fairly late during evolution of chordates and for EWS both the EAD and the RNA-binding domain are strikingly conserved between frogs and mammals with 70% dispersed identity. Conservation within the EAD includes almost all Tyr residues, several Tyr phosphorylation sites and multiple SH2/SH3 binding motifs (Fig. 3) and points to a crucial and specialised biological role in higher animals. Strong conservation in chordates and between different mammalian TET proteins (Fig. 1) together with structural and functionally independence indicate that the EAD can be considered to be a disordered protein domain.[22]

A high proportion of IDRs, such as the EAD, exhibit reiterated sequences. DNA sequence analysis suggests that the EAD was created by expansion (probably via gene conversion events) of a primordial minisatellite encoding the SYGQQS motif. Typical of reiterated sequences in the human proteome, the EAD is rich in Ser, Ala, Gly and Gln. Repeat expansions readily undergo divergent evolution and this is apparent by comparison of different TETs and also within the EAD of EWS (Fig. 3). For EWS mutation of the presumptive SYGQQS minisatellite together with replication slippage can account for the observed substitutions/spacing that result in concentration of contiguous consensus DHRs in the C-terminal region and diverged DHRs with spacing in the N-terminal region (Fig. 3). Interestingly the above regions markedly differ in activity (see later and Fig. 3) and this provides some insight into molecular recognition determinants (see later).

IDRs containing repetitive elements are thought to have evolved in three ways.[23] Type I has multiple repeats that interact with the same target, Type II has diversified repeats that interact with different targets and Type III has novel functions resulting from repeat expansion and/or divergence. Within the above framework our studies indicate that the EAD represents a novel variation of Type III in which repeat evolution has created an extended molecular recognition interface that is not related to the repeat sequence *per se* but rather to overall evolved amino acid composition.[12] Repetitive IDRs often function as flexible linkers or entropic chains that facilitate molecular recognition by other elements. In cases such as the EAD however the repeat sequence forms an extended domain[22] with a direct and autonomous molecular recognition function (albeit as part of a larger protein). Significantly it appears that the function of such extended disordered domains is refractory to mutation because only a small and dispersed proportion of the residues are critical for activity and, critically, the domain is likely to remain unstructured when bound to its target protein(s). This concept is crucial for understanding molecular recognition by the EAD.

EAD TRANSCRIPTIONAL ACTIVITY

In the context of a rudimentary reporter assay in cultured mammalian cells, several properties of the EAD have been established. The intact EAD of ~280 residues is required for full function[24] and contains multiple dispersed elements that synergise to create a potent activation domain.[19,25] The above is to be expected in light of the highly repetitive EAD structure. Small EAD sub-regions containing as few as four DHRs (EAD8-40 [Fig. 8]) functionally cooperate (or synergise) to produce high levels of activity either when linked in cis within a single protein[19,26] or in trans on a promoter containing multiple activator binding sites[26] (see also Fig. 4). Overall there is sound evidence that the EAD harbours reiterated, cooperative and flexible functional elements that are related to DHRs.

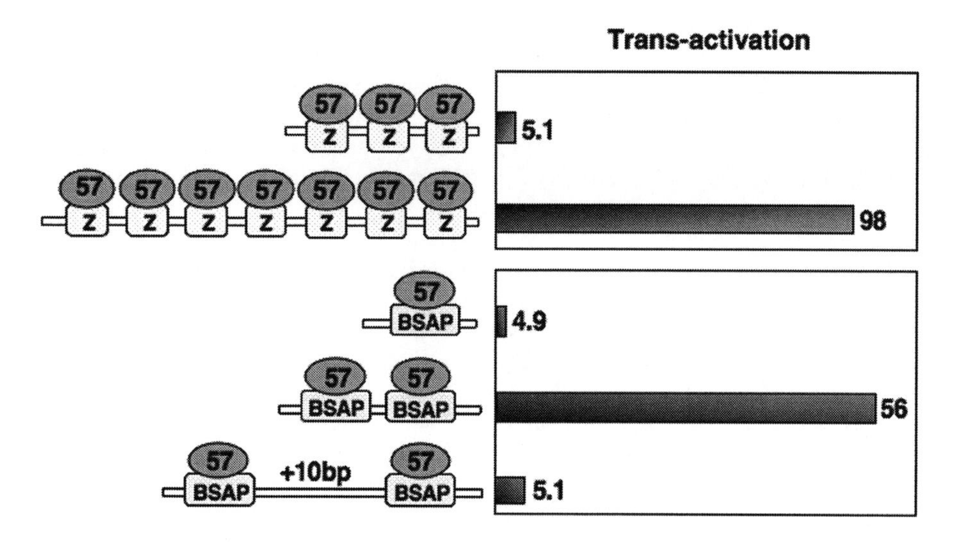

Figure 4. Trans-cooperation by minimal EAD peptides. 57Z protein contains EAD residues 1-57 (harboring six DHR Tyr residues) fused to the EBV Zta dimeric DNA-binding domain. A reporter promoter containing seven Zta binding sites is activated ~20 fold more than one with three Zta sites, demonstrating an ~10 fold synergistic/cooperative effect. A protein containing EAD1-57 that binds to DNA as a monomer (57BSAP) exhibits cooperativity (5-fold) on a two site versus one site BSAP reporter but only when the two BSAP sites are very close together. Distance dependent trans-cooperation by EAD1-57 monomers suggests that the two monomers physically cooperate. The data shown in this figure for 57Z is published in Oncogene (see ref. 26). A color version of this image is available at www.landesbioscience.com/curie.

The ability to assay minimal EAD peptides[26] facilitates analysis of specificity determinants but some constraints apply when extrapolating to the intact EAD. Specifically 57Z protein contains EAD1-57 (including six DHRs) but it binds to DNA as a dimer (through the EBV Zta DNA-binding domain) and might therefore exhibit geometrically constrained synergy between two monomers (and thus dependence on twelve DHRs). Alternatively synergy might simply reflect kinetic as opposed to structural synergy between monomeric EAD1-57 peptides (i.e., a nonlinear response to increasing concentration of promoter bound EAD1-57 peptide). Testing of monomeric activators (obtained by fusion of EAD1-57 to BSAP which binds DNA as a monomer)[27] addresses the above limitation (Fig. 4). Two promoter bound 57BSAP monomers exhibit strong synergy when close to each other but none when separated by just 10 more base pairs (Fig. 4). This raises the possibility that the minimal effective target binding by the EAD may involve more than six DHRs although they need not be covalently linked, again indicating a high degree of flexibility. The above features are incorporated in potential models for molecular recognition by the EAD (Fig. 9).

ANALYSIS OF REPETITIVE EAD SEQUENCES

To characterise repetitive elements in the intact EAD total gene synthesis was exploited for creation of mutant proteins and subsequent testing of transcriptional activation (trans-activation). A well established assay (in the context of EWS/ATF1, Fig. 1) that yields ~250 fold EAD-dependent trans-activation was employed.[12,24] Conservation of DHR Tyr residues (Fig. 3) suggests a critical function and thus extensive Tyr to Ala substitutions were initially tested for the effect on trans-activation (Fig. 5). Since the N-terminal EAD1-176 exhibits higher activity than the remaining C-terminal region[12,19] (see also Fig. 3) mutations were restricted to the N-terminal 176 residues of the EAD. Changing all 17 DHR Tyr to Ala within EAD1-176 (mutant DA, DHR Tyr changed to Ala) abolished activity (<1.6% of wt (Fig. 5). Significantly the 7 nonDHR Tyr residues also contributed similarly to activity.[12]

The mutational burden imposed on DA is quite high and to evaluate the specificity of such gross Tyr to Ala conversions, similar numbers of alternative (nonTyr) residues were changed. QA has all conserved Gln residues at position 4 of the DHR (Fig. 3) substituted by Ala and STA has a total of 16 Ser/Thr residues substituted by Ala. In contrast to DA both QA and STA proteins retained activity, demonstrating that the effect of Tyr to Ala changes is specific (Fig. 5). Additional mutants also established that the changes present in DA do not cause a general protein malfunction or dominant inhibitory effect.[12] Multiple Tyr residues in the EAD (including DHR and nonDHR) are therefore crucial for trans-activation by the EAD.

To scrutinise the Tyr side chain requirements a mutant protein was produced (DF, all DHR Tyr change to Phe) that is equivalent to DA except that Ala is replaced by Phe (Fig. 5). DF protein retained activity (as did a protein in which every Tyr residue in the entire EAD is changed to Phe, see ref. 12) demonstrating that the hydroxyl group of the Tyr side chain is not critical for EAD function. Additional substitutions of Tyr with aromatic, heterocyclic or hydrophobic residues showed that side chains containing an aromatic ring (Tyr, Phe and Trp) supported function while others (Ala, Ile and His) were less active or inactive (Fig. 5 and see also ref. 12). Thus that the aromatic ring of multiple Tyr side chains is critical for transcriptional activity of the EAD.

SEQUENCE INDEPENDENT EAD FUNCTION

DHRs account for 70% of the EAD sequence suggesting that they might be sufficient for EAD activity. Consistent with this a protein called MSP (mutant spacer, in which all the spacers are changed) retained activity (Fig. 5). Similarly a protein called SCR (scrambled) in which the positions of all DHRs present in EAD1-176 were rearranged, also retained activity, indicating that DHRs are functionally interchangeable. The above observations are consistent with the existence of a small repetitive functional element (the DHR) in the EAD. A second possibility however is that EAD function is conferred not by specific peptide sequences but instead by multiple Tyr residues embedded in a permissive overall composition. Three findings support this latter possibility. First, a protein called REV (Fig. 5) has all peptide sequences between consecutive EAD Tyr residues inverted and retains function. Second, the conserved Gln residues in position 4 of DHRs are not required for activity (Fig. 5). Third, all Tyr residues, not just those within DHRs, contribute equally to activity.[12] To examine the effect of EAD mutations in a different context and for a different EFP, the effect on cellular transformation by EWS/Fli1[28] was tested. The relative activity of corresponding EWS/Fli1 and EWS/ATF1 mutants was well correlated[12] indicating that the above findings can be expected to apply to the broader EFP family.

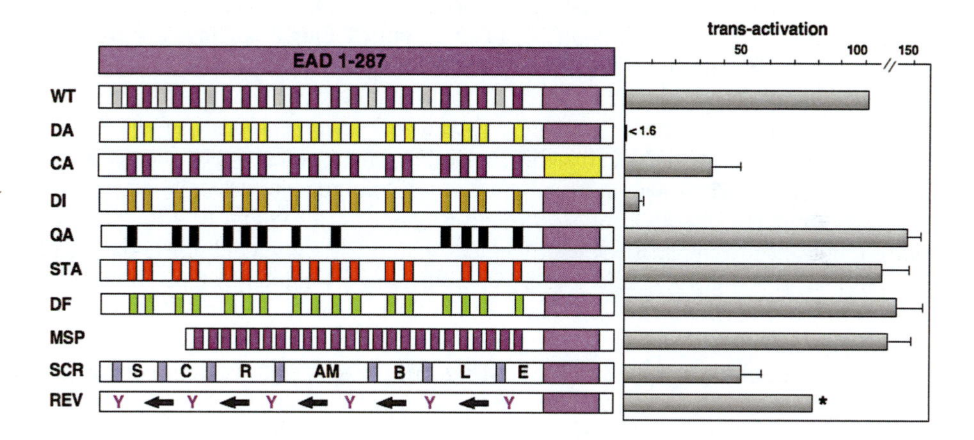

Figure 5. Functional analysis of the EAD. Transcriptional activity of WT and mutant proteins was determined by cotransfection of Jeg3 cells with vectors for EWS/ATF1 and an ATF-dependent CAT reporter.[12,24] EAD mutants: all proteins contain the DNA-binding domain of ATF1 (not shown). WT is an EWS/ATF1 fusion containing the intact EAD (residues 1-287). DHR Tyr residues (purple boxes) and nonDHR Tyr residues (dark gray) are shown. Tyr to Ala changes (DA and CA, yellow), Tyr to Phe (DF, green), Tyr to Ile (DI, brown) Gln to Ala (QA, black) are shown. Ser/Thr to Ala changes (STA, orange) include one change in each DHR within EAD1-176. SCR protein has the positions of all 17 DHRs present in EAD1-176 exchanged in a random manner. REV protein has the peptide sequence between adjacent Tyr residues in EAD1-176 inverted. MSP protein has all spacers (white) between DHRs converted to two residues (either AQ, TT, AP or SG). Data reproduced from: Ng KP et al, Proc Natl Acad Sci USA 2007; 104:479-484;[12] ©2007 National Academy of Sciences.

EAD AMINO ACID COMPOSITION

In addition to the specific role of Tyr residues the overall amino acid composition of the EAD is highly biased and most likely plays a facilitating role in molecular recognition. Accordingly regions of the EAD with a composition reflecting consensus DHRs[19] or synthetic proteins containing only DHRs[19] have very low activity (see peptide B and syn, Fig. 3) whereas enrichment in Ala/Thr and concomitant depletion of Ser significantly increases activity (see peptide A versus B, Fig. 3). The effect of Ser to Ala changes in position 1 of multiple SYGQQS consensus DHRs[25] provides direct evidence that Ala substitution of Ser increases activity (see mutants m1 and m2, Fig. 3). Similarly enrichment of Gln does not appear to be crucial[12] although it should be noted that the Gln-depleted proteins examined retain are high Gln content. In summary the enrichment of particular amino acids (excepting Tyr) reflecting consensus DHR composition is not particularly favourable but Gly and Gln most likely provide a permissive environment and Ala produces a less polar interface compared with Ser. Lack of significant net charge is atypical for IDRs[20] but it should be noted that Ser phosphorylation could compensate for this.

The effect of Pro content for the intact EAD is difficult to assess because Pro depletion results in poor expression (KL. unpublished data). However the most active region of the EAD has only minimal Pro content (Fig. 3) suggesting that Pro is not crucial. High Pro (and Gly) might act indirectly to support a favourable native EAD conformation for example by preventing amyloid type aggregation.[29] The PxxP motif can form a structure called the polyproline II (PPII) helix[30] but three out of four PxxP motifs present in EAD1-176 can be mutated without effect (see MSP, SCR and REV Fig. 5) suggesting that PPII helices do not contribute to EAD activity.[12] To conclude, in addition to the specific Tyr side chain requirement, the restricted amino acid composition of the EAD (reflecting enrichment with Ala, Gly, Gln, Pro and Ser) most probably provides a flexible and polar/neutral environment thus facilitating molecular recognition.

Many natural Tyrosine Enriched Protein Regions (TEPRs, Fig. 6) occur in other transcription factors and share several features with the EAD, including (1) degree of Tyr enrichment; (2) general polar/neutral composition; (3) marked compositional bias and; (4) predicted highly disordered nature. Significantly however, the selected TEPRs are also quite varied in their overall charge and particular amino acid content and exhibit markedly different transcriptional activity (Fig. 6). Perhaps significantly, none of the TEPRs tested are more active than the EAD and some are almost inactive. Overall it is apparent that, in addition to the specific role of Tyr, the overall biochemical nature of the EAD interface is significantly constrained.

EAD POSTTRANSLATIONAL MODIFICATIONS

Several Tyr residues within the EAD are phosphorylation sites and the EAD contains many SH3/SH2 interaction motifs (Fig. 3). Accordingly the EAD exhibits significant Tyr phosphorylation in vivo[31] and both c-Abl and v-Src phosphorylate the EAD and in the latter case modestly augment trans-activation.[32] The above affects may be significant for TET proteins or in some circumstances for EFPs but the finding that a Phe substituted EAD is active[12] (see Fig. 5) rules out a general role for Tyr phosphorylation in EFP function. The EAD has several SH2 (YxxP) and SH3 (PxxP) interaction motifs (Fig. 3). Interaction of SH2 domains with YxxP motifs requires Tyr phosphorylation and EAD/

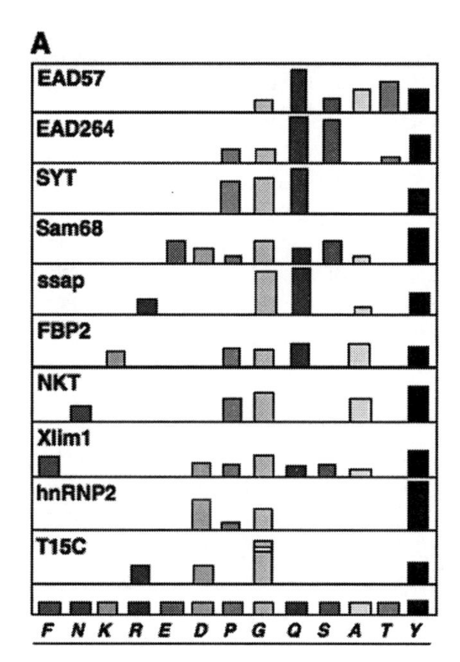

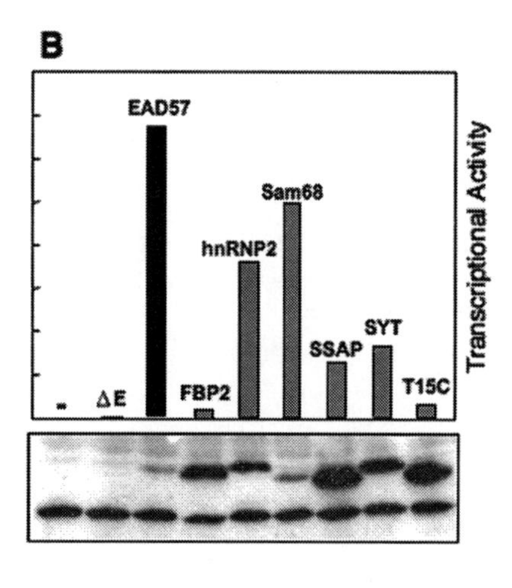

Figure 6. Tyrosine Enriched Protein Regions (TEPRs). Many natural TEPRs exist in other transcription factors and share several features with EAD57 (includes EAD residues 1-57) including size, degree of Tyr enrichment, general polar/neutral composition, marked compositional bias and highly disordered nature. In contrast TEPRs are also quite varied in their overall charge and particular amino acid content. The graph for TEPR composition (A) indicates relative proportions (>5%) of the amino acids shown in the chart (bottom). Transcriptional activity of TEPRs (B) is markedly different (note that most TEPRs are expressed at much higher levels than EAD1-57) indicating that the particular composition of the EAD (excluding Tyr) is required for function. A color version of this image is available at www.landesbioscience.com/curie.

SH2 interactions therefore appear not to be critical for molecular recognition by the EAD (in the context of EFPs).

The EAD is extensively phosphorylated on Ser but not Thr[31,32] in vivo and while not mapped it is certainly possible that Ser phosphorylation impacts molecular recognition. Phosphorylation is also commonly associated with IDRs.[33] Ser is not prevalent in the most active region of the EAD and is enriched in the inactive region (Fig. 3) suggesting that any effect of Ser phosphorylation would be negative. The EAD is also modified by O-GlcNAcylation on Ser/Thr[34] and since no consensus sequence for O-GlcNAcylation has been defined it is possible that several Ser/Thr residues are affected and this might, in turn, modulate molecular recognition by the EAD.

COMPUTATION OF EAD DISORDER

IDRs can be evaluated by various algorithms[35] including Predictors of Natural Disordered Regions (PONDR see ref. 15). PONDR is a set of neural network predictors based on local amino acid composition, flexibility, hydropathy and coordination number. PONDR VL3 is highly sensitive and combines predictions of 30 neural networks for the entire protein sequence, trained using circular dichroism, limited proteolysis and other physical data from more than 150 IDRs.[36] Propensity for order within IDRs is identified by lower PONDR scores and dips in the PONDR curve within generally disordered regions traditionally identify small localised inducible elements that have been referred to Molecular Recognition Elements (MoREs, see ref. 37), Molecular Recognition Features (MoRFs, see ref. 38), Preformed Structural Elements (PSEs; see ref. 39) or primary contact sites.[40]

To assess the potential for induced structure in the EAD (Fig. 7) predictions were performed using the PONDR VL3 predictor (access provided by Molecular Kinetics, Inc. http://www.pondr.com). PONDR VL3 analysis of wt and mutated EAD sequences (functionally tested, Fig. 5) shows that all the proteins are generally disordered (PONDR curves mostly above the 0.5 threshold) but PONDR scores are generally higher (lower order propensity) for Tyr/Ala, Tyr/Ile and Tyr/His changes. In contrast Tyr/Phe changes only marginally shifts the PONDR curve towards disorder. The PONDR curve for other active EAD mutants (QA, STA, SCR and REV) also resembles the wt EAD. REV looks similar to wt EAD (because at a resolution of ~10 residues they are similar) while SCR differs from wt EAD (because particular sub-regions are different) but retains overall propensity for order similar to wt. For alternative disorder predictors (RONN, see ref. 41; IUPred, see ref. 42; FoldIndex, see ref. 43) results for the wt EAD are comparable with PONDR overall.[12] With respect to EAD mutants, RONN and IUPred broadly agree with PONDR but are less sensitive while FoldIndex appears insensitive.[12] In conclusion predictions of order propensity using PONDR are quite sensitive and well correlated with activity for a range of EAD mutants. This strongly suggests that the potential for Tyr-dependent structural perturbations is crucial for EAD activity.

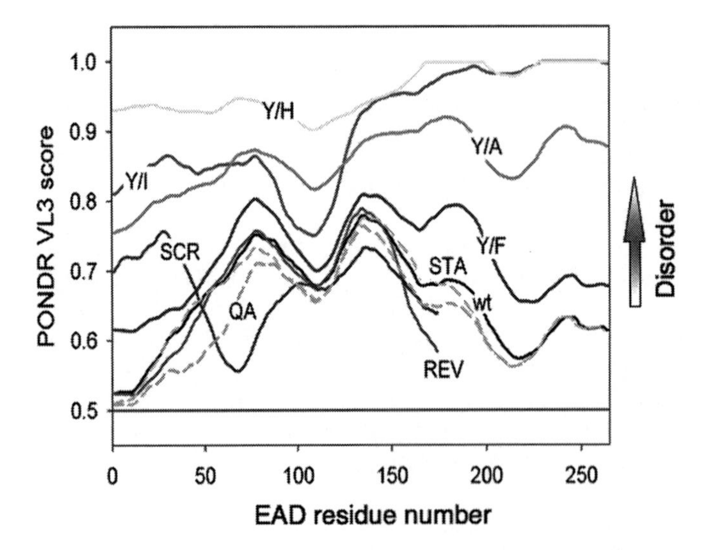

Figure 7. Analysis of WT and EAD mutants by PONDR VL3. Higher PONDR scores reflect propensity for disorder and lower scores propensity for order. For (Y/A), (Y/I), (Y/H) and (Y/F) all EAD Tyr are changed to Ala, Ile, His, or Phe, respectively. Color code is as follows: black, WT EAD; green, Y/A; red, Y/I; yellow, Y/H; blue, Y/F. PONDR plots derived for EAD mutants REV (dark green), SCR (purple), STA (dashed gray) and QA (dashed light blue) also are shown. Reproduced from: Ng KP et al, Proc Natl Acad Sci USA 2007; 104:479-484;[12] ©2007 National Academy of Sciences. A color version of this image is available at www.landesbioscience.com/curie.

MINIMAL FUNCTIONAL ELEMENTS OF THE EAD

The ability to test small EAD sub-regions[26] enables a higher resolution mutational analysis. For a peptide EAD8-40 (Fig. 8) simultaneously changing four Tyr residues to Ala (A1-4) reduced trans-activation to almost background levels (1% of wt), while changing any two Tyr residues (A12, A23 A13, A24 and A14) also greatly reduced trans-activation (average 4% of wt) and changing one Tyr alone (A1, A2, A3 and A4) had a significant but much smaller effect in each case (average of 25% wt). Since the effect of altering any combination of Tyr residues is quantitatively similar this indicates that each Tyr is functioning similarly. In addition, the Tyr residues strongly synergise with each other. In a contrast to the lack of effect of Tyr to Phe for intact EAD[12] partial Phe substitution of Tyr (F12) does moderately reduce activity of EAD8-40. A reasonable interpretation of this is that the function of small trans-cooperating peptides (such as EAD8-40) is not quite optimal and thus allows detection of minor contributions to molecular recognition that do not, however, impact the native EAD.

Similar to the intact EAD[12] there is a positive correlation between activity and order propensity for EAD8-40 (Fig. 8) although the effect of multiple Tyr to Ala changes on order propensity is additive and not cooperative. Tyr to Ala changes result in global decreases in order propensity suggesting that a higher order structure of EAD8-40 may be important for target interaction. A model for molecular recognition by EAD8-40 is shown (Fig. 8) and invokes multiple cooperative contacts with Tyr. In the model proposed, efficient functioning of a minimal EAD peptide (containing only 33 residues and five Tyr residues) implies lack of a requirement for long-range cis-interactions for the intact EAD.

MOLECULAR RECOGNITION BY THE EAD

Together the functional and computational data and theoretical considerations are highly indicative of a "random" model for molecular recognition[44] in which the EAD remains unstructured when bound to its target(s). This mode of interaction applies to several other IDRs, including the T-cell receptor ζchain,[45] self-associating elastin[29,46] and dimerisation of *E. Coli* SOS response protein umuD.[47]

There are two general possibilities for overall EAD structure and these may be dynamic and not mutually exclusive. In one scenario the EAD is an extended and highly flexible rope like structure with multiple Tyr residues accessible for target contact (see models A-D). Alternatively a hydrophobic centre with shielded Tyr residues could produce an interaction surface with other neutral/polar residues exposed to the target (not shown). The latter possibility may not be excluded (note that it also shares similarity with model C) and the ability of the EAD to undergo weak homotypic interactions[48] may support it. However theoretical considerations and some experimental evidence appears not to be in favor. First, in contrast to globular proteins IDRs commonly utilize exposed hydrophobic residues much more than polar residues to contact partners.[49] Second, the severely biased amino acid composition of the EAD (particularly enrichment in Gly, Ala, Gln and Pro) supports flexibility and has low potential for backbone self-interactions required to form a compact core. Third, reducing overall entropy (via intramolecular constraints in the unbound state) would compromise the entropic chain potential central to the random model[44] for target binding. Finally, for small but highly active EAD peptides (EAD8-40, Fig. 8) the high energy barrier to acute backbone folding and would probably be prohibitive.

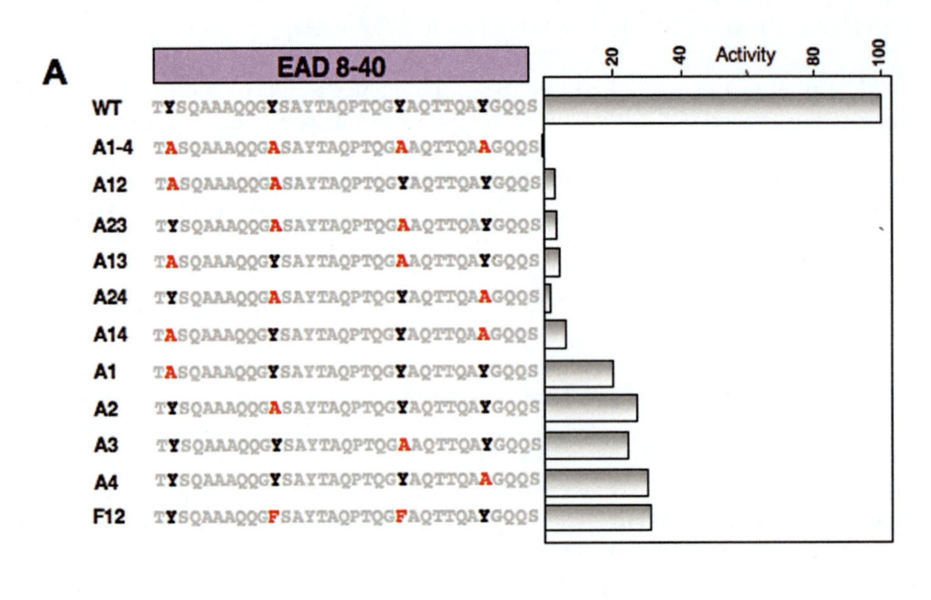

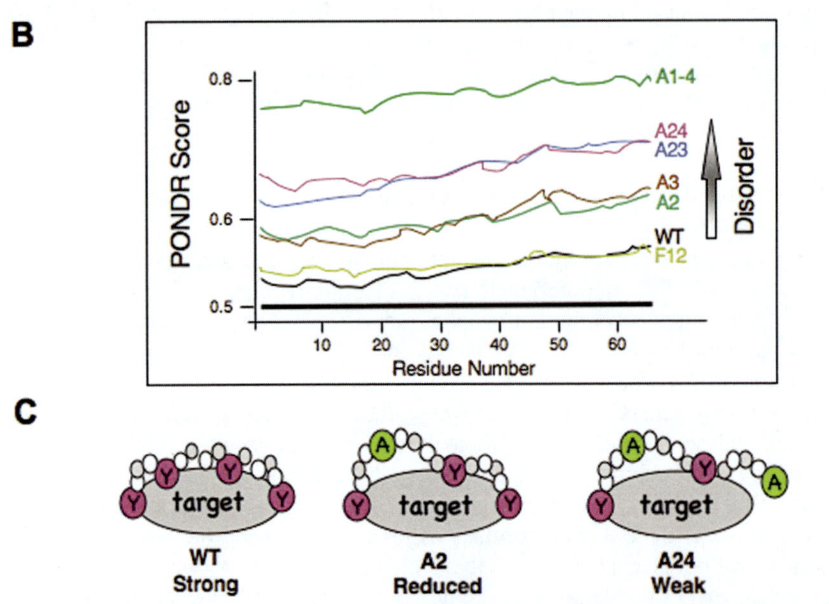

Figure 8. A) Transcriptional activity of a minimal EAD peptide. Activity was determined using the seven-site Zta reporter described in Figure 4 and EAD8-40 fused to the Zta DNA-binding domain. The EAD8-40 sequence is shown (WT) and Tyr to Ala or Phe are indicated in red. The data shown is published.[26] B) PONDR VL3 analysis. Changing one Tyr to Ala (A2 and A3), two Tyr (A23 and A24) or all four Tyr (A1-4) incrementally increases disorder in a global manner. C) Speculative model. Peptides with four Tyr residues are able to effectively recognise the target protein (although because the protein tested is dimeric this might correspond to eight Tyr residues, see Fig. 4). Each Try contributes equally to activity (target recognition) suggesting that individual Tyr residues contact the target. Strong activity (target recognition) involves higher order interactions as reflected by high cooperativity and thus low activity of peptides containing only two out of four Tyr residues. The data shown in this part A of this figure is published in Oncogene (see ref. 26).

OVERALL MODEL

Several potential models of molecular recognition are presented (Fig. 9). Excepting model C these models share the same core feature involving a number of Tyr-dependent contacts embedded in a highly flexible unstructured but chemically/physically permissive environment. The models are distinguished by the number or configuration of the Tyr contacts (model A versus B), the number of contacts on the target (model C) and the number of target proteins (model D).

Strong interactions may be achieved by simply reaching a threshold number of Tyr contacts (model A) or by separate Try clusters with each cluster making an independent contact (model B). The discontinuous interaction in Model B would be dictated by target geometry and resembles clamp type fuzziness[44,50] except that the bound elements and not just the linker remain unstructured. Inherent in each model is the likelihood that a dynamic ensemble of alternative interactions occur (involving subsets of Tyr contacts) for each of different (and numerous) protein targets.

A proportion of residues between Tyr contacts are likely to engage in hydrogen bonding with the target with others being looped out (Fig. 9). In this way the EAD peptide backbone partially follows the contours of the target but this would not correspond to structure in the regular sense. Rather this putative mode of interaction might best be described as "gratuitous shaping" and particularly because the function under scrutiny has not experienced evolutionary guidance. With respect to computation of structure as predicted by PONDR such predictions might coincide with the ability of the EAD to form a diverse shape continuum that does not correspond to any generalised structure. Gratuitous shaping would represent the most extreme form of random fuzziness[44] and would extend the observation that IDR/target interfaces exhibit a greater degree of complementary than for globular proteins.

AFFINITY DETERMINANTS

The molecular recognition models proposed (Fig. 9) invoke cooperative effects of multiple low affinity contacts to produce a strong interaction. Given the length of the EAD, multivalent recognition, lack of sequence constraints and EAD flexibility, the productive interaction surface is probably quite large and this tends to increase infinity.[51] For IDRs that conform to the random model, stabilisation would also be engendered by conformational freedom (of the bound IDR) rather than in entropically penalized preformed structural elements.[49] In the gratuitous shaping scenario only rotational/translational entropy would be lost upon binding but even this might be minimized either due to dynamic changes reflecting closely related alternative interaction surfaces. Besides specific Tyr contacts EAD target interaction is likely to be stabilized by the significant hydrogen bonding capacity of other prevalent amino acids in the EAD. Thus peptide sequences between Tyr residues are also proposed to make partial contacts with the target. Tyr itself frequently participates in intermolecular hydrogen bonding at protein-protein interfaces[52] and this could also contribute to stable interaction.

Cooperative multivalent binding is similar to many classical biomolecular interactions and is also an emerging feature of molecular recognition by IDRs that remain unstructured when bound. Several examples resemble the EAD and might offer instructive guides. In yeast the WD40 domain of Cdc4 interacts with Sic1 via nine suboptimal phosphopeptide motifs

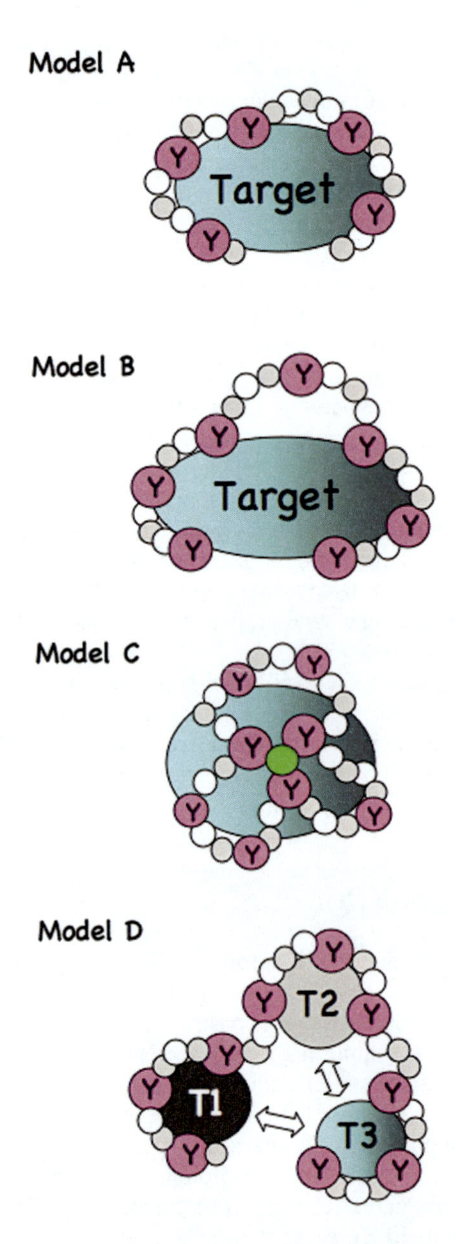

Figure 9. Models for molecular recognition by the EAD. In model A the EAD is shown as a flexible rope like structure with several Tyr residues (pink circles) contacting the target protein. Some of the other residues (white and grey circles) are proposed to weakly contact the target. Model B is variation of model A in which clusters of Tyr residues each form a low affinity recognition element but more than one cluster is required for binding. Model C differs in that multiple Tyr coalesce by cation-π interaction around a single Arg/Lys residue (green circle) in the target with intervening EAD sequences (including other Tyr residues) looped out. For model D the potential mode of target binding is included in models A-C but multiple target proteins (T1-3) are simultaneously bound by the EAD. T1-3 may also interact with each other (indicated by two-way arrows) thus further stabilising complex formation. See text for discussion. A version of model D was published in Lee KAW. Cell Research 2007; 17:286-288.[68] ©2007 the National Academy of Sciences, USA.

(CPDs) and binding is triggered upon reaching a threshold level of CPD phosphorylation.[53,54] Based on the relationship between net charge and binding, a polyelectrostatic model has been proposed to account for Cdc4/Sic1 interaction. A second example is the Arg/Ser-rich (R/S) domain of splicing enhancer factors that can be reconstituted by twenty one RS di-peptides but not by seven.[55] Thirdly, elastin self association is driven by multiple minimal binding motifs/scattered hydrophobic interactions and a small degree of hydrogen bonding with high Pro/Gly content providing flexibility and antagonizing ordered structure.[29]

SPECIFICITY DETERMINANTS

While molecular recognition by other sequence independent IDRs (VP16 TAD, Linker histone CTDs and Prion Sup35p) shares broad similarity with the EAD the specificity determinants are probably distinct. The VP16 TAD has several Phe residues but hydrophobicity rather than the aromatic ring appears to be important.[56] The amino acid composition of linker histone CTDs (which bind the apoptotic protein DFF40) bares little resemblance to the EAD being composed of 75% Lys/Ala and a small proportion of Val as the only hydrophobic residue.[57] The proposed structure for Sup35p and related prions involves β-sheet formation[58] which is not indicated for the EAD.

What is the precise nature of individual Tyr-dependent contacts made by the EAD? These might be related to a class of recognition elements called Linear Motifs or LMs[59,60] that can be small (3-10 residues) but with only a few specificity residues that are generally aromatic (similar to the EAD) or Pro.[59] LMs are often functional as single copies with typically low affinities but multiple cooperative LMs some times produce higher affinity interactions.[59] Unlike the most active regions of the EAD (Fig. 3) LMs tend to be depleted in Ala and Gly and enriched in Pro. In addition there is a remarkable lack of any spatially constrained (relative to Tyr) specificity residues in the EAD since even the relatively conserved Gln in position 2 of the DHR (Fig. 3) is not required for function (Fig. 5 and see ref. 12). Overall it appears that specificity elements within the EAD may not be related to LMs or may represent extremely minimal LMs.

An interaction with high potential as a contact element for the EAD is the electrostatic cation-π interaction between aromatic and basic amino acids.[52,61] Arg is a very common and versatile interface residue and Tyr/Arg pairing is one of the most frequent cation-π interactions at protein–protein interfaces. Furthermore single cation-π interactions are proposed as specificity elements involved in the encounter stage of recognition followed by rapid reorientation of the remaining interface to the optimal binding configuration. The majority of cation-π pairs are also involved in intermolecular hydrogen bonding.[52,61] The above characteristics, namely Tyr specific contacts and multiple hydrogen bonding interactions, fit very well with the proposed models for molecular recognition by the EAD (Fig. 9). Studies of *Aspergillus niger* glucoamylase demonstrate an intriguing alternative interaction model involving cation-π interactions. In the fungal glucoamylase a cluster of four cation-π interactions involving a single Lys residue and four Trp/Tyr residues form a very stable interaction.[52] Such a model involving Tyr clustering and looping out of the intervening flexible sequences is depicted for the EAD (Fig. 9 model C). A feature of this potential mode of recognition is that it could facilitate EAD promiscuity by greatly reducing the complexity (number of contact residues) of the target interface. The disadvantage is that the EAD would become compacted, thus incurring a significant entropic penalty that might antagonise a random mode of interaction.[44]

MULTIPLE EAD TARGET PROTEINS

In Model D efficient and promiscuous action by the EAD involves recruitment of several transcriptional components although the core mode of molecular recognition remains similar to single target models. The available biological, biochemical and computational data point towards the multiple target model as follows. First, the EAD is known to contact a number of transcriptional components including the co-activator CBP,[62,63] multiple TAFs[16] and the RNA PolII sub-units rpb7[16,64,65] and rpb5.[16] Second, similar to other long and malleable IDRs, the EAD most likely interacts with a complex array of proteins as a network hub[9,10] or as a scaffold protein.[11] The EAD may resemble HMGA (of the HMG family of architectural transcription factors) which is highly unstructured[66] and interacts with numerous transcription factors.[66] Third, the EAD shares properties with other TADs which are highly flexible and generally lack structure (PONDR predicts TADs are 73-94% disordered).[66] Current models for TAD function also invoke multiple weak contacts between TADs and their targets, resembling a "molecular Velcro".[67] Considering the above framework, the size, number of potential recognition elements and malleability of the EAD could readily account for the promiscuity and potency of EAD-mediated trans-activation.

CONCLUSION

Future progress can come from finer biochemical definition of the EAD interface using the methodology described.[12] However direct structural assessment of target-bound EAD is a prerequisite for further progress and thus identification of cognate functional EAD partners is the essential next step. The panel of functionally characterised EAD mutants described[12] are indispensable tools for the above task.

The EAD should continue to provide a valuable model for complex extended IDRs with minimal sequence requirement and that employ the "random" mode of interaction.[44] The fact that EFP function is aberrant and not evolutionarily constrained suggests that studies of the EAD in this context may provide unique insights into molecular recognition by IDRs. In addition uncovering the basis for differential EAD function/molecular recognition for EFPs versus native EWS/TETs, might ultimately provide an avenue for development of EFP-specific therapeutic agents.

Notwithstanding the obvious differences between EFPs and TETs it seems inevitable that the core mode of molecular recognition by the EAD (as proposed here) will impact native TET function. It will be of great interest to determine how evolutionarily conserved modifications (such Tyr phosphorylation) or structural impositions (linkage to the TET RNA-binding domain) impact EAD function. The malleable and flexible nature of molecular recognition by the EAD is surely pivotal in allowing EWS to spread its tentacles so broadly within the human proteome interaction network and across the far reaches of mammalian cells.

ACKNOWLEDGEMENTS

I am very grateful to Vladimir Uversky, Keith Dunker and Peter Tompa for insightful discussion and to TOP Gene Technologies Inc., Montreal, for efficient and economical gene synthesis.

REFERENCES

1. Law WJ, Cann KL, Hicks GG. TLS, EWS and TAF15: a model for transcriptional integration of gene expression. Brief Funct Genomic Proteomic 2006; 5:8-14.
2. Tan AY, Manley JL. The TET family of proteins: functions and roles in disease. J Mol Cell Biol 2009; 1-11.
3. Kim J, Pelletier J. Molecular genetics of chromosome translocations involving EWS and related family members, Physiol Genomics 1999; 1:127-138.
4. Arvand A, Denny CT. Biology of EWS/ETS fusions in Ewing's family tumors. Oncogene 2001; 20:5747-5754.
5. Janknecht R. EWS-ETS oncoproteins: The linchpins of Ewing tumors. Gene 2005; 363:1-14.
6. Kovar H, Aryee, D, Zoubek A. The Ewing family of tumors and the search for the Achilles' heel. Curr Opin Oncol 1999; 11:275-284.
7. Li KKC, Lee KAW. Transcriptional activation by the EWS oncogene can be cis-repressed by the EWS RNA-binding domain. J Biol Chem 2000; 275:23053-23058.
8. Alex D, Lee KAW. RGG-boxes of the EWS oncoprotein repress a range of transcriptional activation domains. Nucleic Acids Res 2005; 33:1323-1331.
9. Rual JF, Venkatesan K, Hao T et al. Towards a proteome-scale map of the human protein-protein interaction network, Nature 2005; 437:1173-1178.
10. Haynes C, Oldfield CJ, Ji F et al. Intrinsic disorder is a common feature of hub proteins from four eukaryotic interactomes. PLoS Comput Biol 2006; 2:0890-0901
11. Cortese MS, Uversky VN, Dunker AK. Intrinsic disorder in scaffold proteins: Getting more from less. Prog Biophys and Mol Biol 2008; 98:85-106.
12. Ng KP, Potikyan G, Savene RO et al. Multiple aromatic side chains within a disordered structure are critical for transcription and transforming activity of EWS family oncoproteins. Proc Natl Acad Sci USA 2007; 104:479-484.
13. Ng KP, Li KKC, Lee KAW. In Vitro Activity of the EWS Oncogene Transcriptional Activation Domain. Biochemistry 2009; 48:2849-2857.
14. Bertolotti A, Lutz Y, Heard DJ et al. hTAFII68, a novel RNA/ssDNA-binding protein with homology to the pro-oncoproteins TLS/FUS and EWS is associated with both TFIID and RNA polymerase II. EMBO J 1996; 15:5022-5031.
15. Uversky VN, Oldfield CJ, Dunker AK. Showing your ID: intrinsic disorder as an ID for recognition, regulation and cell signaling. J Mol Recognition 2005; 18:343-384.
16. Bertolotti A, Melot T, Acker J et al. EWS, but notEWS-FLI-1, is associated with both TFIID and RNA polymerase II: interactions between two members of the TET family, EWS and hTAFII68 and subunits of TFIID and RNA polymerase II complexes. Mol Cell Biol 1998; 18:1489-1497.
17. Bertolotti A, Bell B, Tora L. The N-terminal domain of human TAFII68 displays transactivation and oncogenic properties. Oncogene 2000; 18:8000-8010.
18. Rossow KI, Janknecht R. The Ewing's sarcoma gene product functions as a transcriptional activator. Cancer Res 2001; 61:2690-2695.
19. Pan S, Ming KY, Dunn TA et al. The EWS/ATF1 fusion protein contains a dispersed activation domain that functions directly. Oncogene 1998; 16:1625-1631.
20. Uversky VN. Why are "natively unfolded" proteins unstructured under physiological conditions? Proteins 2000; 41:415-427.
21. Uren A, Tcherkasskaya O, Toretsky JA. Recombinant EWS-FLI1 oncoprotein activates transcription. Biochemistry 2004; 43:13579-13589.
22. Tompa P, Fuxreiter M, Oldfield CJ et al. Close encounters of the third kind: disordered domains and the interactions of proteins. BioEssays 2009; 31:328-335.
23. Tompa P. Intrinsically unstructured proteins evolve by repeat expansion. BioEssays 2003; 25:847-855.
24. Brown AD, Lopez-Terrada D, Denny CT et al. Promoters containing ATF-binding sites are de-regulated in tumour-derived cell lines that express the EWS/ATF1 oncogene. Oncogene 1995; 10:1749-1756.
25. Kim J, Lee KAW, Pelletier J. The DNA binding domains of the WT1 tumor suppressor gene product and chimeric EWS/WT1 oncoprotein are functionally distinct. Oncogene 1998; 16:1021-1030.
26. Feng L, Lee KAW. A repetitive element containing a critical tyrosine residue is required for transcriptional activation by the Ewing's sarcoma oncogene. Oncogene 2001; 20:4161-4168.
27. Krajewski W, Lee KAW. A monomeric derivative of the cellular transcription factor CREB functions as a constitutive activator. Mol Cell Biol 1994; 14:7204-7210.
28. Lessnick SL, Braun BS, Denny CT et al. Multiple domains mediate transformation by the Ewing's sarcoma EWS/FLI-1 fusion gene. Oncogene 1995; 10:423-431.
29. Rauscher S, Baud S, Miao M et al. Proline and glycine control protein self-organization into elastomeric or amyloid fibrils. Structure 2006; 14:1667-1676.
30. Li SS-C. Specificity and versatility of SH3 and other proline-recognition domains: structural basis and implications for cellular signal transduction. Biochem J 2005; 390:641-653.

31. Kim J, Lee JM, Branton PE et al. Modification of EWS/WT1 functional properties by phosphorylation. Proc Natl Acad Sci USA 1999; 96:14300-14305.
32. Kim J, Lee JM, Branton PE et al. Modulation of EWS/WT1 activity by the v-Scr protein tyrosine kinase. FEBS Lett 2000; 474:121-128.
33. Iakoucheva LM, Radivojac P, Brown CJ et al. The importance of intrinsic disorder for protein phosphorylation. Nucleic Acids Res 2004; 32:1037-1049.
34. Bachmaier R, Aryee DNT, Jug G et al. O-GlcNAcylation is involved in the transcriptional activity of EWS-FLI1 in Ewing's sarcoma. Oncogene 2009; 28:1280-1284.
35. Ferron F, Longhi S, Canard B. A practical overview of protein disorder prediction methods. Proteins 2006; 65:1-14.
36. Obradovic Z, Peng K, Vucetic S et al. Predicting intrinsic disorder from amino acid sequence. Proteins 2003; 53:Suppl 6, 566-572.
37. Oldfield CJ, Cheng Y, Cortese MS et al. Coupled folding and binding with alpha-helix-forming molecular recognition elements. Biochemistry 2005; 44:12454-12470.
38. Mohan A, Oldfield CJ, Radivojac P et al. Analysis of molecular recognition features (MoRFs). J Mol Biol 2006; 362:1043-1059.
39. Fuxreiter M, Simon I, Friedrich P. Preformed structural elements feature in partner recognition by intrinsically unstructured proteins. J Mol Biol 2004; 338:1015-1026.
40. Csizmok V, Bokor M, Banki P et al. Primary contact sites in intrinsically unstructured proteins: the case of calpastatin and microtubule-associated protein. Biochemistry 2005; 15:3955-3964.
41. Yang ZR, Thomson R, McNeil P et al. RONN: the bio-basis function neural network technique applied to the detection of natively disordered regions in proteins. Bioinformatics 2005; 21:3369-3376
42. Dosztanyi Z, Csizmok V, Tompa. IUPred: web server for the prediction of intrinsically unstructured regions of proteins based on estimated energy content. Bioinformatics 2005; 21:3433-3434.
43. Prilusky J, Felder CE, Zeev-Ben-Mordehai T et al. FoldIndex: a simple tool to predict whether a given protein sequence is intrinsically unfolded. Bioinformatics 2005; 21:3435-3438.
44. Tompa P, Fuxreiter M. Fuzzy complexes: polymorphism and structural disorder in protein-protein interactions. Trends in Biochem Sci 2007; 33:1-8.
45. Sigalov AB, Kim, WM, Saline M et al. The intrinsically disordered cytoplasmic domain of the T-cell receptor z chain binds to the Nef protein of simian immunodeficiency virus without a disorder-to-order transition. Biochemistry 2008; 47:12942-12944.
46. Pometun MS, Chekmenev EY, Wittebort RJ. Quantitative observation of backbone disorder in native elastin. J Biol Chem 2004; 279:7982-7987.
47. Simon SM, Sousa FJR, Mohana-Borges R et al. Regulation of Escherichia coli SOS mutagenesis by dimeric intrinsically disordered umuD gene products Proc Natl Acad Sci USA 2008; 105:1152-1157.
48. Spahn l, Siligan C, Bachmaier R et al. Homotypic and heterotypic interactions of EWS, FLI1 and their oncogenic fusion protein. Oncogene 2003; 22:6819-6829.
49. Meszaros BP, Tompa P, Simon I et al. Molecular principles of the interactions of disordered proteins. J Mol Biol 2007; 372:549-561.
50. Hazy E, Tompa P. Limitations of induced folding in molecular recognition by intrinsically disordered proteins. ChemPhysChem 2009; 10:1415-1419.
51. Nooren IMA, Thornton JM. Structural characterisation and functional significance of transient protein-protein interactions. J Mol Biol 2003; 325:991-1018.
52. Gallivan JP, Dougherty DA. 1999 Cation-pi interactions in structural biology. Proc Natl Acad Sci USA 1999; 96:9459-9464.
53. Nash P, Tang X, Orlicky S et al. Multisite phosphorylation of the CDK inhibitor sets a threshold for the onset of DNA replication. Nature 2001; 414:514-521.
54. Borg M, Mittag T, Pawson T et al. Polyelectrostatic interactions of disordered ligands suggest a physical basis for ultrasensitivity. Proc Natl Acad Sci USA 2007; 104:9650-9655.
55. Philipps D, Celotto AM, Wang Q et al. Arginine/serine repeats are sufficient to constitute a splicing activation domain. Nucleic Acids Research 2003; 31:6502-6508.
56. Sullivan SM, Horn PJ, Olson VA et al. Mutational analysis of a transcriptional activation region of the VP16 protein of herpes simplex virus. Nucleic Acids Res 1998; 26:4487-4496.
57. Hansen JC, Xu L, Ross ED et al. Intrinsic protein disorder, amino acid composition and histone terminal domains. J Biol Chem 2006; 281:1853-1856.
58. Ross ED, Edskes HK, Terry MJ et al. Primary sequence independence for prion formation. Proc Natl Acad Sci USA 2005; 102:12825-12830.
59. Neduva V, Russell RB. Linear motifs: evolutionary interaction switches. FEBS Lett. 2005; 579:3342-3345
60. Fuxreiter M, Tompa P, Simon I. Local structural disorder imparts plasticity on linear motifs. Bioinformatics 2007; 23:950-956.
61. Crowley PB, Golovin A. Cation–pi interactions in protein–protein interfaces. Proteins 2005; 59:231-239.

62. Fujimura Y, Siddique H, Leo L. EWS-ATF-1 chimeric protein in soft tissue clear cell sarcoma associates with CREB-binding protein and interferes with p53-mediated trans-activation function. Oncogene 2001; 20:6653-6659.
63. Araya N, Hirota K, Shimamoto Y et al. Cooperative interaction of EWS with CREB-binding protein selectively activates hepatocyte nuclear factor 4-mediated transcription. J Biol Chem 2003; 278:5427-5432.
64. Petermann R, Mossier BM, Aryee DNT et al. Oncogenic EWS-Fli1 interacts with hsRPB7, a subunit of human RNA polymerase II. Oncogene 1998; 17:603-610.
65. Zhou H, Lee KAW. An hsRPB4/7-dependent yeast assay for trans-activation by the EWS oncogene. Oncogene 2001; 20:1519-1524.
66. Liu J, Perumal NB, Oldfield CJ et al. Intrinsic disorder in transcription factors. Biochemistry 2006; 45:6873-6888.
67. Melcher K. The strength of acidic activation domains correlates with their affinity for both transcriptional and non-transcriptional protein. J Mol Biol 2000; 301:1097-1112.
68. Lee KAW. Ewing's family oncoproteins: drunk, disorderly and in search of partners. Cell Research 2007; 17:286-288.

CHAPTER 8

THE MEASLES VIRUS N_{TAIL}-XD COMPLEX: An Illustrative Example of Fuzziness

Sonia Longhi

Architecture et Fonction des Macromolécules Biologiques, Universités d'Aix-Marseille I et II, Marseille, France
Email: Sonia.Longhi@afmb.univ-mrs.fr

Abstract: In this chapter, I focus on the biochemical and structural characterization of the complex between the intrinsically disordered C-terminal domain of the measles virus nucleoprotein (N_{TAIL}) and the C-terminal X domain (XD) of the viral phosphoprotein (P). I summarize the main experimental data available so far pointing out the prevalently disordered nature of N_{TAIL} even after complex formation and the role of the flexible C-terminal appendage in the binding reaction. I finally discuss the possible functional role of these residual disordered regions within the complex in terms of their ability to capture other regulatory, binding partners.

INTRODUCTION

The nonsegmented, single-stranded RNA genome of measles virus (MeV) is encapsidated by the nucleoprotein (N) within a helical nucleocapsid that is the substrate used by the viral polymerase for transcription and replication. The viral polymerase consists of the large (L) protein and of the phosphoprotein (P), with this latter being an essential polymerase cofactor in that it recruits the L protein onto the nucleocapsid template (for reviews on transcription and replication see refs. 1-4) (Fig. 1A).

In the course of the structural and functional characterization of MeV replicative complex proteins, we discovered that the N and P proteins contain long disordered regions that possess the sequence and biochemical features that typify intrinsically disordered proteins (IDPs).[5-14] Intrinsically disordered proteins (IDPs) are functional proteins that lack highly populated secondary and tertiary structure under physiological conditions in the

Fuzziness: Structural Disorder in Protein Complexes, edited by Monika Fuxreiter and Peter Tompa.

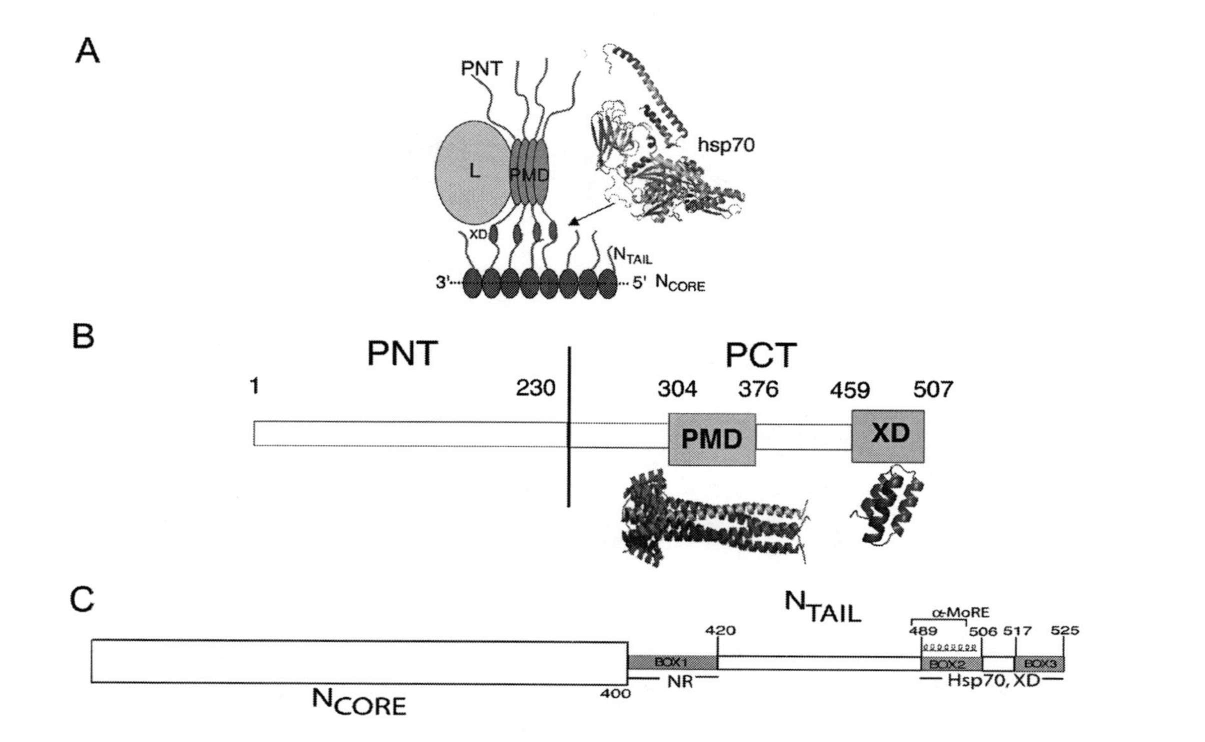

Figure 1. A) Schematic representation of the polymerase complex (L-P) bound to the nucleocapsid template. The RNA genome is represented by a dotted line. The disordered N_{TAIL} ($N^{401-525}$) and PNT (P^{1-230}) regions are represented by lines. P is depicted as tetrameric by analogy with the closely related Sendai virus P.[82] The segment connecting PMD and XD is represented as disordered according to references 6 and 7. The L protein is shown bound to a P tetramer through the multimerization domain of P (P,$^{304-375}$ PMD). A cartoon representation of hsp70 (pdb codes 1YUW and 1UDO) is also shown. The arrow indicates the ability of hsp70 to compete out XD for binding to N_{TAIL} (see text). B) Modular organization of the P protein, where globular and disordered regions are represented by large and narrow boxes, respectively. The crystal structures of MeV XD (pdb code 1OKS)[48] and of the closely related Sendai virus PMD (pdb code 1EZJ)[82] are also shown. Structures were drawn using Pymol.[83] C) Modular organization of the N protein, where the disordered N_{TAIL} domain is represented by a narrow box. The three N_{TAIL} boxes conserved in members belonging to the same genus of MeV[43] are shown (grey), as are the α-MoRE and the location of the NR, hsp70 and XD binding sites.

absence of a partner and rather exist as dynamic ensembles of interconverting conformers (for recent reviews on intrinsically disordered proteins see refs.15-18.)

Indeed, the P protein was shown to have a modular organization, being composed of alternating structured and disordered regions, with the N-terminal domain (PNT, $P^{1\text{-}230}$) being the largest disordered domain[5,7] (see Fig. 1B).

Like the P protein, the N protein has a mixed structural nature, being composed of a structured N-terminal domain, N_{CORE} ($N^{1\text{-}400}$), containing all the regions necessary for self-assembly and RNA-binding[19] and of a disordered C-terminal domain N_{TAIL} ($N^{401\text{-}525}$)[6] that protrudes from the globular body of N_{CORE} and that is exposed at the surface of the viral nucleocapsid[19] (Fig. 1A).

One of the functional advantages of disorder is related to an increased plasticity that enables disordered regions to bind to numerous structurally distinct targets.[20-22] Hence, intrinsic disorder is a distinctive and common feature of "hub" proteins, with disorder serving as a determinant of protein interactivity.[20-22]

The disordered nature of N_{TAIL} and its exposure at the surface of the viral nucleocapsid confer to this domain a large flexibility enabling the establishment of a broad molecular partnership with various viral and cellular proteins. Indeed, beyond serving as a tethering anchor for the P-L polymerase complex[6,23-25] (see Fig. 1A), N_{TAIL} interacts with the matrix protein[26] and with cellular proteins, including the major inducible heat shock protein 70 kDa (hsp70),[27,28] the interferon regulatory factor 3 (IRF3),[29,30] the cell protein responsible for the nuclear export of N,[31] as well as cell receptors involved in MeV-induced immunosuppression.[32-35] Likewise, the PNT domain has been reported to interact with N[36] and cellular proteins.[37]

THE DISORDERED N_{TAIL} DOMAIN AND ITS INTERACTION WITH XD

Computational, biochemical and spectroscopic analyses showed that N_{TAIL} belongs to the family of IDPs.[6] Although N_{TAIL} is primarily unfolded in solution, it nevertheless retains a certain degree of compactness based upon its Stokes radius (27 Å) and ellipticity values at 200 and 222 nm.[6] Altogether, these characteristics indicate that N_{TAIL} is a premolten globule,[6,8] i.e., it has a conformational state intermediate between a random coil and a molten globule.[38,39] In solution, premolten globules possess a certain degree of residual compactness due to the presence of residual and fluctuating secondary and/or tertiary structures. It has been proposed that the residual intramolecular interactions that typify the premolten globule state may enable a more efficient start of the folding process induced by a partner by lowering the entropic cost of the folding-coupled-to-binding process.[40-42]

That N_{TAIL} does indeed undergo induced folding was documented by far-UV circular dichroism (CD) studies, where N_{TAIL} was shown to undergo an α-helical transition in the presence of the C-terminal region of P (PCT, $P^{231\text{-}507}$).[6] Using computational approaches, an α-helical Molecular Recognition Element (α-MoRE, $N^{488\text{-}499}$) has been identified within one (namely Box2) out of three N_{TAIL} regions (referred to as Box1, Box2 and Box3) that are conserved within viruses belonging to the same genus of MeV[43] (see Fig. 1C). MoREs are short, order-prone regions within IDPs that have a certain propensity to bind to a partner and thereby to undergo induced folding.[44-47] The role of the α-MoRE in binding to P and in the α-helical induced folding was further confirmed by spectroscopic and biochemical experiments carried out on a truncated N_{TAIL} form devoid of the 489-525 region, where removal of Box2 was shown to both impair the ability of N_{TAIL} to bind to

P and reduce the gain of α-helicity in the presence of the secondary structure stabilizer 2,2,2 trifluoro ethanol (TFE).[8]

The PCT region responsible for the interaction with and induced folding of N_{TAIL} has been mapped to the C-terminal X domain of P (XD, $P^{459-507}$) and the crystal structure of this domain has been solved[48] (see Fig. 1B). A model of the interaction between the triple α-helical bundle of XD and the α-MoRE of N_{TAIL} was then proposed[48] and thereafter experimentally validated[49] by the determination of the crystal structure of a chimeric construct consisting of XD and of $N^{486-504}$ (Fig. 2). In both the proposed and experimentally observed structure of the complex, N_{TAIL} is embedded in a large hydrophobic cleft delimited by helices α2 and α3 of XD thus leading to a pseudo-four helix arrangement that occurs frequently in nature (Fig. 2A). Burying of hydrophobic residues of the α-MoRE in the hydrophobic cleft of XD (see Fig. 2B) is thought to provide the driving force to induce the folding of the α-MoRE, in agreement with the findings by Meszaros and coworkers who reported that the binding interfaces of protein complexes involving IDPs are often enriched in hydrophobic residues.[50]

Small angle X-ray scattering (SAXS) studies provided a low-resolution model of the N_{TAIL}-XD complex, which showed that most of N_{TAIL} (residues 401-488) remains disordered within the complex (Fig. 3).[9] As such, the N_{TAIL}-XD complex provides an illustrative example of "fuzziness," where this term has been recently coined by Tompa and Fuxreiter to designate the persistence of conspicuous regions of disorder, often playing a functional role in binding, within protein complexes implicating IDPs.[51]

In the N_{TAIL}-XD complex, the lack of a protruding appendage corresponding to the extreme C-terminus of N_{TAIL} suggests that, beyond Box2, Box3 could also be involved

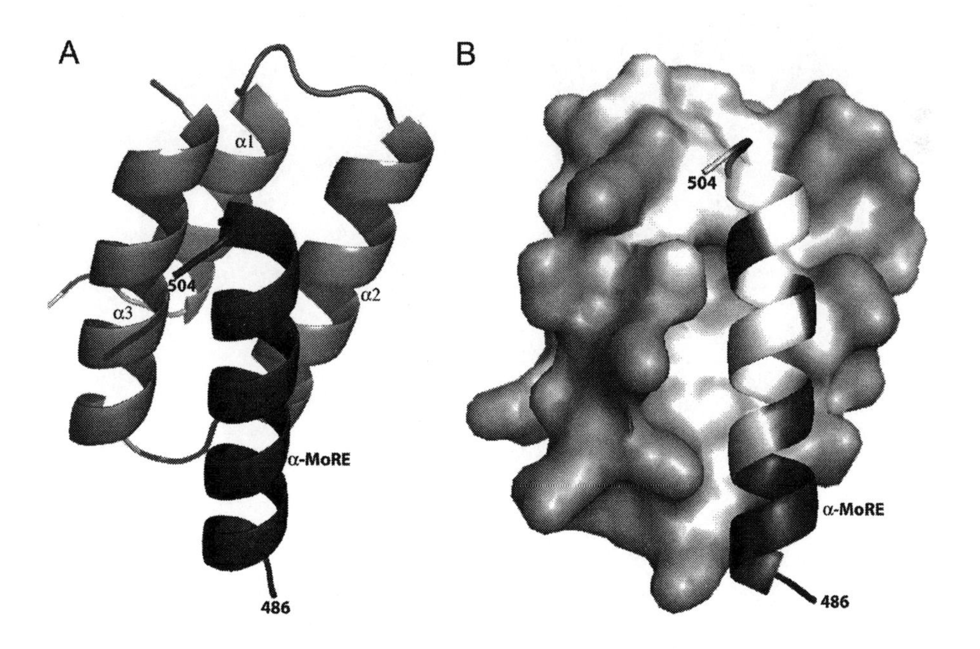

Figure 2. Crystal structure of the chimeric construct between XD and the α-MoRE of N_{TAIL} ($N^{486-504}$, dark grey, ribbon representation) (pdb code 1T6O).[49] XD is shown by a ribbon (A) or surface (B) representation. In panel B, the hydrophobic residues of XD and of the α-MoRE are shown in white.

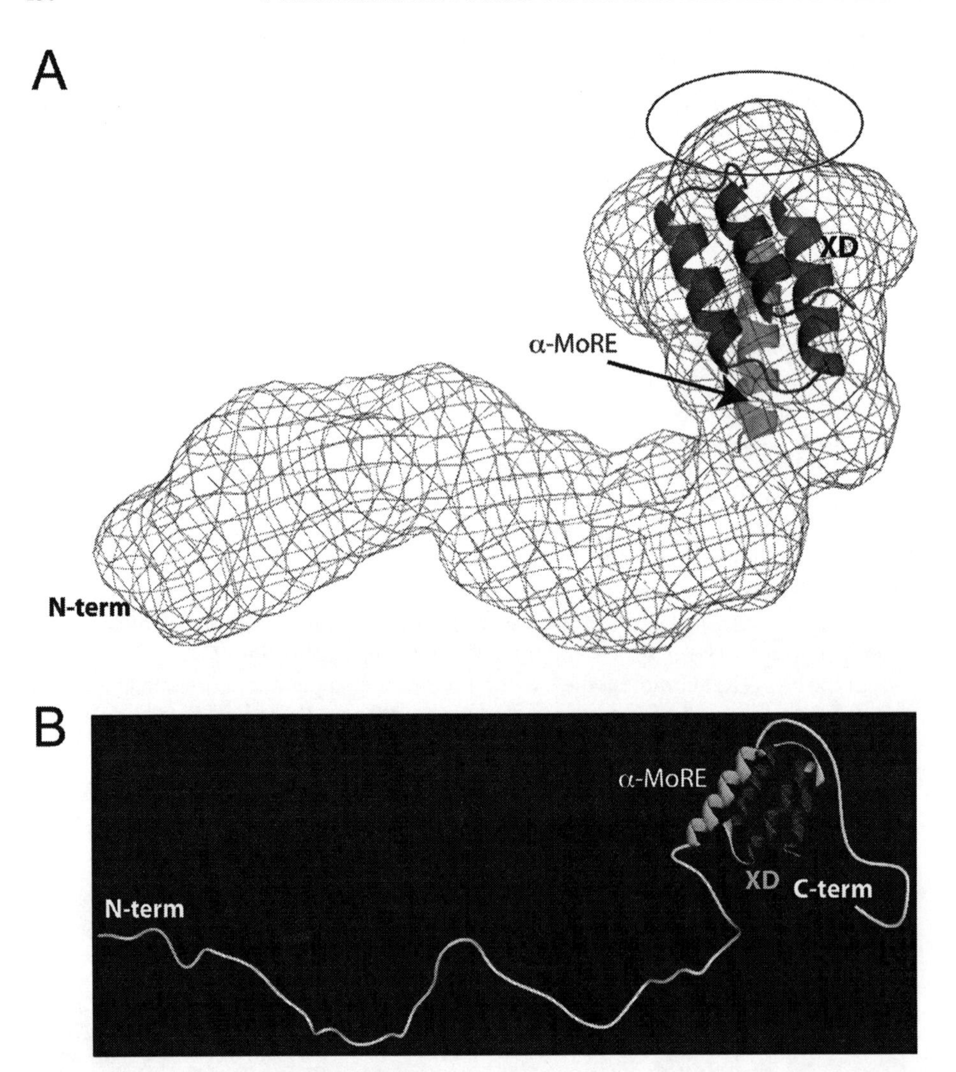

Figure 3. A) Global shape of the N_{TAIL}-XD complex as derived by small angle X-ray scattering studies.[9] The circle points to the lack of a protruding shape from the globular body of the model. The crystal structure of the chimera between XD and the N_{TAIL} region encompassing residues 486-504 (pdb code 1T6O)[49] is shown. The picture was drawn using Pymol.[83] B) Low resolution model of the N_{TAIL}-XD complex showing that (i) the 401-488 region of N_{TAIL} is disordered and exposed to the solvent, (ii) the α-MoRE is packed against XD and (iii) the C-terminus of N_{TAIL} (Box3) does not protrude from the globular part of the model. Data were taken and modified from reference 9.

in binding to XD.[9] To directly assess the possible implication of the C-terminus of N_{TAIL} in binding to XD, truncated N_{TAIL} constructs have been designed and purified from the soluble fraction of *E. coli*[9] (Fig. 4). Far-UV CD studies showed that Box2 plays a key role in the α-helical transition triggered by either XD or TFE, contrary to Box3 that does not affect the folding potential of N_{TAIL}.[9] In agreement, heteronuclear NMR studies (HN-NMR) showed that while the addition of XD to ^{15}N-labeled N_{TAIL} (or to $N_{TAIL\Delta 3}$,

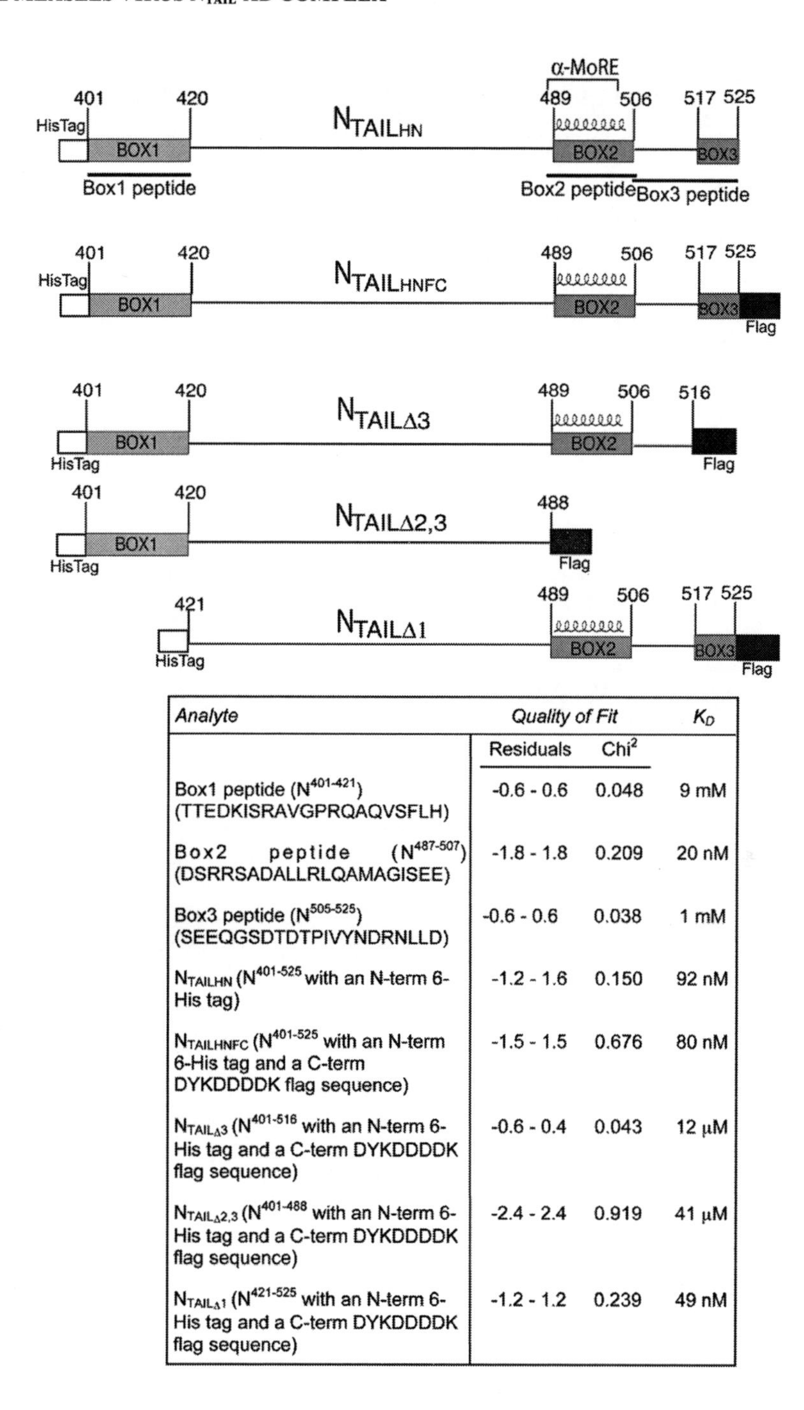

Analyte	*Quality of Fit*		K_D
	Residuals	Chi^2	
Box1 peptide ($N^{401\text{-}421}$) (TTEDKISRAVGPRQAQVSFLH)	-0.6 - 0.6	0.048	9 mM
Box2 peptide ($N^{487\text{-}507}$) (DSRRSADALLRLQAMAGISEE)	-1.8 - 1.8	0.209	20 nM
Box3 peptide ($N^{505\text{-}525}$) (SEEQGSDTDTPIVYNDRNLLD)	-0.6 - 0.6	0.038	1 mM
N_{TAILHN} ($N^{401\text{-}525}$ with an N-term 6-His tag)	-1.2 - 1.6	0.150	92 nM
$N_{TAILHNFC}$ ($N^{401\text{-}525}$ with an N-term 6-His tag and a C-term DYKDDDDK flag sequence)	-1.5 - 1.5	0.676	80 nM
$N_{TAIL\Delta 3}$ ($N^{401\text{-}516}$ with an N-term 6-His tag and a C-term DYKDDDDK flag sequence)	-0.6 - 0.4	0.043	12 μM
$N_{TAIL\Delta 2,3}$ ($N^{401\text{-}488}$ with an N-term 6-His tag and a C-term DYKDDDDK flag sequence)	-2.4 - 2.4	0.919	41 μM
$N_{TAIL\Delta 1}$ ($N^{421\text{-}525}$ with an N-term 6-His tag and a C-term DYKDDDDK flag sequence)	-1.2 - 1.2	0.239	49 nM

Figure 4. Schematic representation of the truncated N_{TAIL} constructs used in SPR studies. The location of the peptides used in these studies is indicated. The white and black boxes represent the hexahistidine tag and the flag, respectively. The inset shows the calculated equilibrium dissociation constants (K_D) between XD and N_{TAIL} proteins and peptides using SPR. Data were taken from references 9 and 14.

i.e., N_{TAIL} lacking Box3) triggered α-helical folding of Box2, it does not promote gain of regular secondary structure within Box3 although this latter undergoes a significant magnetic perturbation[9,52] (Fig. 5).

In further support of a role of Box3 in binding to XD, surface plasmon resonance (SPR) studies showed that removal of either Box3 alone or Box2 *plus* Box3 results in a strong increase (three orders of magnitude) in the equilibrium dissociation constant,

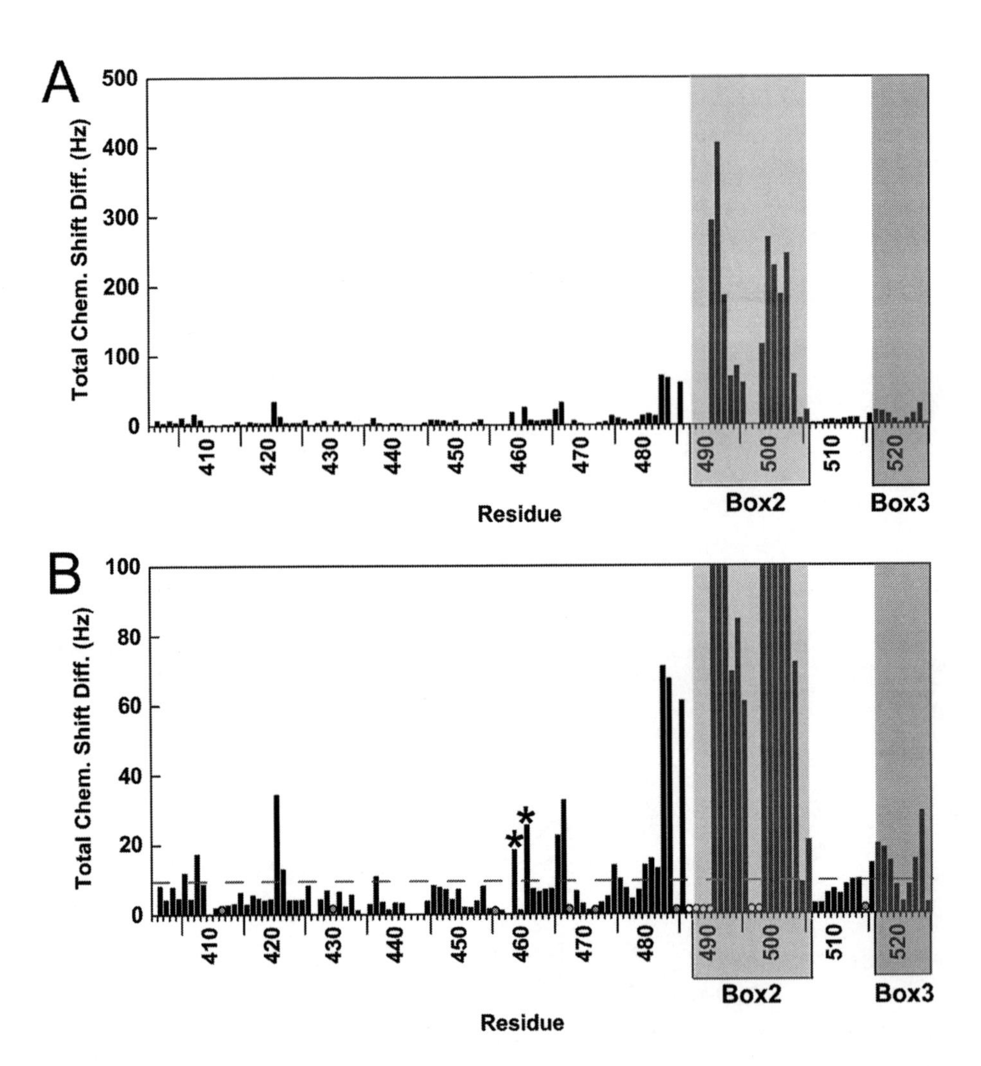

Figure 5. A) Total ^{1}H and ^{15}N chemical shift differences in Hz for free and bound forms of N_{TAIL}. The N_{TAIL} concentration in both free and bound sample was 0.3 mM, while that of XD was 0.6 mM. In panel (B), the scale has been reduced so as to highlight smaller, yet significant, chemical shift changes. Dashed grey line is at 2 times the digital resolution (10 Hz). Stars show residues in the N-terminal half of N_{TAIL} that have resonance overlap with residues in the C-terminal half (where 459 overlaps with 518 and 461 overlaps with 516). Grey circles correspond to proline residues and empty circles show the position of unassigned resonances in the N_{TAIL}-XD complex that probably have large chemical shifts changes. Box2 and Box3 regions are shaded. Data were modified from reference 52.

with the K_D increasing from 80 nM to either 12 μM or 41 μM, respectively[9] (see inset in Fig. 4). Conversely and as expected from the low-resolution model inferred from SAXS data and from CD studies carried out on $N_{TAIL\Delta1}$ (i.e., N_{TAIL} devoid of Box1) in the presence of XD, SPR studies showed that Box1 does not contribute to binding to XD (see inset in Fig. 4). When synthetic peptides mimicking Box1, Box2 and Box3 were used, Box2 peptide ($N^{487\text{-}507}$), was found to display an affinity for XD (K_D of 20 nM) that was similar to that between XD and N_{TAIL} (K_D of 80 nM) consistent with a role of Box2 as the primary binding site (see inset in Fig. 4). Surprisingly however, Box3 peptide ($N^{505\text{-}525}$) exhibits an insignificant affinity for XD (K_D of approximately 1 mM) (see inset in Fig. 4). In the same vein, HN-NMR experiments using ^{15}N-labeled XD pointed out lack of magnetic perturbation within this latter upon addition of unlabeled Box3 peptide, consistent with the lack of stable contacts between XD and Box3.[53] The discrepancy between the data obtained with N_{TAIL} truncated proteins and with peptides could be accounted for by assuming that Box3 would act only in the context of N_{TAIL} and not in isolation. Consistent with this hypothesis, both N_{TAIL} and $N_{TAIL\Delta3}$ (i.e., N_{TAIL} devoid of Box3) trigger slightly more pronounced chemical shift variations within ^{15}N-labeled XD than Box2 peptide alone, indicating that the region downstream Box2 (either Box3 itself or the region connecting Box2 to Box3) contributes to binding to XD, although only when acting in concert with Box2.[53] Thus, according to this model, Box3 and Box2 would be functionally coupled in the binding of N_{TAIL} to XD: the burying of the hydrophobic side of the α-MoRE in the hydrophobic cleft formed by helices α2 and α3 of XD could provide the primary driving force in the N_{TAIL}-XD interaction, with Box3 acting to further stabilize the bound conformation.

In view of unraveling the precise boundaries of the N_{TAIL} region undergoing α-helical folding, as well as the impact of XD binding on Box3, the N_{TAIL}-XD interaction has been also investigated by using site-directed spin-labeling (SDSL) electron paramagnetic resonance (EPR) spectroscopy. The basic strategy of SDSL involves the introduction of a paramagnetic nitroxide side chain through covalent modification of a selected protein site. This is usually accomplished by cysteine-substitution mutagenesis, followed by covalent modification of the unique sulfydryl group with a selective nitroxide reagent, such as the methanethiosulphonate derivative (for a review see ref. 54). From the EPR spectral shape of a spin-labeled protein one can extract information in terms of radical mobility, which reflects the local mobility of residues in the proximity of the radical. Variations in the radical mobility can therefore be monitored in the presence of partners, ligands, or organic solvents.

Fourteen single-site N_{TAIL} cysteine variants were designed, purified and labeled, thus enabling grafting of a nitroxide paramagnetic probe on 12 sites scattered in the 488-525 region and on two sites located outside the reported region of interaction with XD[55,56] (see diamonds in Fig. 6). EPR spectra were then recorded in the presence of either TFE or XD in both 0 and 30% sucrose.

Different regions of N_{TAIL} were shown to contribute to a different extent to the binding to XD: while the mobility of the spin labels grafted at positions 407 and 460 was unaffected upon addition of XD, that of spin labels grafted within the 488-502 and the 505-522 regions was severely and moderately reduced, respectively[56] (Fig. 6). Furthermore, EPR experiments in the presence of 30% sucrose (i.e., under conditions in which the intrinsic motion of the protein becomes negligible with respect to the intrinsic motion of the spin label), allowed precise mapping of the N_{TAIL} region undergoing α-helical folding to residues 488-502.[56] The drop in the mobility of the 505-522 region upon binding to

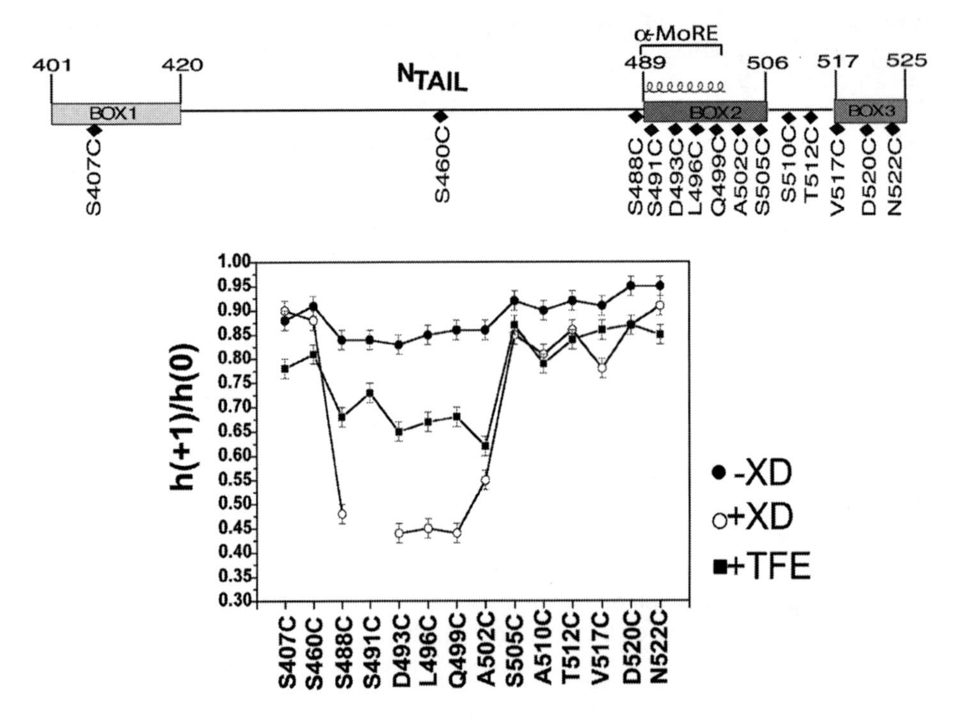

Figure 6. Mobility (expressed as h(+1)/h(0) ratios, see ref. 54) of the spin-labeled N_{TAIL} proteins free and in the presence of either saturating amounts of XD or 20% TFE as a function of spin-label position. Note that the h(+1)/h(0) ratio of the spin-labeled S491C variant was not indicated, as it is not a reliable indicator of the mobility of this spin label because of a highly restricted mobility (see ref. 56). The schematic representation of the N_{TAIL} variants is shown on the top. Modified from references 55 and 56.

XD was shown to be comparable to that observed in the presence of TFE[56] (Fig. 6). This observation suggests that the restrained mobility that the Box3 region experiences upon binding to XD is due neither to a steric hindrance exerted by XD nor to a direct interaction with XD and rather only arises from α-helical folding of the neighboring Box2.

The mobility of the 488-502 region was found to be restrained even in the absence of the partner (see Fig. 6 and refs. 56,57), a behavior that could be accounted for by the existence of a transiently populated folded state. That the N_{TAIL} region spanning residues 491-499 adopts an α-helical conformation in about 50% of the conformers sampled by unbound N_{TAIL}, has been recently experimentally confirmed by HN-NMR.[52]

These findings are in agreement with previous reports that showed that the conformational space of MoREs[45] in the unbound state is often restricted by their inherent conformational propensities. Noteworthy, the lack of a rigid 3D structure is thought to allow IDPs to establish interactions that are characterized by a high specificity and a low affinity: while the former is ensured by the very large surface area that is generally buried in complexes involving IDPs,[58] the low affinity arises from an unfavorable entropic contribution associated to the disorder-to-order transition.[39,59-66] In practice however, the extent of the entropic penalty is tightly related to the extent of conformational sampling of the prerecognition state, i.e., on the degree to which MoREs are preconfigured in solution prior to binding: the occurrence of a partly preconfigured MoREs in the unbound state in fact reduces the entropic cost of binding thereby enhancing affinity.[40-42,46,47,67] As such, IDPs exhibit a wide binding diversity, with some of them binding their partners with strong affinities. This is for instance the case of the N_{TAIL}-XD complex, whose rather high affinity (K_D of 80 nM) probably arises from a reduced entropic penalty related to the occurrence of the partly preformed MoRE prior to binding.

Although a strong N_{TAIL}-XD affinity explains the relatively long half life (well over 6 hours) of active P-L transcriptase complexes tethered on the nucleocapsid template,[68] as well as the ability to readily purify nucleocapsid-P complexes using rather stringent techniques such as CsCl isopycnic density centrifugation,[69-72] a stable XD-N_{TAIL} complex would be expected to hinder the processive movement of P along the nucleocapsid template. In agreement with this model, the elongation rate of MeV polymerase was found to be rather slow (three nucleotides/s)[68] and N_{TAIL} amino acid substitutions that lower the affinity towards XD result in enhanced transcription and replication levels, as well as in increased polymerase rate (Oglesbee, Gerlier and Longhi, unpublished data).

Nevertheless, despite the rather strong affinity of the N_{TAIL}-XD binding reaction, EPR equilibrium displacement experiments showed that the XD-induced folding of N_{TAIL} is a reversible phenomenon.[55,56] These results, beyond representing the first experimental evidence indicating that N_{TAIL} adopts its original premolten globule conformation after dissociation from XD, point out the dynamic nature of the association and dissociation of the N_{TAIL}-XD couple. This latter point is particularly relevant taking into consideration that the contact between XD and N_{TAIL} within the replicative complex has to be dynamically made and broken to allow the polymerase to progress along the nucleocapsid template during both transcription and replication.[13,14] As we will see in the next paragraph, modulation of the strength of the N_{TAIL}-XD interaction can also rely on the intervention of cellular cofactors that can compete out XD by binding to residual disordered regions of N_{TAIL} in the N_{TAIL}-XD complex.

FUNCTIONAL ROLE OF RESIDUAL DISORDER WITHIN THE N_{TAIL}-XD COMPLEX

What is the functional role of the flexible N_{TAIL} appendages within the N_{TAIL}-XD complex? The prevalently disordered nature of N_{TAIL} even after complex formation may serve as a platform for the capture of other binding partners.

Flow cytofluorimetry studies carried out on truncated forms of N_{TAIL} allowed mapping to Box1 the N_{TAIL} region responsible for the interaction with the cellular receptor NR, where this interaction occurs in the extracellular compartment following apoptosis of infected cells.[33,34] Given the rather high affinity between N_{TAIL} and XD, it is conceivable that released nucleocapsids are decorated by P (and P-L) complexes. The considerable flexibility of the 401-480 N_{TAIL} region even in the P-bound form, would allow the nucleocapsid to bind to NR without the requirement for XD to dissociate. In agreement, flow cytofluorimetry studies showed that neither hsp70 nor XD are able to competitively inhibit binding of N_{TAIL} to NR.[35]

In the same vein, the C-terminus of N_{TAIL} has been shown to retain a certain flexibility in the complex, despite the reduction in its conformational freedom arising from the XD-induced α-helical folding of the neighboring Box2. Strikingly, although the C-terminus of N_{TAIL} does not establish stable contacts with XD, it was shown to play an important regulatory role in transcription and replication. Indeed, not only Box3 was shown to affect the affinity of XD for N_{TAIL} (see inset in Fig. 4 and ref. 9), but also was it found to inhibit viral transcription and replication.[27]

Thus, Box3 would dynamically control the strength of the N_{TAIL}-XD interaction, by stabilizing the complex probably through several weak, nonspecific contacts with XD. Removal of Box3 would reduce the affinity of N_{TAIL} for XD which would stimulate transcription and replication by promoting successive cycle of binding and release of the polymerase that are essential to polymerase movement along the nucleocapsid template. Modulation of N_{TAIL}-XD binding affinity could also be dictated by interactions between N_{TAIL} and cellular and/or viral cofactors that could act by modulating the strength of the interaction between the polymerase complex and the nucleocapsid template. Indeed, in MeV, viral transcription and replication are enhanced by hsp70 and this stimulation relies on interaction with N_{TAIL}.[27,28,73-78] Two binding sites for hsp70 have been identified within N_{TAIL}: while high affinity binding (K_D of 10 nM) is supported by the α-MoRE,[28] a second low-affinity binding site is present within Box3.[27,77] As hsp70 was shown to competitively inhibit binding of XD to N_{TAIL},[28] hsp70-dependent stimulation of viral transcription and replication has been proposed to be due to a reduced stability of P-N_{TAIL} complexes (see Fig. 1A). This reduction would rely on competition between hsp70 and XD for binding to N_{TAIL} through (i) competition for binding to the α-MoRE (and this would occur at low hsp70 concentrations) and (ii) neutralization of the contribution of the C-terminus of N_{TAIL} to the formation of a stable P-N_{TAIL} complex (and this would occur in the context of elevated cellular levels of hsp70).[28]

CONCLUSION

In conclusion, using a panel of various physico-chemical approaches, the interaction between N_{TAIL} and XD was shown to lead to the formation of a "fuzzy" complex, in which the region upstream the α-MoRE remains disordered and the C-terminus retains a considerable flexibility. Complex formation implies the stabilization of the helical conformation of the α-MoRE, which is otherwise only transiently populated in the unbound form. The occurrence of a transiently populated α-helix even in the absence of the partner, suggests that the molecular mechanism governing the folding of N_{TAIL} induced by XD could rely on conformer selection (i.e., selection by the partner of a pre-existing conformation).[79,80] Recent data based on a quantitative analysis of NMR titration studies[52] suggest however that the binding reaction may also imply a binding intermediate in the form of a weak, nonspecific encounter complex and hence may also occur through a "fly casting" mechanism[81] (Fig. 7).

Stabilization of the helical conformation of the α-MoRE is also accompanied by a reduction in the mobility of the downstream region. The lower flexibility of the region downstream Box2 is not due to gain of α-helicity, nor can it be ascribed to a restrained motion due to the presence of XD or to the establishment of stable contacts with this latter. Rather, it likely arises from a gain of rigidity brought by α-helical folding of the neighboring Box2 region that results in a reduced Box3 conformational sampling. The reduced conformational freedom of Box3 may favor the establishment of weak, nonspecific contacts with XD (Fig. 7). At present, the exact role that Box3 plays in the stabilization of the N_{TAIL}-XD complex remains to be unraveled. Indeed, if recent data

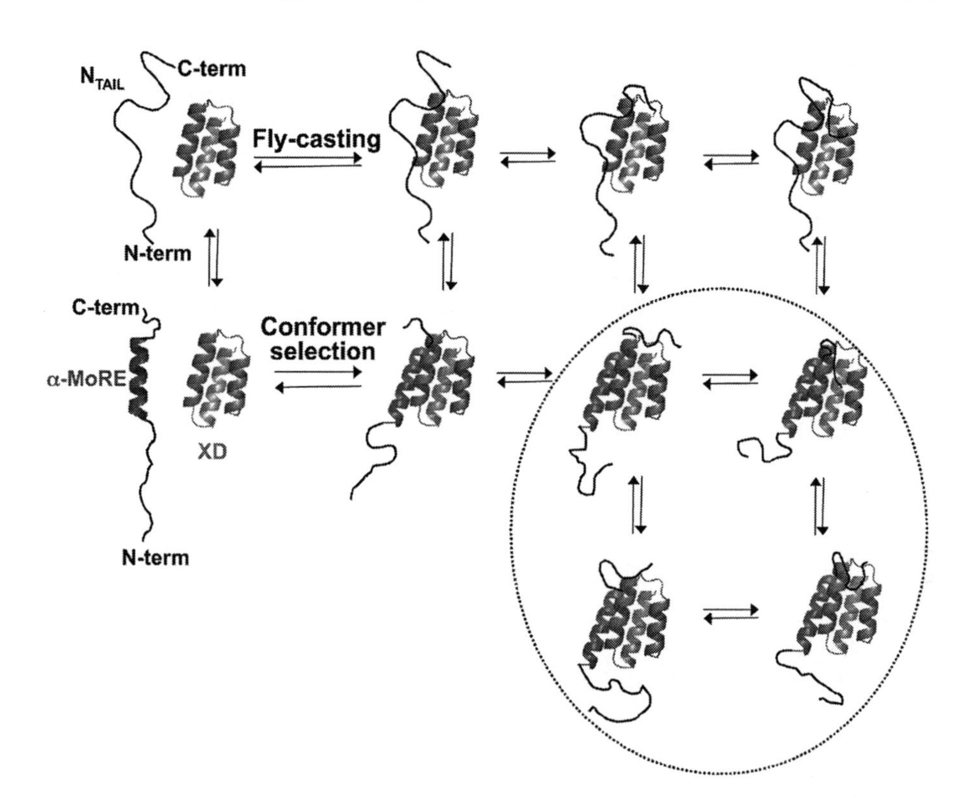

Figure 7. Model for N_{TAIL}-XD complex formation that utilizes both conformer selection and nonspecific encounter complex formation (modified from ref. 52).[52] The helix, corresponding to the primary binding site for XD, is partly preformed in the absence of XD. In the nonspecific binding model, a weak encounter complex may also form between N_{TAIL} and XD. This encounter complex is converted to a tightly bound complex by the folding of the α-MoRE. It is unclear if this folding occurs during the lifetime of the encounter complex. In the conformer selection model, the preformed helix interacts with XD to form a tightly bound complex. In both cases, following α-helical folding of Box2, Box3 becomes more rigid. The four conformers that are contoured by a dotted line schematically represent the final stage in complex formation, which consists of an ensemble of conformers in which Box3 has a reduced conformational freedom that may favor the establishment of weak, nonspecific contacts with XD.

clearly indicate that transient long-range tertiary contacts between Box2 and Box3 are unlikely,[52] it is still unclear whether Box3 contributes to binding to XD through weak (transient) nonspecific contacts with this latter or rather through another unknown mechanism. The possibility that Box3 may act by reducing the entropic penalty of the binding-coupled-to-folding process[51] seems to be unlikely: indeed replacement of the Box3 region by an irrelevant, 8-residues long Flag sequence (DYKDDDDK) led to a dramatic drop (two orders of magnitude) in the affinity towards XD, contrary to what would be expected for a nonbinding region contributing to the overall entropy of the system (*cfr.* N_{TAILHN} and $N_{TAIL\Delta3}$ in inset of Fig. 4).

Irrespective of the mechanism by which Box3 participates to binding to XD, the occurrence of this flexible appendage within the complex provides a scaffold for the capture of other, regulatory partners, such as hsp70.

ACKNOWLEDGEMENTS

I wish to thank all the persons who contributed to the works herein described. In particular, within the AFMB laboratory, I would like to thank Jean-Marie Bourhis, Benjamin Morin, Stéphanie Costanzo, Sabrina Rouger, Elodie Liquière, Bruno Canard, Kenth Johansson, David Karlin, François Ferron, Véronique Receveur-Brechot, Hervé Darbon, Cédric Bernard, Valérie Campanacci and Christian Cambillau. I also thank my coworkers Michael Oglesbee (Ohio State University, USA), Keith Dunker (Indiana University, USA), Denis Gerlier (LabVirPAth, Lyon, France), Hélène Valentin and Chantal Rabourdin-Combe (INSERM, Lyon, France), Gary Daughdrill (University of South Florida), Martin Blackledge and Malene Ringkjobing-Jensen (IBS, Grenoble, France), André Fournel, Valérie Belle and Bruno Guigliarelli (BIP, CNRS, Marseille, France). I wish to express my gratitude to Frédéric Carrière (EIPL, CNRS, Marseille, France) for having introduced me to EPR spectroscopy and for constant support. I also want to thank Vladimir Uversky and Denis Gerlier for stimulating discussions. The studies mentioned in this chapter were carried out with the financial support of the European Commission, program RTD, QLK2-CT2001-01225, "Towards the design of new potent antiviral drugs: structure-function analysis of *Paramyxoviridae* polymerase," of the Agence Nationale de la Recherche, specific program "Microbiologie et Immunologie," ANR-05-MIIM-035-02, "Structure and disorder of measles virus nucleoprotein: molecular partnership and functional impact" and of the National Institute of Neurological Disorders and Stroke, specific program "The cellular stress response in viral encephalitis," R01 NS031693-11A2.

REFERENCES

1. Longhi S, Canard B. Mécanismes de transcription et de réplication des Paramyxoviridae. Virologie 1999; 3:227-240.
2. Lamb RA, Kolakofsky D. Paramyxoviridae: the viruses and their replication. In: Fields BN, Knipe DM, Howley PM, eds. Fields Virology. 4th ed. Philadelphia, PA: Lippincott-Raven 2001:1305-1340.
3. Albertini AAV, Schoehn G, Ruigrok RW. Structures impliquées dans la réplication et la transcription des virus à ARN non segmentés de sens négatif. Virologie 2005; 9:83-92.
4. Roux L. Dans le génome des Paramyxovirinae, les promoteurs et leurs activités sont façonnés par la « règle de six. Virologie 2005; 9:19-34.
5. Karlin D, Longhi S, Receveur V et al. The N-terminal domain of the phosphoprotein of morbilliviruses belongs to the natively unfolded class of proteins. Virology 2002; 296:251-262.
6. Longhi S, Receveur-Brechot V, Karlin D et al. The C-terminal domain of the measles virus nucleoprotein is intrinsically disordered and folds upon binding to the C-terminal moiety of the phosphoprotein. J Biol Chem 2003; 278:18638-18648.
7. Karlin D, Ferron F, Canard B et al. Structural disorder and modular organization in Paramyxovirinae N and P. J Gen Virol 2003; 84:3239-3252.
8. Bourhis J, Johansson K, Receveur-Bréchot V et al. The C-terminal domain of measles virus nucleoprotein belongs to the class of intrinsically disordered proteins that fold upon binding to their pohysiological partner. Virus Research 2004; 99:157-167.
9. Bourhis JM, Receveur-Bréchot V, Oglesbee M et al. The intrinsically disordered C-terminal domain of the measles virus nucleoprotein interacts with the C-terminal domain of the phosphoprotein via two distinct sites and remains predominantly unfolded. Protein Sci 2005; 14:1975-1992.
10. Bourhis JM, Canard B, Longhi S. Désordre structural au sein du complexe réplicatif du virus de la rougeole: implications fonctionnelles. Virologie 2005; 9:367-383.
11. Bourhis JM, Canard B, Longhi S. Structural disorder within the replicative complex of measles virus: functional implications. Virology 2006; 344:94-110.
12. Bourhis JM, Longhi S. Measles virus nucleoprotein: structural organization and functional role of the intrinsically disordered C-terminal domain. In: Longhi S, ed. Measles virus nucleoprotein. Hauppage, NY: Nova Publishers Inc., 2007:1-35.

13. Longhi S. Nucleocapsid Structure and Function. Curr Top Microbiol Immunol 2009; 329:103-128.
14. Longhi S, Oglesbee M. Structural disorder within the measles virus nucleoprotein and phosphoprotein. Protein and Peptide Letters 2010; 17:961-978.
15. Radivojac P, Iakoucheva LM, Oldfield CJ et al. Intrinsic disorder and functional proteomics. Biophys J 2007; 92:1439-1456.
16. Dunker AK, Oldfield CJ, Meng J et al. The unfoldomics decade: an update on intrinsically disordered proteins. BMC Genomics 2008; 9 Suppl 2:S1.
17. Dunker AK, Silman I, Uversky VN et al. Function and structure of inherently disordered proteins. Curr Opin Struct Biol 2008; 18:756-764.
18. Uversky VN. The mysterious unfoldome: structureless, underappreciated, yet vital part of any given proteome. J Biomed Biotechnol 2010; 2010:568068.
19. Karlin D, Longhi S, Canard B. Substitution of two residues in the measles virus nucleoprotein results in an impaired self-association. Virology 2002; 302:420-432.
20. Dunker AK, Cortese MS, Romero P et al. Flexible nets. Febs J 2005; 272:5129-5148.
21. Uversky VN, Oldfield CJ, Dunker AK. Showing your ID: intrinsic disorder as an ID for recognition, regulation and cell signaling. J Mol Recognit 2005; 18:343-384.
22. Haynes C, Oldfield CJ, Ji F et al. Intrinsic disorder is a common feature of hub proteins from four eukaryotic interactomes. PLoS Comput Biol 2006; 2:e100.
23. Liston P, Batal R, DiFlumeri C et al. Protein interaction domains of the measles virus nucleocapsid protein (NP). Arch Virol 1997; 142:305-321.
24. Bankamp B, Horikami SM, Thompson PD et al. Domains of the measles virus N protein required for binding to P protein and self-assembly. Virology 1996; 216:272-277.
25. Kingston RL, Walter AB, Gay LS. Characterization of nucleocapsid binding by the measles and the mumps virus phosphoprotein. J Virol 2004; 78:8615-8629.
26. Iwasaki M, Takeda M, Shirogane Y et al. The matrix protein of measles virus regulates viral RNA synthesis and assembly by interacting with the nucleocapsid protein. J Virol 2009.
27. Zhang X, Glendening C, Linke H et al. Identification and characterization of a regulatory domain on the carboxyl terminus of the measles virus nucleocapsid protein. J Virol 2002; 76:8737-8746.
28. Zhang X, Bourhis JM, Longhi S et al. Hsp72 recognizes a P binding motif in the measles virus N protein C-terminus. Virology 2005; 337:162-174.
29. tenOever BR, Servant MJ, Grandvaux N et al. Recognition of the Measles Virus Nucleocapsid as a Mechanism of IRF-3 Activation. J Virol 2002; 76:3659-3669.
30. Colombo M, Bourhis JM, Chamontin C et al. The interaction between the measles virus nucleoprotein and the Interferon Regulator Factor 3 relies on a specific cellular environment. Virol J 2009; 6:59.
31. Sato H, Masuda M, Miura R et al. Morbillivirus nucleoprotein possesses a novel nuclear localization signal and a CRM1-independent nuclear export signal. Virology 2006; 352:121-130.
32. Marie JC, Kehren J, Trescol-Biemont MC et al. Mechanism of measles virus-induced suppression of inflammatory immune responses. Immunity 2001; 14:69-79.
33. Laine D, Trescol-Biémont M, Longhi S et al. Measles virus nucleoprotein binds to a novel cell surface receptor distinct from FcgRII via its C-terminal domain: role in MV-induced immunosuppression. J Virol 2003; 77:11332-11346.
34. Laine D, Bourhis J, Longhi S et al. Measles virus nucleoprotein induces cell proliferation arrest and apoptosis through N_{TAIL}/NR and N_{CORE}/FcgRIIB1 interactions, respectively. J Gen Virol 2005; 86:1771-1784.
35. Laine D, Vidalain P, Gahnnam A et al. Interaction of measles virus nucleoprotein with cell surface receptors: impact on cell biology and immune response. In: Longhi S, ed. Measles virus nucleoprotein. Hauppage, NY: Nova Publishers Inc., 2007:113-152.
36. Chen M, Cortay JC, Gerlier D. Measles virus protein interactions in yeast: new findings and caveats. Virus Res 2003; 98:123-129.
37. Liston P, DiFlumeri C, Briedis DJ. Protein interactions entered into by the measles virus P, V and C proteins. Virus Res 1995; 38:241-259.
38. Uversky VN. Natively unfolded proteins: a point where biology waits for physics. Protein Sci 2002; 11:739-756.
39. Dunker AK, Lawson JD, Brown CJ et al. Intrinsically disordered protein. J Mol Graph Model 2001; 19:26-59.
40. Tompa P. Intrinsically unstructured proteins. Trends Biochem Sci 2002; 27:527-533.
41. Fuxreiter M, Simon I, Friedrich P et al. Preformed structural elements feature in partner recognition by intrinsically unstructured proteins. J Mol Biol 2004; 338:1015-1026.
42. Lacy ER, Filippov I, Lewis WS et al. P27 binds cyclin-CDK complexes through a sequential mechanism involving binding-induced protein folding. Nat Struct Mol Biol 2004; 11:358-364.
43. Diallo A, Barrett T, Barbron M et al. Cloning of the nucleocapsid protein gene of peste-des-petits-ruminants virus: relationship to other morbilliviruses. J Gen Virol 1994; 75 (Pt 1):233-237.
44. Garner E, Romero P, Dunker AK et al. Predicting Binding Regions within Disordered Proteins. Genome Inform Ser Workshop Genome Inform 1999; 10:41-50.

45. Oldfield CJ, Cheng Y, Cortese MS et al. Coupled Folding and Binding with alpha-Helix-Forming Molecular Recognition Elements. Biochemistry 2005; 44:12454-12470.
46. Mohan A, Oldfield CJ, Radivojac P et al. Analysis of Molecular Recognition Features (MoRFs). J Mol Biol 2006; 362:1043-1059.
47. Vacic V, Oldfield CJ, Mohan A et al. Characterization of molecular recognition features, MoRFs and their binding partners. J Proteome Res 2007; 6:2351-2366.
48. Johansson K, Bourhis JM, Campanacci V et al. Crystal structure of the measles virus phosphoprotein domain responsible for the induced folding of the C-terminal domain of the nucleoprotein. J Biol Chem 2003; 278:44567-44573.
49. Kingston RL, Hamel DJ, Gay LS et al. Structural basis for the attachment of a paramyxoviral polymerase to its template. Proc Natl Acad Sci USA 2004; 101:8301-8306.
50. Meszaros B, Tompa P, Simon I et al. Molecular principles of the interactions of disordered proteins. J Mol Biol 2007; 372:549-561.
51. Tompa P, Fuxreiter M. Fuzzy complexes: polymorphism and structural disorder in protein-protein interactions. Trends Biochem Sci 2008; 33:2-8.
52. Gely S, Lowry DF, Bernard C et al. Solution structure of the C-terminal X domain of the measles virus phosphoprotein and interaction with the intrinsically disordered C-terminal domain of the nucleoprotein J Mol Recognit 2010; 23:435-447.
53. Bernard C, Gely S, Bourhis JM et al. Interaction between the C-terminal domains of N and P proteins of measles virus investigated by NMR. FEBS Lett 2009; 583:1084-1089.
54. Belle V, Rouger S, Costanzo S et al. Site-directed spin labeling EPR spectroscopy. In: Uversky VN, Longhi S, eds. Instrumental analysis of intrinsically disordered proteins: assessing structure and conformation. Hoboken, New Jersey John Wiley and Sons, 2010.
55. Morin B, Bourhis JM, Belle V et al. Assessing induced folding of an intrinsically disordered protein by site-directed spin-labeling EPR spectroscopy. J Phys Chem B 2006; 110:20596-20608.
56. Belle V, Rouger S, Costanzo S et al. Mapping alpha-helical induced folding within the intrinsically disordered C-terminal domain of the measles virus nucleoprotein by site-directed spin-labeling EPR spectroscopy. Proteins: Structure, Function and Bioinformatics 2008; 73:973-988.
57. Kavalenka A, Urbancic I, Belle V et al. Conformational analysis of the partially disordered measles virus N_{TAIL}-XD complex by SDSL EPR spectroscopy. Biophys J 2010; 98:1055-1064.
58. Tompa P. The functional benefits of disorder. J Mol Structure (Theochem) 2003; 666-667:361-371.
59. Dunker AK, Garner E, Guilliot S et al. Protein disorder and the evolution of molecular recognition: theory, predictions and observations. Pac Symp Biocomput 1998; 3:473-484.
60. Wright PE, Dyson HJ. Intrinsically unstructured proteins: re-assessing the protein structure-function paradigm. J Mol Biol 1999; 293:321-331.
61. Dunker AK, Obradovic Z. The protein trinity—linking function and disorder. Nat Biotechnol 2001; 19:805-806.
62. Dunker AK, Brown CJ, Obradovic Z. Identification and functions of usefully disordered proteins. Adv Protein Chem 2002; 62:25-49.
63. Uversky VN, Li J, Souillac P et al. Biophysical properties of the synucleins and their propensities to fibrillate: inhibition of alpha-synuclein assembly by beta- and gamma- synucleins. J Biol Chem 2002; 25:25.
64. Gunasekaran K, Tsai CJ, Kumar S et al. Extended disordered proteins: targeting function with less scaffold. Trends Biochem Sci 2003; 28:81-85.
65. Fink AL. Natively unfolded proteins. Curr Opin Struct Biol 2005; 15:35-41.
66. Dyson HJ, Wright PE. Intrinsically unstructured proteins and their functions. Nat Rev Mol Cell Biol 2005; 6:197-208.
67. Sivakolundu SG, Bashford D, Kriwacki RW. Disordered p27Kip1 exhibits intrinsic structure resembling the Cdk2/cyclin A-bound conformation. J Mol Biol 2005; 353:1118-1128.
68. Plumet S, Duprex WP, Gerlier D. Dynamics of viral RNA synthesis during measles virus infection. J Virol 2005; 79:6900-6908.
69. Oglesbee M, Tatalick L, Rice J et al. Isolation and characterization of canine distemper virus nucleocapsid variants. J Gen Virol 1989; 70 (Pt 9):2409-2419.
70. Robbins SJ, Bussell RH. Structural phosphoproteins associated with purified measles virions and cytoplasmic nucleocapsids. Intervirology 1979; 12:96-102.
71. Robbins SJ, Bussell RH, Rapp F. Isolation and partial characterization of two forms of cytoplasmic nucleocapsids from measles virus-infected cells. J Gen Virol 1980; 47:301-310.
72. Stallcup KC, Wechsler SL, Fields BN. Purification of measles virus and characterization of subviral components. J Virol 1979; 30:166-176.
73. Vasconcelos D, Norrby E, Oglesbee M. The cellular stress response increases measles virus-induced cytopathic effect. J Gen Virol 1998; 79:1769-1773.
74. Vasconcelos DY, Cai XH, Oglesbee MJ. Constitutive overexpression of the major inducible 70 kDa heat shock protein mediates large plaque formation by measles virus. J Gen Virol 1998; 79:2239-2247.

75. Oglesbee MJ, Kenney H, Kenney T et al. Enhanced production of morbillivirus gene-specific RNAs following induction of the cellular stress response in stable persistent infection. Virology 1993; 192:556-567.
76. Oglesbee MJ, Liu Z, Kenney H et al. The highly inducible member of the 70 kDa family of heat shock proteins increases canine distemper virus polymerase activity. J Gen Virol 1996; 77:2125-2135.
77. Carsillo T, Zhang X, Vasconcelos D et al. A single codon in the nucleocapsid protein C terminus contributes to in vitro and in vivo fitness of Edmonston measles virus. J Virol 2006; 80:2904-2912.
78. Oglesbee M. Nucleocapsid protein interactions with the major inducible 70 kDa heat shock protein. In: Longhi S, ed. Measles virus nucleoprotein. Hauppage, NY: Nova Publishers Inc., 2007:53-98.
79. Tsai CD, Ma B, Kumar S et al. Protein folding: binding of conformationally fluctuating building blocks via population selection. Crit Rev Biochem Mol Biol 2001; 36:399-433.
80. Tsai CJ, Ma B, Sham YY et al. Structured disorder and conformational selection. Proteins: Structure, Function and Bioinformatics 2001; 44:418-427.
81. Shoemaker BA, Portman JJ, Wolynes PG. Speeding molecular recognition by using the folding funnel: the fly-casting mechanism. Proc Natl Acad Sci USA 2000; 97:8868-8873.
82. Tarbouriech N, Curran J, Ruigrok RW et al. Tetrameric coiled coil domain of Sendai virus phosphoprotein. Nat Struct Biol 2000; 7:777-781.
83. DeLano WL. The PyMOL molecular graphics system Proteins: Structure, Function and Bioinformatics 2002; 30:442-454.

CHAPTER 9

FUZZINESS IN THE CORE OF THE HUMAN PATHOGENIC VIRUSES HCV AND HIV

Roland Ivanyi-Nagy[1] and Jean-Luc Darlix*[,2]

[1]Molecular Parasitology Group, The Weatherall Institute of Molecular Medicine, University of Oxford, Oxford, UK; [2]LaboRetro, Unité de Virologie Humaine INSERM 758, IFR 128, ENS de Lyon, Lyon, France
**Corresponding Author: Jean-Luc Darlix—Email: jldarlix@ens-lyon.fr*

Abstract: Nucleocapsid proteins are the molecular jacks-of-all-trades of small RNA viruses because they play pivotal roles in viral genomic RNA selection and packaging, regulate genome replication and virus budding and at the same time orchestrate a complex, dynamic interaction network with host cell proteins contributing to viral persistence and pathogenecity. These promiscuous interactions are made possible by the intrinsic flexibility of viral nucleocapsid proteins, facilitating either simultaneous or sequential binding to a plethora of structurally unrelated substrates, resulting in flexible, ever-changing multiprotein, RNA-protein and lipid-protein complexes during the viral replicative cycle. In this chapter, we examine the flexibility and multifunctionality of the assemblages formed by the nucleocapsid proteins of two important human pathogens, hepatitis C virus and human immunodeficiency virus.

INTRODUCTION

An abundance of proteins containing long intrinsically unstructured regions has recently been recognized as a common feature in the proteomes of eukaryotic organisms.[1-3] Flexible protein regions play important roles in molecular recognition, signal transduction and protein-protein and protein-RNA interactions, while they are relatively rare in enzymes.[4,5] As a consequence, intrinsically unstructured proteins (IUPs) or protein domains are especially widespread among highly connected ("hub") proteins in cellular interaction networks.[6,7] This notion is particularly pertinent for RNA viruses, whose coding capacity is limited by the low-fidelity replication associated with RNA- or DNA-dependent RNA polymerases. Indeed, only a handful of proteins account for the

Fuzziness: Structural Disorder in Protein Complexes, edited by Monika Fuxreiter and Peter Tompa.

replication, assembly and spread of these viruses, at the same time orchestrating subversion of varied cellular processes and fighting elimination by the host's restriction and immune responses. Although viral proteins with enzymatic activity—similarly to their cellular counterparts—are usually folded in a well-defined three-dimensional structure, a large number of regulatory and structural proteins belong to the IUP class. Well-characterized examples in pathogenic human viruses include the nucleocapsid (NCp7), Tat and Vif proteins of HIV-1, the nucleoprotein (N) and phosphoprotein (P) of measles virus and nucleocapsid proteins of flavi- and coronaviruses.[8-16] As discussed in detail below, the core protein of hepatitis C virus and the nucleocapsid protein of HIV-1 show a number of analogous features, chaperoning RNA-RNA interactions and orchestrating dynamic multi-component interaction networks regulating viral replication, assembly and budding. Thus flexible regions in such isolated viral proteins and probably in fuzzy replication complexes play central roles in the viral life cycle.

THE CORE PROTEIN OF HEPATITIS C VIRUS

Hepatitis C virus (HCV) is a small, enveloped virus belonging to the *Flaviviridae* family of positive-stranded RNA viruses, together with other clinically important, emerging human pathogens, including West Nile virus and the dengue viruses.[17] More than 120 million people, corresponding to a world-wide prevalence of ~2%, are chronically infected with HCV,[18] presenting an ever-growing health and socio-economic concern despite declining infection rates.[19] Hepatitis C accounts for substantial morbidity and mortality due to its propensity to establish a chronic infection, resulting in progressive liver disease associated with life-threatening sequelae, including liver cirrhosis, steatosis and hepatocellular carcinoma.[20] In addition to liver pathology, a variety of extrahepatic disorders, affecting the endocrine, nervous and immune systems, are linked with chronic infection.[21,22]

The genome of HCV contains a single long open reading frame (ORF), flanked by highly conserved and structured untranslated regions (UTRs) that constitute *cis*-acting RNA elements regulating viral translation and replication (reviewed in refs. 23,24). Translation of the ORF yields a precursor polyprotein of ~3000 amino acids, which is co and posttranslationally processed by cellular and viral proteases into at least 10 viral proteins (Fig. 1A). The structural proteins of HCV comprise the core (capsid) protein and the envelope glycoproteins E1 and E2, which are followed by the small viroporin p7 and the nonstructural proteins and enzymes (NS2 to NS5B), involved in genome replication (Fig. 1A).

The core (capsid) protein, located at the N-terminal region of the polyprotein precursor, is released in its mature form by the sequential action of the host-encoded signal peptidase (SP) and signal peptide peptidase (SPP) enzymes.[25,26] The mature core protein consists of two distinct domains distinguished by markedly different amino acid compositions and hydrophobicity profiles (Fig. 1B).[27] The N-terminal domain D1 contains the majority of the core basic residues, arranged in three highly charged amino acid clusters. The isolated D1 domain preserves the RNA binding[28] and RNA chaperoning activities[29] of the full length protein, as well as its capacity to form virus-like particles in vitro in the presence of structured RNA molecules.[30,31] All these varied functions are carried out in the absence of a well-defined three-dimensional structure of D1, as shown by a variety of biophysical methods, including proteinase digestion,[32] circular dichroism (CD) spectroscopy[33,34] and

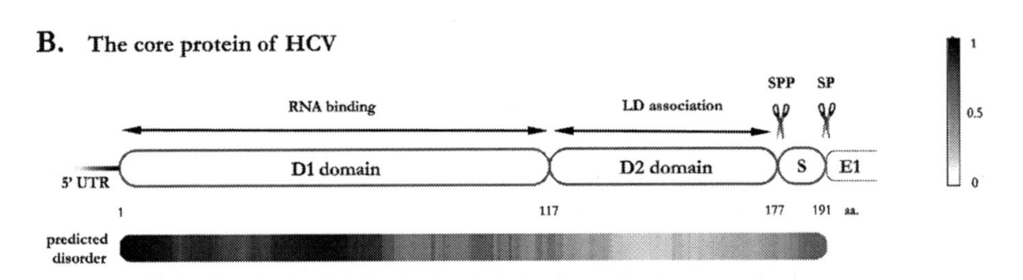

Figure 1. A) Genome organization of hepatitis C virus. The ORF encoding the viral polyprotein is flanked by highly conserved and structured untranslated regions. The structural proteins are released from the precursor by cellular proteases (black scissors), while the nonstructural region is processed by virus-encoded enzymes (gray scissors). B) Domain organization of hepatitis C virus core protein. The mature core protein consists of an RNA-binding and a lipid-binding domain (D1 and D2, respectively). The heat map illustrates the predicted disorder in core protein, with highly flexible segments in black and well-folded domains in white. Computer prediction of disordered regions was obtained using the DisProt VL3-H predictor.[118] An amino acid with a disorder score above or equal to 0.5 is considered to be in a disordered environment, while below 0.5 in an ordered environment.

chemical shift indexing,[35] confirmed functionally by deletion analysis.[36] The C-terminal region (D2 domain) of core protein is mostly hydrophobic and it serves as a targeting signal, mediating the attachment of core to cellular lipid droplets—the proposed sites for the interaction of structural proteins with the viral replication complex—leading to nucleocapsid assembly.[27,37]

VIRAL PARTICLE MORPHOGENESIS: CRYSTAL GROWTH VS. 'FUZZINESS'

Viral particle assembly is often envisaged as a straightforward process of nucleation followed by self-assembly, completely determined and governed by the same, simple laws of physics as inanimate crystal formation. Although this is certainly an over-simplification, it might apply as a working model for simple, non-enveloped viruses. Indeed, in the case of tobacco mosaic virus (TMV), the example regularly used in textbooks to illustrate viral assembly, the rod-shaped particle is assembled in a well-defined succession of intermediate complexes from 2130 subunits of the coat protein binding to the viral RNA.[38,39] Importantly, purified TMV coat protein and the viral genomic RNA spontaneously assemble in vitro under appropriate conditions to form fully infectious viral particles.[40]

However, virion morphogenesis in most clinically important human viruses is an infinitely more complex and dynamic process. Since nucleocapsid proteins of enveloped viruses establish the link between viral replication and egress, they have to interact with components of the viral replication machinery, as well as with viral and cellular proteins involved in the budding process. Homo-oligo/multimerization of capsid proteins, as well as their sequential interaction with cellular membranes, the viral genomic RNA and the envelope glycoproteins are essential for the formation of infectious viruses. Thus, in sharp contrast to crystal formation, HCV core functions in a highly regulated and dynamic fashion, organizing a sophisticated network of interactions with viral proteins and components of various cellular pathways (Fig. 2). Although information on the dynamics, exact composition, stoichiometry and structure of the resulting protein-RNA, protein-protein and protein-lipid complexes is extremely limited at present, there is accumulating evidence suggesting that the intrinsic flexibility of the core protein, partially retained in the various complexes, is important for its functions.[33] Thus, a model based on a succession of dynamic, fuzzy complexes[41] might provide a better description of HCV capsid function and assembly than the simplistic nucleation/self-assembly paradigm.

HCV CORE PROTEIN IN VIRAL PARTICLE FORMATION

Owing to its propensity to form heterogeneous, high molecular weight aggregates in aqueous solutions, the structure of the full-length core protein is still unresolved.[34] Far-UV CD measurements on core have shown a single ellipticity minimum at ~200 nm, suggesting that it is mostly unstructured in its unliganded form, with less than 10% alpha-helical content.[33] However, core protein is thought to undergo (partial) disorder-to-order transitions in the course of the sequential binding reactions during its intracellular targeting and nucleocapsid formation.

From its synthesis to virion budding, core protein is associated with cellular membrane structures, trafficking along the lipid bilayer from the endoplasmic reticulum to lipid droplets.[42] Membrane association is mediated by the formation of two putative short amphipathic α-helices in the D2 domain of core, separated by a flexible hydrophobic loop region.[42] Lipid binding induces conformational changes not only in the D2 domain but leads to partial α-helical folding of the N-terminal region, as evidenced by CD measurements on core in membrane-mimetic environments.[34] In addition, membrane binding and protein conformation may be further regulated by posttranslational modifications, as palmitoylation of cys-172 was reported to be required for proper intracellular localization of core and infectious virion production.[43] Membrane-induced folding is also important for the stabilization of core protein, since mutants deficient in membrane association—and thus probably largely unfolded—are rapidly degraded by the proteasome.[42]

The D1 domain of core is additionally stabilized upon RNA binding, as evidenced by its increased resistance to proteinase digestion.[32] Since RNA binding and nucleocapsid formation are intimately interrelated in HCV, particle formation has so far precluded the structural characterization of the RNA-induced folding reaction.[35] Besides coating and protecting the viral genomic RNA, core protein is also endowed with nucleic acid chaperone activities, i.e., it can promote RNA structural rearrangements, leading to the formation of the most stable RNA structure.[29,33,36] RNA binding and chaperoning are carried out by the intrinsically unstructured N-terminal region of core protein and probably involve mutual induced folding of the interacting partners.[32,33] RNA chaperones in diverse virus

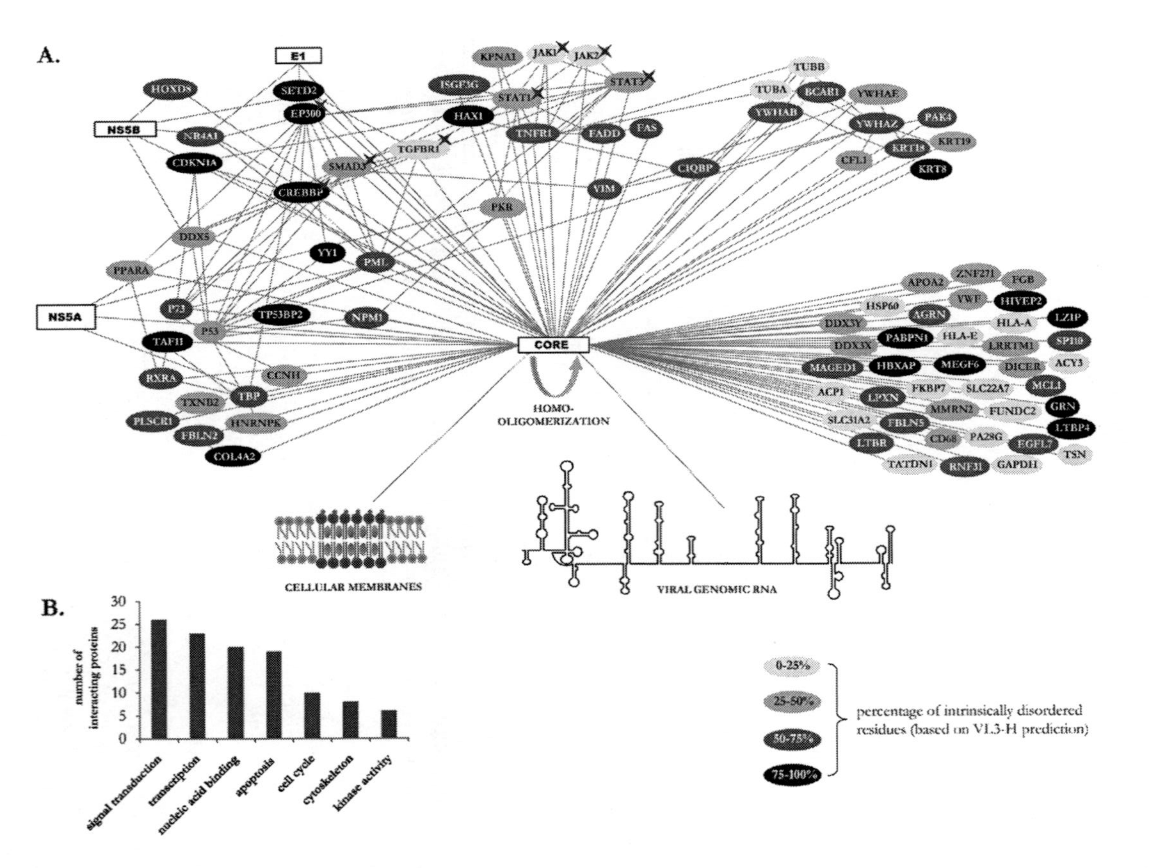

Figure 2. A) The interaction network of HCV core protein. Cellular proteins interacting with HCV core were collected from reference 58, the VirHostNet website,[59] and through extensive literature review. Disorder prediction for the proteins was carried out using the DisProt VL3-H predictor.[118] Important proteins of the IJT network are indicated by an asterisk. B) GO terms frequently associated with core-interacting proteins.

families, including retroviruses, coronaviruses and several distinct members of flaviviruses employ disordered regions to modulate RNA structure.[44,45] Indeed, intrinsic disorder is prevalent in both viral and cellular RNA chaperones and might even be a prerequisite of ATP-independent chaperone action, as implied by the 'entropy exchange model' of Tompa and Csermely.[46]

Will the successive conformational changes in HCV core protein upon membrane- and RNA-binding lead to the formation of an ordered viral nucleocapsid shell? The organization of HCV nucleocapsid remains elusive to date. However, in related flaviviruses, cryo-electron microscopy based reconstruction of immature and mature particle structures suggests that the nucleocapsid is poorly ordered or that its symmetry does not match that of the outer glycoprotein layer.[47-49] A disordered nucleocapsid organization has also been suggested for bovine viral diarrhoea virus, a member of the closely related pestivirus genus.[50] These studies suggest the possibility that fuzziness may be a conserved feature in the nucleocapsids of all three genera of the *Flaviviridae* family.

CORE PROTEIN: A PROMISCUOUS DATE HUB WITH A CENTRAL ROLE IN HCV PATHOGENESIS

Transgenic mice expressing the core protein of HCV develop progressive steatosis, insulin resistance and hepatocellular carcinoma,[51–54] the hallmarks of chronic infection in human patients. Although the observed phenotype may depend on the genetic background of the mouse strains and on the expression constructs used,[55] core protein expression undoubtedly accounts for varied and serious physio-pathological manifestations. This is achieved by perturbation of a large number of cellular processes, including regulation of cellular transcription, signal transduction, apoptosis, lipid metabolism and immunomodulation (Figs. 2A, 2B; reviewed in refs. 56, 57). Yeast two-hybrid screens and co-immunoprecipitation experiments have identified close to hundred cellular proteins directly interacting with HCV core (Fig. 2A).[58,59] Of particular relevance for HCV pathogenesis is the major perturbation of the insulin-JAK/STAT-TGFβ (IJT) network through direct interactions between core and a number of proteins participating in and connecting the signalling cascades (Fig. 2A),[58] ultimately resulting in insulin resistance and fibrogenesis in chronic infection.[60] Interestingly, the same pathways were shown to be activated in patients who do not achieve sustained virological response upon interferon therapy,[61] linking the effect of core protein to treatment failure. Another key mechanism mediating the action of core is its binding to nuclear hormone receptors, including retinoid X receptor (RXRα) and peroxisome proliferator-activated receptor (PPARα), leading to transcriptional regulation of a number of genes involved in cellular lipid metabolism, cell differentiation and proliferation.[62] These changes in the gene expression pattern, ultimately dependent on the upstream regulator proteasome activator PA28γ, are essential for HCV-associated insulin resistance, steatosis and carcinogenesis.[63-65]

As expected, most of the mapped core-cellular protein interactions involve the N-terminal, intrinsically unstructured region of core protein, which provides a large, accessible surface area and might adopt different conformations to allow high specificity/low affinity interactions with multiple, structurally heterogeneous substrates.[66] Indeed, highly connected (hub) proteins in eukaryotic interactomes are characterized by a higher amount of disorder than proteins with fewer interaction partners.[6,7] In addition to constituting a major hub itself, core protein tends to interact with cellular hubs and with proteins that function as key connectors of different pathways (bottlenecks),[58] a shared feature of many proteins from viral and bacterial pathogens.[67] Interestingly, a number of core-interacting human proteins, like the histone acetyltransferase EP300, the p53 tumour suppressor protein or the double-stranded RNA-activeted kinase PKR, are targeted by a large number of pathogenic viruses, underlying their importance in the subversion of cellular processes and/or immune evasion.[67]

HIV-1 AND ITS NUCLEOCAPSID PROTEIN

The human immunodeficiency virus Type 1 (HIV-1) is a small, enveloped virus belonging to the *Retroviridae* family, which possesses unique features such as two copies of the positive strand genomic RNA[68,69] together with the viral enzymes reverse transcriptase (RT), integrase (IN) and protease (PR).[70-73] HIV-1 is a member of the Lentivirus genus together with other pathogenic viruses such as HIV-2, FIV (feline immunodeficiency virus), BIV (bovine immunodeficiency virus) and Visna/CAEV (caprine arthritis encephalitis virus). HIV-1 is the causative agent of the acquired immunodeficiency syndrome (AIDS) and according to WHO estimates has caused the death of about 25 million since its discovery some 28 years ago (ref. 74 and references therein). HIV-1 is mainly transmitted through sexual contacts and by contaminated syringes among intravenous drug users. At present about 35 million persons world-wide are living with the AIDS virus and unfortunately only a small proportion of them benefit from therapeutic treatments among which are the highly active antiretroviral therapies (HAART) targeting the viral enzymes RT, IN and PR.[75] Although HAART proved to be very effective by causing a nearly complete reduction of the virus load in the blood, none of the available treatments can yet cure the disease and eliminate the virus from infected persons.

The HIV-1 genome contains nine open reading frames (ORF) coding for the virion structural proteins (Gag), enzymes (Pol) and envelope glycoproteins (Env) and for viral factors notably those required for proviral DNA transcription (Tat) and viral RNA export and translation (Rev). The other regulatory factors include Nef which modulates the immune responses targeting virus-infected cells, Vif which counteracts the host restriction factor APOBEC 3G, Vpr involved in the nuclear import of the newly made viral DNA and Vpu acting at the level of virus budding. The genomic RNA is flanked by 5' and 3' untranslated regions (UTR), that are highly structured and constitute *cis*-acting genetic elements playing key roles throughout the virus replication cycle,[76,77] notably during reverse transcription of the genomic RNA by RT assisted by the nucleocapsid protein (NC) and in genomic RNA dimerization and packaging in the course of virion formation[69,78,79] (reviewed in ref. 80).

NC protein is located at the C-terminus of the Gag polyprotein precursor, which is released in its mature form called NCp7, together with the matrix Map17, capsid Cap24 and p6 proteins, by the sequential cleavage of Gag by the viral protease during virus morphogenesis in the infected cell (Fig. 3). The nucleocapsid structure is in the interior of the mature globular viral particle of about 115 nm in diameter and is made up of approximately 1500 molecules of NCp7 coating the genomic RNA of positive sense which alike other retroviral genomes is in a dimeric 60S complex (see refs. 79,81,82 and references therein). In addition, 80-100 molecules of the viral enzymes RT, IN and PR are present in the virion nucleocapsid together with 100-200 molecules of the viral protein Vpr (reviewed in ref. 79).

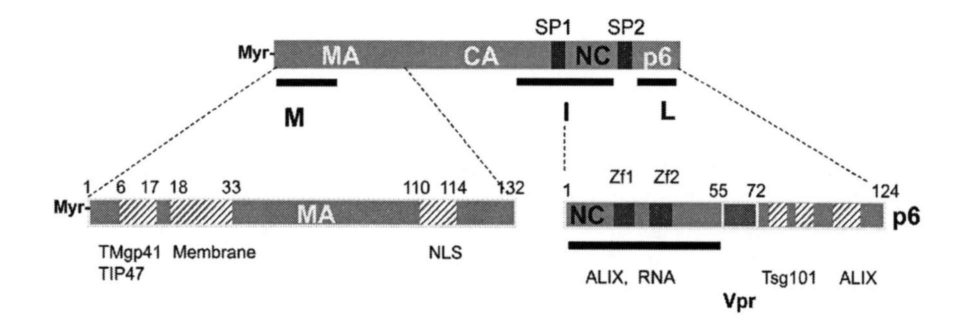

Figure 3. Signals important for HIV-1 assembly in the MA and NC regions of GAG. Simple scheme showing signals in the matrix (MA) and nucleocapsid (NC) regions important for the nucleation of Gag polyprotein assembly in infected cells. 1-The M-domain or plasma membrane-binding domain includes a posttranslational modification corresponding to a myristate linked to glycine residue at position 2 and the downstream 32 residues of MA sequence, notably a stretch of basic residues. The M-domain targets the Pr55Gag molecules to microdomains of the plasma membrane enriched in cholesterol, glycolipids and sphingolipids, called lipid rafts.[119,120] The M-domain also recruits the viral envelope during assembly by interacting with the cellular factor TIP47.[121] 2-The interacting I-domain directs Gag oligomer formation via Gag-RNA and Gag-Gag interactions where the RNA is essential to start assembly via a nucleation process. It overlaps the C-terminal subdomain (CTD) of the CA structural domain and NC domain with its two zinc fingers (Zf) and the flanking basic residues. The NC zinc fingers, which form a hydrophobic plateau (see text), are major determinants for Gag-NC directed packaging of the genomic RNA in a dimeric form.[70,78-82,122] The Zfs are also important signals for vesicular trafficking and virus budding *via* interactions with ALIX[107-109] as well as for the recruitment of Vpr[123] and the overall architecture of the nucleocapsid in mature virions (see Fig. 4 and ref. 112). 3-The nucleocapsid NCp7 of HIV-1 is a 72-residue basic protein in its mature form, with two zinc fingers of the CCHC type flanked by basic residues. NC is essential for virus particle assembly and the selection, dimerization and packaging of the genomic RNA. Later NC assists RT throughout the reverse transcription reaction (reviewed in refs. 78,79,81,82,86,124). 4-The p6 late L-domain is required for the release of newly formed viral particles and has been mapped to the conserved proline-rich (PT/SAPP) motif in p6 (reviewed in ref. 120). Molecular and cellular studies showed that the L-domain recruits the cellular factor TSG101 to the virus assembly site at the plasma membrane. TSG101, which is the product of a tumor susceptibility gene, is a homologue of ubiquitin-conjugating (E2) enzyme involved in the vacuolar protein sorting machinery. TSG101 interacts with the p6 PTAPP motif, causing the recruitment of ESCRT-I (endosomal sorting complex required for transport-I) and associated factors to the site of assembly and budding (refs. 125,126 reviewed in ref. 120). Another cellular endosomal sorting factor, called ALIX, interacts with the L domain and plays a significant role in the release of retroviral particles.[107-109] 5-SP1 and SP2 are the spacer peptides 1 and 2 which regulate, at least in part, the binding of NC to RNA and Gag processing by the viral protease (PR), respectively.[127,128]

Moreover, molecules of cellular origin are incorporated in the virion nucleocapsid such as ribosomal RNAs, actin and tRNAs, notably primer tRNALys,3.[83]

HIV-1 NCp7 is a small basic protein characterized by two copies of a highly conserved CCHC zinc finger motif (ZF), both being flanked by basic residues. Each ZF motif coordinates one zinc ion with high affinity via CCHC—Zn^{2+} interactions, which act as the main driving force to provide a well-defined three dimensional structure to this very small peptidic domain[84,85] (reviewed in ref. 79). The proximal and distal ZFs are linked together by the basic sequence 29-RAPRKKG, which may induce a close proximity of the two ZFs due to the proline kink and contribute to the formation of a hydrophobic plateau.[8,9,84] In contrast, the N- and C-terminal domains of NCp7 appear to adopt a flexible conformation, which can be viewed as instrinsically disordered (Fig. 3) according to NMR analyses. The role of NC in HIV-1 replication has been the subject of a large number of investigations and, taken together, findings demonstrate that both the structured ZFs with the hydrophobic platform and the flexible basic sequences are absolutely required for virus formation and infectivity as well as nucleocapsid structure. [79,86-88]

Similarly to HCV core protein, HIV-1 NC belongs to a growing class of nucleic acid binding proteins (NABP) endowed with RNA chaperoning properties. These proteins called RNA chaperones are ubiquitous and abundant in all living organisms and viruses where they play essential roles such as regulation of gene transcription and RNA translation in cells and genome replication and virion morphogenesis in RNA viruses.[89]

HIV-1 PARTICLE MORPHOGENESIS, THE ROLE OF GAG-NC AND FUZZINESS

As mentioned above (cf. crystal growth *vs* fuzziness), viral particle formation in enveloped viruses, including HIV-1, is much different from crystal growth governed by simple laws of physics. There are two distinct series of players in HIV particle assembly in infected cells, the viral components among which Gag is the major one, on the one hand and host machineries hijacked by the virus, on the other hand.[90]

The HIV-1 Gag polyprotein precursor of 55 kDa is formed of structural domains and contains specific signals necessary and sufficient for viral particle assembly and thus is the major viral player in the process of virus assembly, from protein synthesis to viral particle budding (reviewed in refs. 91, 92 and references therein). In infected cells, the full length viral RNA has two fates, either directed to function as a mRNA for Gag and Gag-Pol synthesis by the host translational machinery,[93] or as a viral genome to be selected and packaged into assembling particles. NC, as part of the newly made Gag precursor, may actually bind its own mRNA and in turn act as a switch from viral RNA translation to particle assembly (Fig. 4).[79] Yet the different fates of the genomic RNA may well coexist with a translation-packaging balance varying according to the levels of viral DNA transcription and RNA translation, both regulated by Tat which is another essential viral factor member of the intrinsically unstructured protein family (IUPs).[94]

How Does Gag-NC Recognize the Genomic RNA to Nucleate Assembly?

This molecular recognition is most probably directed by NC binding to stem-loop structures (SL) in the 5' UTR of the viral RNA, encompassing the internal ribosome entry signal (IRES) which pilots viral RNA translation.[93] But NC can also bind any sequence along the genomic RNA showing only moderate sequence specificity for the 5' SL structures in vitro.[78,79] These findings favour the notion that the nucleation event proceeds *via* different modes of Gag-NC binding to the genomic RNA. In addition, Gag-NC binding causes dimerization of the RNA genome by chaperoning RNA-RNA interactions via the dimer linkage structure in the 5'UTR,[80] meaning that two genomic RNAs should be recognized at a time by Gag-NC. Then the dimeric genome functions as a nucleation platform or a

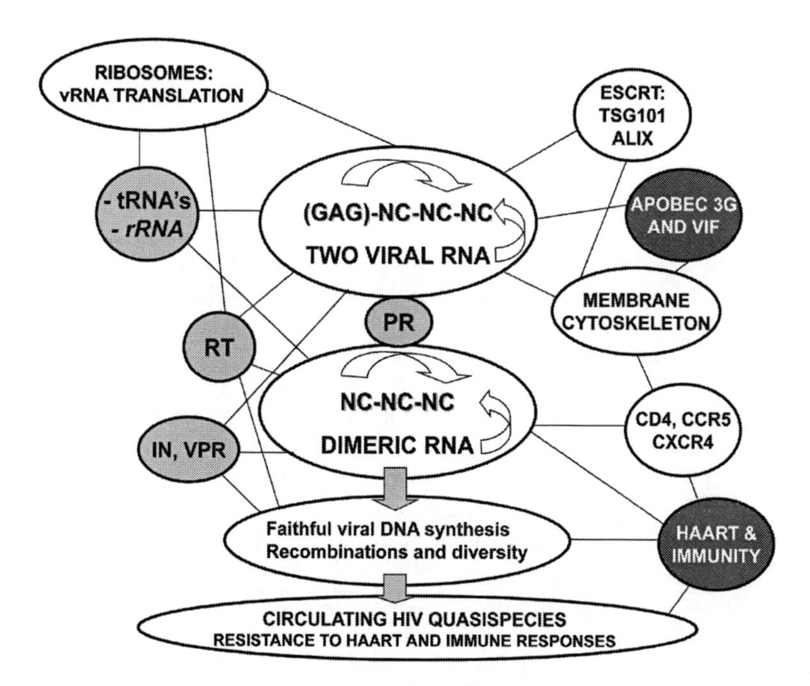

Figure 4. Dynamic molecular interactions during HIV assembly. This simple scheme illustrates essential molecular interactions taking place in the course of virus assembly. Although genetic and biochemical studies have been able to draw a rather well-defined picture of each individual interaction, overall understanding of these interactions, their dynamics and the structure of Gag-NC/RNA and NC/RNA nucleoprotein complexes is probably more than vague. The newly made full length viral RNA is translated by an IRES mechanism to produce the Gag and Gag-Pol polyprotein precursors (reviewed in ref. 93). As shown here, Gag-NC recruits the viral RNA in a *cis*-acting manner to nucleate assembly. At the very end of assembly the viral protease (PR) processes Gag and Gag-Pol polyproteins in the form of oligomers in the particle. This maturation reaction seems to be facilitated by Gag-NC binding to the viral RNA,[128] and in turn activates core condensation[129] and RTion.[124] NC oligomers coat the dimeric genome in the particle and assist RT throughout viral DNA synthesis and subsequently its integration by the viral enzyme IN into the genome of the newly infected cells (for more detail see text). White ovals represent viral and cellular machineries pertaining to viral particle formation in infected cells. Light gray circles and ovals show viral and cellular molecules essential for the formation of infectious virions. Dark gray ovals point to molecules that can combat HIV-1 replication and dissemination (reviewed in refs. 130,131), notably HAART treatments.[75,132,133]

scaffold[95,96] to recruit more Gag-NC molecules *via* RNA-Gag and Gag-Gag interactions.[96] At the same time, NC can recognize cellular RNA sequences, which are in a vast excess over that of the viral RNA in infected cells. In fact, cellular tRNAs and ribosomal RNAs represent roughly half of all RNAs incorporated in virions.[78,79]

All in all, the start of virus assembly corresponds to a nucleation event, where NC in the Gag precursor has a major contribution in recruiting the genomic RNA. This essential reaction conducted by Gag-NC seems to proceed via several modes, namely specific and nonspecific binding to the viral RNA which chaperones its dimerization and nonspecific binding to cellular RNAs, as well as different routes, in *cis* and in *trans*, in a fuzzy manner. Such a flexible pathway early in assembly is fundamental to the formation of viruses with two distinct viral genomes, that will fuel the genetic variability of HIV-1 upon recombination during the reverse transcription reaction in newly infected cells (Fig. 4) (see refs. 68,69; reviewed in refs. 70,79 and references therein). Furthermore, this flexible pathway is currently used to generate hundreds of Lentiviral vectors for gene transfer in cells and animals as well as gene therapy (GT) trials.[97,98]

Gag-NC/RNA Interactions and the Myristyl Switch

An important consequence of the nucleation event is the targeting of the Gag-NC/RNA ribonucleoprotein complex to cellular membranes. In fact, the N-terminal matrix domain (MA) contains a myristate (myr), which amidifies the N-terminal glycine residue of MA. The myristyl group is buried in a cavity of MA in the Gag monomer and undergoes a conformational switch once Gag forms oligomeric structures upon RNA binding directed by NC.[99,100] This in turn allows the Gag RNP to be targeted to cellular membranes, either the plasma membrane of infected TCD4+ cells or the endosomal/MVB membrane of macrophages.[101-103] Subsequently, interactions between the Gag RNP and cellular membranes, notably the phosphoinositide phosphatidylinositol (4,5) bisphosphate ($PI(4,5)P_2$), are stabilized by N-terminal basic clusters of Matrix.[104,105]

At the end of the assembly process, viral particles are released from infected cells by budding. To this end Gag engages components of the ESCRT (endosomal sorting complex required for transport) pathway, such as the cellular proteins TSG101 and ALIX, via late-assembly signals present in Gag-p6 (reviewed in refs. 106,107; see also references therein). Recruitment of the ESCRT machinery by Gag is also ensured by interactions between ALIX and Gag-NC.[108,109]

Taken together, the findings briefly summarized above clearly indicate that the NC domain plays key roles in orchestrating Gag assembly from the nucleation event to particle budding.

ROLE OF HIV-1 NC, OLIGOMER FORMATION AND FUZZINESS

The multiple roles of NC in particle morphogenesis are most probably accomplished through the action of NC oligomers, which can be easily visualized in vitro by electron microscopy.[110] In these oligomeric structures of different sizes and forms, extensive molecular interactions such as NC-RNA, NC-NC and RNA-RNA, are taking place as indicated by biochemical and genetic analyses.[78,79,82] These NC-RNA structures appear to be porous because oligonucleotides complementary to the RNA sequence are getting access to the RNA and are hybridized to the RNA by the NC annealing activity (reviewed in refs. 78,79). Along this line, the heterogeneity and dynamics of these nucleocapsid structures are probably essential to promote other molecular interactions, notably with the cellular factor ALIX to facilitate virus budding[108,109] and with RT so that the enzyme gains access to the RNA and performs cDNA synthesis with high efficiency.[110-112] Taken together, these findings favor the notion that NC-RNA ribonucleoprotein structures resemble fuzzy assemblages essential to perform diverse biological functions.

The highly dynamic nature of such fuzzy molecular assemblages is illustrated by investigations on the functionality of viral RNA-NC complexes mimicking HIV-1 nucleocapsid (ref. 110; reviewed in ref. 78). Under these conditions, reverse transcription (RTion) of a viral RNA representing the 5' and 3' UTR sequences of the genome was found to experience the two strand transfers necessary to generate the long terminal repeats (LTR) allowing the reaction to go to completion. In these reconstituted nucleocapsids, NC oligomers are playing multiple roles to ensure a faithful and efficient viral DNA synthesis by chaperoning the DNA strand transfers and, at the same time, providing the RT enzyme with excision-repair activities.[113] At the end of the RTion reaction, NC-oligomers bind the newly made viral DNA, which contributes to maintain its integrity and facilitates the integration reaction according to in vitro and in vivo data.[111,114]

The functionality of such fuzzy nucleocapsid assemblages is also illustrated by the dimeric nature of the genomic RNA, which provides the basis for multiple forced and unforced recombination reactions during vDNA synthesis by RT.[115] These recombinations are chaperoned by NC and fulfill two functions, formation of a complete viral DNA by RT in conditions where each RNA monomer contains nicks and reassortments of specific genetic traits when the two monomers are different, which contributes to a high level of genetic diversity resulting in a quasispecies population in HIV-1 infected persons.

CONCLUSION

In conclusion, the fuzzy assemblage of NC oligomers evenly coating the genomic RNA, which interact with RT and IN, forms the viral replication machinery. This fuzzy macromolecular assemblage is essential to ensure faithful and efficient DNA synthesis and integration and thus HIV-1 replication and at the same time to permit sufficient genetic diversity for the virus to escape HAART treatments and specific immune responses. Such a vaguely ordered fuzzy replication machinery probably holds true also for the alpharetroviruses (e.g., avian leukosis virus) and gammaretroviruses (e.g., murine leukemia viruses) that are widespread in birds and rodents, respectively.[79,80]

On a more general basis, intrinsic disorder may confer numerous functional advantages to viral nucleocapsid proteins, providing the driving force for RNA structural rearrangements upon RNA chaperoning during genome replication, as well as facilitating nucleocapsid assembly and promiscuous interactions with a large number of cellular target molecules. A complex array of simultaneous and sequential interactions with other viral constituents, cellular lipids and proteins may induce various amounts (and possibly type) of structure in the context of macromolecular assemblies. Determination of the structural changes in the dynamic interaction networks is of pivotal importance in order to identify potential druggable sites in nucleocapsid proteins or in the interacting protein network.[116,117]

ACKNOWLEDGEMENTS

Thanks are due to INSERM, ANRS, Sidaction and FINOVI (France) and Europe (FP6) for their support to HIV NC and HCV Core Research programs).

REFERENCES

1. Ward JJ, Sodhi JS, McGuffin LJ et al. Prediction and functional analysis of native disorder in proteins from the three kingdoms of life. J Mol Biol 2004; 337(3):635-645.
2. Dunker AK, Obradovic Z, Romero P et al. Intrinsic protein disorder in complete genomes. Genome Inform Ser Workshop Genome Inform 2000; 11:161-171.
3. Mohan A, Sullivan WJ Jr., Radivojac P et al. Intrinsic disorder in pathogenic and nonpathogenic microbes: discovering and analyzing the unfoldomes of early-branching eukaryotes. Mol Biosyst 2008; 4(4):328-340.
4. Xie H, Vucetic S, Iakoucheva LM et al. Functional anthology of intrinsic disorder. 3. Ligands, posttranslational modifications and diseases associated with intrinsically disordered proteins. J Proteome Res 2007;6(5):1917-1932.
5. Xie H, Vucetic S, Iakoucheva LM et al. Functional anthology of intrinsic disorder. 1. Biological processes and functions of proteins with long disordered regions. J Proteome Res 2007; 6(5):1882-1898.
6. Dosztanyi Z, Chen J, Dunker AK et al. Disorder and sequence repeats in hub proteins and their implications for network evolution. J Proteome Res 2006; 5(11):2985-2995.
7. Haynes C, Oldfield CJ, Ji F et al. Intrinsic disorder is a common feature of hub proteins from four eukaryotic interactomes. PLoS Comput Biol 2006; 2(8):e100.
8. Morellet N, de Rocquigny H, Mely Y et al. Conformational behaviour of the active and inactive forms of the nucleocapsid NCp7 of HIV-1 studied by 1H NMR. J Mol Biol 1994;235(1):287-301.
9. Morellet N, Jullian N, De Rocquigny H et al. Determination of the structure of the nucleocapsid protein NCp7 from the human immunodeficiency virus type 1 by 1H NMR. EMBO J 1992; 11(8):3059-3065.
10. Goh GK, Dunker AK, Uversky VN. Protein intrinsic disorder toolbox for comparative analysis of viral proteins. BMC Genomics 2008; 9 Suppl 2:S4.
11. Longhi S, Oglesbee M. Structural disorder within the measles virus nucleoprotein and phosphoprotein. Protein Pept Lett 2010; 17(8):961-978.
12. Reingewertz TH, Shalev DE, Friedler A. Structural disorder in the HIV-1 Vif protein and interaction-dependent gain of structure. Protein Pept Lett 2010; 17(8):988-998.
13. Shojania S, O'Neil JD. Intrinsic disorder and function of the HIV-1 Tat protein. Protein Pept Lett 17(8):999-1011.
14. Ivanyi-Nagy R, Darlix JL. Intrinsic disorder in the core proteins of flaviviruses. Protein Pept Lett 2010; 17(8):1019-1025.
15. Chang CK, Hsu YL, Chang YH et al. Multiple nucleic acid binding sites and intrinsic disorder of severe acute respiratory syndrome coronavirus nucleocapsid protein: implications for ribonucleocapsid protein packaging. J Virol 2009; 83(5):2255-2264.
16. Chang CK, Sue SC, Yu TH et al. Modular organization of SARS coronavirus nucleocapsid protein. J Biomed Sci 2006; 13(1):59-72.
17. Lindenbach B, Thiel HJ, Rice CM. Flaviviridae: The Viruses and Their Replication. Fields Virology. Philadelphia: Lippincott-Raven Publishers; 2007:1101-1152.
18. Shepard CW, Finelli L, Alter MJ. Global epidemiology of hepatitis C virus infection. Lancet Infect Dis 2005; 5(9):558-567.
19. Lavanchy D. The global burden of hepatitis C. Liver Int 2009;29 Suppl 1:74-81.
20. Hoofnagle JH. Course and outcome of hepatitis C. Hepatology 2002; 36(5 Suppl 1):S21-29.
21. Galossi A, Guarisco R, Bellis L et al. Extra hepatic manifestations of chronic HCV infection. J Gastrointestin Liver Dis 2007; 16(1):65-73.
22. Acharya JN, Pacheco VH. Neurologic complications of hepatitis C. Neurologist 2008; 14(3):151-156.
23. Bartenschlager R, Frese M, Pietschmann T. Novel insights into hepatitis C virus replication and persistence. Adv Virus Res 2004; 63:71-180.
24. Moradpour D, Penin F, Rice CM. Replication of hepatitis C virus. Nat Rev Microbiol 2007; 5(6):453-463.
25. Hijikata M, Kato N, Ootsuyama Y et al. Gene mapping of the putative structural region of the hepatitis C virus genome by in vitro processing analysis. Proc Natl Acad Sci USA 1991; 88(13):5547-5551.
26. McLauchlan J, Lemberg MK, Hope G et al. Intramembrane proteolysis promotes trafficking of hepatitis C virus core protein to lipid droplets. EMBO J 2002; 21(15):3980-3988.
27. Hope RG, McLauchlan J. Sequence motifs required for lipid droplet association and protein stability are unique to the hepatitis C virus core protein. J Gen Virol 2000; 81(Pt 8):1913-1925.
28. Santolini E, Migliaccio G, La Monica N. Biosynthesis and biochemical properties of the hepatitis C virus core protein. J Virol 1994; 68(6):3631-3641.
29. Cristofari G, Ivanyi-Nagy R, Gabus C et al. The hepatitis C virus Core protein is a potent nucleic acid chaperone that directs dimerization of the viral (+) strand RNA in vitro. Nucleic Acids Res 2004; 32(8):2623-2631.
30. Majeau N, Gagne V, Boivin A et al. The N-terminal half of the core protein of hepatitis C virus is sufficient for nucleocapsid formation. J Gen Virol 2004; 85(Pt 4):971-981.

31. Kunkel M, Lorinczi M, Rijnbrand R et al. Self-assembly of nucleocapsid-like particles from recombinant hepatitis C virus core protein. J Virol 2001; 75(5):2119-2129.
32. Kunkel M, Watowich SJ. Conformational changes accompanying self-assembly of the hepatitis C virus core protein. Virology 2002; 294(2):239-245.
33. Ivanyi-Nagy R, Lavergne JP, Gabus C et al. RNA chaperoning and intrinsic disorder in the core proteins of Flaviviridae. Nucleic Acids Res 2008; 36(3):712-725.
34. Boulant S, Vanbelle C, Ebel C et al. Hepatitis C virus core protein is a dimeric alpha-helical protein exhibiting membrane protein features. J Virol 2005; 79(17):11353-11365.
35. Duvignaud JB, Savard C, Fromentin R et al. Structure and dynamics of the N-terminal half of hepatitis C virus core protein: an intrinsically unstructured protein. Biochem Biophys Res Commun 2009;378(1):27-31.
36. Ivanyi-Nagy R, Kanevsky I, Gabus C et al. Analysis of hepatitis C virus RNA dimerization and core-RNA interactions. Nucleic Acids Res 2006; 34(9):2618-2633.
37. Miyanari Y, Atsuzawa K, Usuda N et al. The lipid droplet is an important organelle for hepatitis C virus production. Nat Cell Biol 2007; 9(9):1089-1097.
38. Namba K, Stubbs G. Structure of tobacco mosaic virus at 3.6 A resolution: implications for assembly. Science 1986; 231(4744):1401-1406.
39. Butler PJ. Self-assembly of tobacco mosaic virus: the role of an intermediate aggregate in generating both specificity and speed. Philos Trans R Soc Lond B Biol Sci 1999; 354(1383):537-550.
40. Fraenkel-Conrat H, Williams RC. Reconstitution of Active Tobacco Mosaic Virus from Its Inactive Protein and Nucleic Acid Components. Proc Natl Acad Sci USA 1955; 41(10):690-698.
41. Tompa P, Fuxreiter M. Fuzzy complexes: polymorphism and structural disorder in protein-protein interactions. Trends Biochem Sci 2008; 33(1):2-8.
42. Boulant S, Montserret R, Hope RG et al. Structural determinants that target the hepatitis C virus core protein to lipid droplets. J Biol Chem 2006; 281(31):22236-22247.
43. Majeau N, Fromentin R, Savard C et al. Palmitoylation of hepatitis C virus core protein is important for virion production. J Biol Chem 2009.
44. Ivanyi-Nagy R, Davidovic L, Khandjian EW et al. Disordered RNA chaperone proteins: from functions to disease. Cell Mol Life Sci 2005; 62(13):1409-1417.
45. Zuniga S, Sola I, Cruz JL et al. Role of RNA chaperones in virus replication. Virus Res 2009; 139(2):253-266.
46. Tompa P, Csermely P. The role of structural disorder in the function of RNA and protein chaperones. FASEB J 2004; 18(11):1169-1175.
47. Kuhn RJ, Zhang W, Rossmann MG et al. Structure of dengue virus: implications for flavivirus organization, maturation and fusion. Cell 2002; 108(5):717-725.
48. Mukhopadhyay S, Kim BS, Chipman PR et al. Structure of West Nile virus. Science 2003; 302(5643):248.
49. Zhang Y, Corver J, Chipman PR et al. Structures of immature flavivirus particles. EMBO J 2003; 22(11):2604-2613.
50. Murray CL, Marcotrigiano J, Rice CM. Bovine viral diarrhea virus core is an intrinsically disordered protein that binds RNA. J Virol 2008; 82(3):1294-1304.
51. Moriya K, Fujie H, Shintani Y et al. The core protein of hepatitis C virus induces hepatocellular carcinoma in transgenic mice. Nat Med 1998; 4(9):1065-1067.
52. Moriya K, Yotsuyanagi H, Shintani Y et al. Hepatitis C virus core protein induces hepatic steatosis in transgenic mice. J Gen Virol 1997; 78 (Pt 7):1527-1531.
53. Lerat H, Honda M, Beard MR et al. Steatosis and liver cancer in transgenic mice expressing the structural and nonstructural proteins of hepatitis C virus. Gastroenterology 2002; 122(2):352-365.
54. Shintani Y, Fujie H, Miyoshi H et al. Hepatitis C virus infection and diabetes: direct involvement of the virus in the development of insulin resistance. Gastroenterology 2004; 126(3):840-848.
55. Liang TJ, Heller T. Pathogenesis of hepatitis C-associated hepatocellular carcinoma. Gastroenterology 2004; 127(5 Suppl 1):S62-71.
56. McLauchlan J. Properties of the hepatitis C virus core protein: a structural protein that modulates cellular processes. J Viral Hepat 2000; 7(1):2-14.
57. Ray RB, Ray R. Hepatitis C virus core protein: intriguing properties and functional relevance. FEMS Microbiol Lett 2001; 202(2):149-156.
58. de Chassey B, Navratil V, Tafforeau L et al. Hepatitis C virus infection protein network. Mol Syst Biol 2008; 4:230.
59. Navratil V, de Chassey B, Meyniel L et al. VirHostNet: a knowledge base for the management and the analysis of proteome-wide virus-host interaction networks. Nucleic Acids Res 2009; 37(Database issue):D661-668.
60. Del Campo JA, Romero-Gomez M. Steatosis and insulin resistance in hepatitis C: a way out for the virus? World J Gastroenterol 2009; 15(40):5014-5019.
61. Birerdinc A, Afendy A, Stepanova M et al. Functional pathway analysis of genes associated with response to treatment for chronic hepatitis C. J Viral Hepat 2009.
62. Koike K, Moriya K, Matsuura Y. Animal models for hepatitis C and related liver disease. Hepatol Res 40(1):69-82.

63. Miyamoto H, Moriishi K, Moriya K et al. Involvement of the PA28gamma-dependent pathway in insulin resistance induced by hepatitis C virus core protein. J Virol 2007; 81(4):1727-1735.
64. Moriishi K, Mochizuki R, Moriya K et al. Critical role of PA28gamma in hepatitis C virus-associated steatogenesis and hepatocarcinogenesis. Proc Natl Acad Sci USA 2007; 104(5):1661-1666.
65. Moriishi K, Okabayashi T, Nakai K et al. Proteasome activator PA28gamma-dependent nuclear retention and degradation of hepatitis C virus core protein. J Virol 2003; 77(19):10237-10249.
66. Kriwacki RW, Hengst L, Tennant L et al. Structural studies of p21Waf1/Cip1/Sdi1 in the free and Cdk2-bound state: conformational disorder mediates binding diversity. Proc Natl Acad Sci USA 1996; 93(21):11504-11509.
67. Dyer MD, Murali TM, Sobral BW. The landscape of human proteins interacting with viruses and other pathogens. PLoS Pathog 2008; 4(2):e32.
68. Temin HM. Sex and recombination in retroviruses. Trends Genet 1991; 7(3):71-74.
69. Chen J, Nikolaitchik O, Singh J et al. High efficiency of HIV-1 genomic RNA packaging and heterozygote formation revealed by single virion analysis. Proc Natl Acad Sci USA 2009; 106(32):13535-13540.
70. Coffin JM. Structure, replication and recombination of retrovirus genomes: some unifying hypotheses. J Gen Virol 1979; 42(1):1-26.
71. Coffin JM. Retroviridae and their replication. In: Fields B, Knipe D, Chanock R, eds. Virology Vol I. New York: Raven Press; 1990:1437-1500.
72. Gilboa E, Mitra SW, Goff S et al. A detailed model of reverse transcription and tests of crucial aspects. Cell 1979; 18(1):93-100.
73. Lewinski MK, Bushman FD. Retroviral DNA integration—mechanism and consequences. Adv Genet 2005; 55:147-181.
74. Wainberg MA, Jeang KT. 25 years of HIV-1 research—progress and perspectives. BMC Med 2008; 6:31.
75. Broder S. The development of antiretroviral therapy and its impact on the HIV-1/AIDS pandemic. Antiviral Res 85(1):1-18.
76. Jones KA. HIV trans-activation and transcription control mechanisms. New Biol 1989; 1(2):127-135.
77. Yilmaz A, Bolinger C, Boris-Lawrie K. Retrovirus translation initiation: Issues and hypotheses derived from study of HIV-1. Curr HIV Res 2006; 4(2):131-139.
78. Darlix JL, Garrido JL, Morellet N et al. Properties, functions and drug targeting of the multifunctional nucleocapsid protein of the human immunodeficiency virus. Adv Pharmacol 2007; 55:299-346.
79. Darlix JL, Lapadat-Tapolsky M, de Rocquigny H et al. First glimpses at structure-function relationships of the nucleocapsid protein of retroviruses. J Mol Biol 1995; 254(4):523-537.
80. Berkhout B. Structure and function of the human immunodeficiency virus leader RNA. Prog Nucleic Acid Res Mol Biol 1996; 54:1-34.
81. Levin JG, Guo J, Rouzina I et al. Nucleic acid chaperone activity of HIV-1 nucleocapsid protein: critical role in reverse transcription and molecular mechanism. Prog Nucleic Acid Res Mol Biol 2005; 80:217-286.
82. Rein A, Henderson LE, Levin JG. Nucleic-acid-chaperone activity of retroviral nucleocapsid proteins: significance for viral replication. Trends Biochem Sci 1998; 23(8):297-301.
83. Kleiman L, Halwani R, Javanbakht H. The selective packaging and annealing of primer tRNALys3 in HIV-1. Curr HIV Res 2004; 2(2):163-175.
84. De Guzman RN, Wu ZR, Stalling CC et al. Structure of the HIV-1 nucleocapsid protein bound to the SL3 psi-RNA recognition element. Science 1998; 279(5349):384-388.
85. de Rocquigny H, Shvadchak V, Avilov S et al. Targeting the viral nucleocapsid protein in anti-HIV-1 therapy. Mini Rev Med Chem 2008; 8(1):24-35.
86. Thomas JA, Gorelick RJ. Nucleocapsid protein function in early infection processes. Virus Res 2008; 134(1-2):39-63.
87. Grigorov B, Decimo D, Smagulova F et al. Intracellular HIV-1 Gag localization is impaired by mutations in the nucleocapsid zinc fingers. Retrovirology 2007; 4:54.
88. Tanchou V, Decimo D, Pechoux C et al. Role of the N-terminal zinc finger of human immunodeficiency virus type 1 nucleocapsid protein in virus structure and replication. J Virol 1998; 72(5):4442-4447.
89. Cristofari G, Darlix JL. The ubiquitous nature of RNA chaperone proteins. Prog Nucleic Acid Res Mol Biol 2002; 72:223-268.
90. Goff SP. Host factors exploited by retroviruses. Nat Rev Microbiol 2007; 5(4):253-263.
91. Freed EO. HIV-1 gag proteins: diverse functions in the virus life cycle. Virology 1998; 251(1):1-15.
92. Freed EO. Viral late domains. J Virol 2002; 76(10):4679-4687.
93. Balvay L, Lopez Lastra M, Sargueil B, Darlix JL, Ohlmann T. Translational control of retroviruses. Nat Rev Microbiol 2007; 5(2):128-140.
94. Kuciak M, Gabus C, Ivanyi-Nagy R et al. The HIV-1 transcriptional activator Tat has potent nucleic acid chaperoning activities in vitro. Nucleic Acids Res 2008; 36(10):3389-3400.
95. Cimarelli A, Darlix JL. Assembling the human immunodeficiency virus type 1. Cell Mol Life Sci 2002; 59(7):1166-1184.

96. Ganser-Pornillos BK, Yeager M, Sundquist WI. The structural biology of HIV assembly. Curr Opin Struct Biol 2008; 18(2):203-217.
97. Mangeot PE, Negre D, Dubois B et al. Development of minimal lentivirus vectors derived from simian immunodeficiency virus (SIVmac251) and their use for gene transfer into human dendritic cells. J Virol 2000; 74(18):8307-8315.
98. Negre D, Duisit G, Mangeot PE et al. Lentiviral vectors derived from simian immunodeficiency virus. Curr Top Microbiol Immunol 2002; 261:53-74.
99. Saad JS, Loeliger E, Luncsford P et al. Point mutations in the HIV-1 matrix protein turn off the myristyl switch. J Mol Biol 2007; 366(2):574-585.
100. Saad JS, Miller J, Tai J et al. Structural basis for targeting HIV-1 Gag proteins to the plasma membrane for virus assembly. Proc Natl Acad Sci USA 2006; 103(30):11364-11369.
101. Raposo G, Moore M, Innes D et al. Human macrophages accumulate HIV-1 particles in MHC II compartments. Traffic 2002; 3(10):718-729.
102. Grigorov B, Arcanger F, Roingeard P et al. Assembly of infectious HIV-1 in human epithelial and T-lymphoblastic cell lines. J Mol Biol 2006; 359(4):848-862.
103. Ono A, Freed EO. Cell-type-dependent targeting of human immunodeficiency virus type 1 assembly to the plasma membrane and the multivesicular body. J Virol 2004; 78(3):1552-1563.
104. Ono A, Freed EO. Role of lipid rafts in virus replication. Adv Virus Res 2005; 64:311-358.
105. Hamard-Peron E, Juillard F, Saad JS et al. Targeting of murine leukemia virus gag to the plasma membrane is mediated by PI(4,5)P2/PS and a polybasic region in the matrix. J Virol 84(1):503-515.
106. Demirov DG, Freed EO. Retrovirus budding. Virus Res 2004; 106(2):87-102.
107. Morita E, Sundquist WI. Retrovirus budding. Annu Rev Cell Dev Biol 2004; 20:395-425.
108. Popov S, Popova E, Inoue M et al. Divergent Bro1 domains share the capacity to bind human immunodeficiency virus type 1 nucleocapsid and to enhance virus-like particle production. J Virol 2009; 83(14):7185-7193.
109. Fujii K, Munshi UM, Ablan SD et al. Functional role of Alix in HIV-1 replication. Virology 2009; 391(2):284-292.
110. Gabus C, Auxilien S, Pechoux C et al. The prion protein has DNA strand transfer properties similar to retroviral nucleocapsid protein. J Mol Biol 2001; 307(4):1011-1021.
111. Buckman JS, Bosche WJ, Gorelick RJ. Human immunodeficiency virus type 1 nucleocapsid zn(2+) fingers are required for efficient reverse transcription, initial integration processes and protection of newly synthesized viral DNA. J Virol 2003; 77(2):1469-1480.
112. Berthoux L, Pechoux C, Ottmann M et al. Mutations in the N-terminal domain of human immunodeficiency virus type 1 nucleocapsid protein affect virion core structure and proviral DNA synthesis. J Virol 1997; 71(9):6973-6981.
113. Bampi C, Bibillo A, Wendeler M et al. Nucleotide excision repair and template-independent addition by HIV-1 reverse transcriptase in the presence of nucleocapsid protein. J Biol Chem 2006; 281(17):11736-11743.
114. Carteau S, Gorelick RJ, Bushman FD. Coupled integration of human immunodeficiency virus type 1 cDNA ends by purified integrase in vitro: stimulation by the viral nucleocapsid protein. J Virol 1999; 73(8):6670-6679.
115. Hu WS, Temin HM. Genetic consequences of packaging two RNA genomes in one retroviral particle: pseudodiploidy and high rate of genetic recombination. Proc Natl Acad Sci USA 1990; 87(4):1556-1560.
116. Cheng Y, LeGall T, Oldfield CJ et al. Rational drug design via intrinsically disordered protein. Trends Biotechnol 2006; 24(10):435-442.
117. Csermely P, Agoston V, Pongor S. The efficiency of multi-target drugs: the network approach might help drug design. Trends Pharmacol Sci 2005; 26(4):178-182.
118. Obradovic Z, Peng K, Vucetic S et al. Predicting intrinsic disorder from amino acid sequence. Proteins 2003; 53 Suppl 6:566-572.
119. Freed EO. HIV-1 Gag: flipped out for PI(4,5)P(2). Proc Natl Acad Sci USA 2006; 103(30):11101-11102.
120. Adamson CS, Freed EO. Human immunodeficiency virus type 1 assembly, release and maturation. Adv Pharmacol 2007; 55:347-387.
121. Lopez-Verges S, Camus G, Blot G et al. Tail-interacting protein TIP47 is a connector between Gag and Env and is required for Env incorporation into HIV-1 virions. Proc Natl Acad Sci USA 2006; 103(40):14947-14952.
122. Paillart JC, Shehu-Xhilaga M, Marquet R et al. Dimerization of retroviral RNA genomes: an inseparable pair. Nat Rev Microbiol 2004; 2(6):461-472.
123. Lavallee C, Yao XJ, Ladha A et al. Requirement of the Pr55gag precursor for incorporation of the Vpr product into human immunodeficiency virus type 1 viral particles. J Virol 1994; 68(3):1926-1934.
124. Mougel M, Houzet L, Darlix JL. When is it time for reverse transcription to start and go? Retrovirology 2009; 6:24.
125. Garrus JE, von Schwedler UK, Pornillos OW et al. Tsg101 and the vacuolar protein sorting pathway are essential for HIV-1 budding. Cell 2001; 107(1):55-65.
126. VerPlank L, Bouamr F, LaGrassa TJ et al. Tsg101, a homologue of ubiquitin-conjugating (E2) enzymes, binds the L domain in HIV type 1 Pr55(Gag). Proc Natl Acad Sci USA 2001; 98(14):7724-7729.

127. Pettit SC, Lindquist JN, Kaplan AH et al. Processing sites in the human immunodeficiency virus type 1 (HIV-1) Gag-Pro-Pol precursor are cleaved by the viral protease at different rates. Retrovirology 2005; 2:66.
128. Pettit SC, Moody MD, Wehbie RS et al. The p2 domain of human immunodeficiency virus type 1 Gag regulates sequential proteolytic processing and is required to produce fully infectious virions. J Virol 1994; 68(12):8017-8027.
129. Fu W, Dang Q, Nagashima K et al. Effects of Gag mutation and processing on retroviral dimeric RNA maturation. J Virol 2006; 80(3):1242-1249.
130. Wolf D, Goff SP. Host restriction factors blocking retroviral replication. Annu Rev Genet 2008; 42:143-163.
131. Goncalves J, Santa-Marta M. HIV-1 Vif and APOBEC3G: multiple roads to one goal. Retrovirology 2004; 1:28.
132. Martinez-Picado J, Martinez MA. HIV-1 reverse transcriptase inhibitor resistance mutations and fitness: a view from the clinic and ex vivo. Virus Res 2008; 134(1-2):104-123.
133. Wlodawer A, Vondrasek J. Inhibitors of HIV-1 protease: a major success of structure-assisted drug design. Annu Rev Biophys Biomol Struct 1998; 27:249-284.

CHAPTER 10

STRUCTURAL DISORDER AND PROTEIN ELASTICITY

Sarah Rauscher and Régis Pomès*

Molecular Structure and Function, Hospital for Sick Children, Canada; Department of Biochemistry, University of Toronto, Canada

**Corresponding Author: Régis Pomès—Email: regis.pomes@sickkids.ca*

Abstract: An emerging class of disordered proteins underlies the elasticity of many biological tissues. Elastomeric proteins are essential to the function of biological machinery as diverse as the human arterial wall, the capture spiral of spider webs and the jumping mechanism of fleas. In this chapter, we review what is known about the molecular basis and the functional role of structural disorder in protein elasticity. In general, the elastic recoil of proteins is due to a combination of internal energy and entropy. In rubber-like elastomeric proteins, the dominant driving force is the increased entropy of the relaxed state relative to the stretched state. Aggregates of these proteins are intrinsically disordered or fuzzy, with high polypeptide chain entropy. We focus our discussion on the sequence, structure and function of five rubber-like elastomeric proteins, elastin, resilin, spider silk, abductin and ColP. Although we group these disordered elastomers together into one class of proteins, they exhibit a broad range of sequence motifs, mechanical properties and biological functions. Understanding how sequence modulates both disorder and elasticity will help advance the rational design of elastic biomaterials such as artificial skin and vascular grafts.

INTRODUCTION

Elasticity is the intrinsic ability of a material to return to its original shape after being deformed by an external force.[1] Elastic recoil is a property of many different materials, familiar examples of which include steel, rubber, silicon and lycra. Flexible elastomers, such as rubber, are stretched or compressed with minimal force. Extensible elastomers, including lycra, can be stretched significantly before rupturing. Resilient

---

Fuzziness: Structural Disorder in Protein Complexes, edited by Monika Fuxreiter and Peter Tompa.

elastic materials are often used as components in mechanical devices because of their ability to undergo stretching and relaxation reversibly, dissipating minimal elastic energy as heat.[1-3] Thus, different elastomers are suited to different functions because of their unique set of elastic properties.

Accordingly, elastomeric proteins fulfill essential roles in species throughout the animal kingdom. In vertebrates, elastin is responsible for the elastic recoil of arteries, skin, lung alveoli and uterine tissue.[4,5] Elastin's mechanical properties are remarkably similar to those of resilin, an elastic insect protein.[1] Diverse insect tissues use resilin deposits to store elastic energy, including the wing joints of dragonflies[6] and the jumping mechanism of fleas.[7] Also found in arthropods, spider silks are a class of elastomeric materials with a wide range of elastic properties; rigid and strong silks are used for encasing eggs and restraining trapped prey, while flexible silks are used to construct webs.[8] Molluscs (scallops and mussels) have two well-characterized elastomeric proteins: abductin and ColP. Abductin forms the flexible hinge of the scallop's shell. When the scallop is 'swimming', the opening and closing motion of the shell propels the scallop through water.[9-11] Other molluscs, including mussels, are stationary and require elastomeric threads to tether their shells to underwater surfaces. The elastomeric protein ColP permits these threads to be stretched by the force of tides without breaking.[12] Taken together, these examples demonstrate that the biological roles of elastomeric proteins are remarkably diverse. Accordingly, elastomeric proteins exhibit a wide range of elastic properties: dragline spider silk is one of the toughest materials ever discovered, resilin is more resilient than the best synthetic rubbers,[13] and elastin exhibits remarkable durability.[14]

The wide range of elastic mechanical properties of biological tissues is dictated by the molecular structure of their constituent elastomeric proteins. Because of their desirable mechanical properties, the study of elastomeric proteins is motivated by their potential use in biomedical engineering and materials science. A detailed understanding of the sequence, structure and function of these proteins provides a framework for the rational design of novel biomaterials.

Importantly, some elastomeric proteins have well-defined secondary and tertiary structures, while others are intrinsically disordered. This dichotomy in structural tendencies is embodied by collagen and elastin. Although both are elastomeric proteins, they have very different structural properties. Collagen is the protein responsible for the strength and elasticity of tendon. In fact, collagen has more than 10 times the elastic energy storage capacity of steel.[1] Collagen has been shown by X-ray crystallography to adopt a highly ordered, triple helix structure.[15] By contrast, elastin's structure is characterized by a high degree of conformational disorder,[16,17] which makes it flexible and easily stretched.[18] Due to intrinsic differences in the degree of structural order of elastin and collagen, very little force is required to stretch blood vessels compared to the force required to stretch tendon. Other examples of elastomeric proteins with well-defined molecular structures include spectrin, keratin,[19,20] and a protein recently discovered in the egg capsule of the marine snail.[21] Structurally-ordered elastomeric proteins have been reviewed in detail elsewhere[19,20] and are beyond the scope of the present discussion. In this chapter, we focus on intrinsically disordered, "fuzzy" elastomeric proteins.

The purpose of this chapter is to review what is currently known about the molecular basis for the elastic properties of rubber-like elastomeric proteins. We introduce the relationship between intrinsic disorder and elasticity in the following section, with a brief background on rubber-like elasticity and its associated mechanical properties. We then provide a detailed description of elastomeric proteins that require structural disorder to function, including elastin, resilin, spider silk, abductin and ColP. Finally, we review the essential sequence features of these proteins and we present an emerging unified model of the sequence, structure and function of disordered elastomeric proteins.

ELASTICITY AND ELASTIC MECHANICAL PROPERTIES

Disorder and Elasticity

Elastic materials exhibit a broad spectrum of mechanical properties due to fundamental differences in their molecular mechanisms of elasticity. The driving force of elastic recoil, f, is the sum of two contributions: an entropic component, f_s, and an internal energy component, f_e:[3,22]

$$f = f_s + f_e.$$

Changes in internal energy occur when an applied force distorts the material's underlying molecular structure. In this case, the driving force for elastic recoil arises from the tendency of the molecular structure to return to the state of lowest potential energy upon removal of the external force. Stiff materials, like steel, store elastic energy in changes in internal energy (i.e., $f_e > f_s$). In contrast, entropic elastomers, like rubber, store elastic energy in the difference in entropy between the stretched and relaxed states (i.e., $f_s > f_e$).[3] Entropic elastomers have a disordered molecular structure (Fig. 1). Because there are many more ways of arranging a recoiled polymer chain than a stretched one, stretching a disordered polymer lowers the chain entropy, which is restored upon release of the strain. Compressing an entropic elastomer has the same effect as stretching: in both cases, the entropy of the relaxed state is higher than that of the deformed state, resulting in elastic recoil. In summary, there are two general mechanisms of elastic recoil:

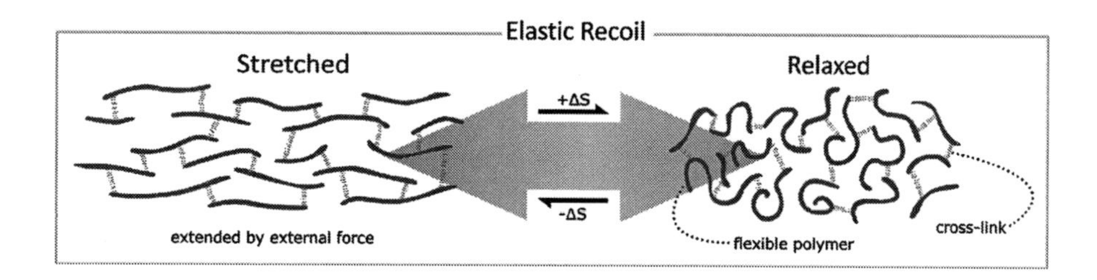

Figure 1. Entropy-driven elastic recoil. The driving force for elastic recoil in a rubber-like elastomer is the increased entropy of the relaxed state relative to the stretched state. The relaxed state has higher entropy because there are many more ways of arranging a collapsed polymer chain than a stretched polymer chain. The effect of an external force is to decrease the entropy, which is recovered when the force is removed and the material recoils to its relaxed state. Cross-links are essential to this mechanism of elastic recoil because they prevent the chains from sliding past each other during stretching. A color version of this figure is available online at www.landesbioscience.com/curie.

(1) due to internal energy, in a structurally-ordered elastomer, and (2) due to entropy, in a structurally-disordered elastomer.

Rubber-Like Elasticity

The molecular driving force of rubber-like elasticity is the increased entropy of the relaxed state relative to the stretched state (Fig. 1).[1] Due to their entropy-driven elastic restoring force, rubber-like materials exhibit near-perfect recovery of stored elastic energy following deformation.[2,3] The term rubber-like refers only to the elastic properties of a material, and does not imply that the chemical composition is similar to that of natural rubber.[6] However, it is important to note that there are three molecular characteristics common to rubber-like materials: (1) sufficient polymer chain length, (2) high chain flexibility, and (3) the presence of interchain cross-links.[2,23] Chain length and flexibility are necessary for elastic recoil: long and flexible polymer chains have many energetically-accessible spatial configurations, and the vast majority of these configurations are compact, resulting in the difference in entropy between the stretched and relaxed states. In turn, covalent or noncovalent cross-links maintain the structural integrity of the network by preventing the polymer chains from being pulled apart during extension.[2,23,24] Thus, in order for proteins to exhibit rubber-like elasticity, their amino acid sequence must encode a sufficiently flexible structure to give rise to entropy-driven elastic recoil, and must contain amino acid residues capable of forming interchain cross-links. The mechanical properties of rubber-like elastomeric proteins are modulated by the flexibility of the disordered regions, the nature of the cross-links and the spacing between cross-links.

Experimentally, rubber-like materials are identified by their unique thermoelastic behaviour: when held at a constant force, a rubber-like elastomer shrinks with increasing temperature.[2] Similarly, if held at constant extension, its elastic restoring force increases in proportion to the temperature.[25] Underlying both of these thermoelastic properties is the enhancement of molecular motion at higher temperature, which results in an increased tendency of the system to populate states of higher entropy.[3,23] Raising the temperature increases the contribution of entropy to the free energy of the system. In addition to being entropically-favourable, the relaxed state of an elastomer has a lower potential energy than the stretched state due to stabilizing interactions between polymer chains, such as hydrogen bonds. The relative contributions of entropy (f_s) and potential energy (f_e) to the elastic restoring force can be determined using thermoelastic (force-temperature) measurements.[2] In natural rubber, 82% of the elastic restoring force is due to entropy and 18% is due to energetic stabilization of the relaxed state.[26] Similarly, the internal energy component of the elastic force is 13% for the synthetic rubber polybutadiene (used in tires).[26]

Measures of Elasticity

In order to fully understand the biological role of elastomeric proteins, and furthermore, to use them effectively in novel biomaterials, measurements of their elastic mechanical properties are required. These measurements are typically performed on a small biomaterial sample using a specialized mechanical testing apparatus.[20] Mechanical (force-deformation) tests produce stress-strain curves. Stress is the applied force normalized by the cross-sectional area of the material (in units of pascals, Pa). An applied stress induces a strain, which is the change in length of the material normalized by the initial length.[14,20] A typical stress-strain curve for a rubber-like elastomeric protein is shown in Figure 2, along with illustrations of several elastic mechanical properties.[1,2]

Two types of measurements are performed to measure mechanical properties: (1) extending the elastomer until it ruptures (Fig. 2A), and (2) allowing the elastomer to relax before it reaches its breaking point (Fig. 2B). The stress and strain at the point of rupture are a measure of the material's strength and extensibility, respectively. Integrating the area under the stress-strain curve is a measure of the work needed to perform a given deformation. The work required to rupture the material is a measure of its toughness. If the elastomer is stretched and allowed to return to its relaxed state, the hysteresis between the two stress-strain curves is a measure of the elastic energy lost to heat. The corresponding mechanical property is resilience, which is the initial work minus the lost heat, normalized by the initial work done to strain the material.[27] An elastomer's stiffness is a measure of how easily it is deformed; stiffness is quantified by the elastic modulus, which is the slope in the linear regime of the stress-strain curve.[3] Taken together, the strength, extensibility, toughness, resilience and elastic modulus provide a description of the elastic behaviour of a material.

RUBBER-LIKE ELASTOMERIC PROTEINS

Here, we review what is currently known about the sequence features, structural characteristics, mechanical properties and biological roles of five rubber-like elastomeric proteins: elastin, resilin, spider silk, abductin and ColP. In addition, we summarize recent advances in other elastomeric proteins.

Thermoelastic measurements have been performed on elastin, resilin, abductin and hydrated major ampullate spider silk. The internal energy components of the recoil force of elastin and major ampullate spider silk are 26%[3,18,28] and 14%,[26] respectively. On the basis of these measurements, the thermoelastic behaviour of rubber-like elastomeric proteins is consistent with entropic elasticity.[9,25,26] It is essential to note that elastomeric

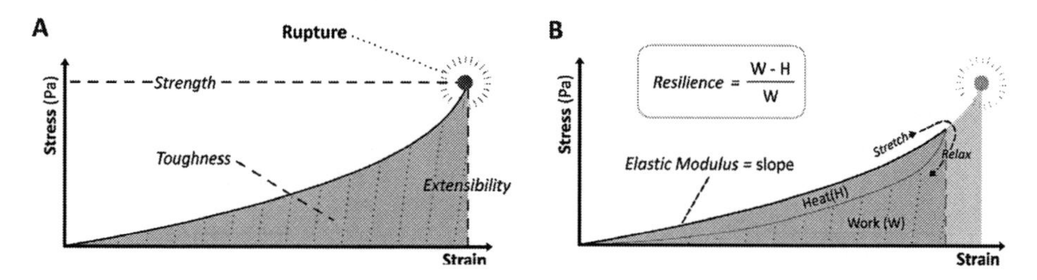

Figure 2. Measurements of elastic mechanical properties. Shown here are stress-strain curves for a rubber-like elastomer in two types of mechanical tests. A) In the first type of experiment, an elastomer is stretched past its breaking point. This experiment measures strength, which is the stress at the rupture point, and extensibility, which is the strain at the rupture point. Toughness is the work required to rupture the material, which is the area below the stress-strain curve. B) In the second type of experiment, the material is allowed to return to its relaxed state without being stretched to its rupture point. The elastic modulus is the slope in the linear regime of the stress-strain curve. The work done to stretch the material is the area under the top curve (W). When the material is allowed to relax, it follows a stress-strain curve below the original curve. The area between the stretch curve and the relax curve is the elastic energy lost to heat (H). Resilience is the difference between W and H, normalized by W.

proteins are rubber-like only when hydrated or in polar solvents.[25] Water is thought to act as a 'plasticizer' by forming direct interactions with the polypeptidic backbone, resulting in elastic mechanical properties.[16]

Measurements of mechanical properties for the elastomeric proteins described in this chapter are provided in Table 1. Elastomeric proteins exhibit a remarkable diversity of elastic properties, which may differ by orders of magnitude. Different biological roles demand different combinations of extensibility, resilience, strength and stiffness, which are all ultimately determined by structural properties. The elastic properties of proteins are modulated by the primary sequence, the domain organization and the spacing between adjacent cross-links; these features are reviewed below.

Elastin

Human life is entirely dependent on the elastic properties of elastin. Together with other structural proteins, elastin forms the fabric of extensible tissues, including skin, blood vessels and elastic ligaments, and provides the elasticity required for proper physiological function.[29] Elastin is a major component of large arteries; bovine aorta is composed of nearly 50% elastin.[30] The aorta expands when the heart contracts (during systole), and recoils elastically when the heart refills with blood (during diastole).[31,32] In the walls of the aorta, elastin functions in tandem with collagen to produce a "J-shaped" stress-strain curve.[32] Elastin is responsible for the initial low stiffness region of the curve, while collagen confers increased stiffness at higher strains. Thus, collagen provides the strength required to prevent rupture due to high blood pressure, while the resilience and extensibility of the aorta imparted by elastin minimize the energetic demands on the heart and ensure smooth blood flow to tissues throughout the body.[31,32]

In addition to extensibility and resilience, elastin possesses remarkable durability: once laid down in tissue during development, elastin does not turn over at an appreciable rate.[14] In order to sustain a lifetime of breaths and heartbeats, elastin must therefore undergo billions of stretching-relaxation cycles without damage or permanent deformation. Unlike elastin in blood vessels and lungs, elastin in the uterus is degraded and replaced during adulthood.[33] In order to accommodate the rapid growth and motion of the fetus, the uterus requires significant extensibility. Accordingly, during pregnancy, uterine elastin content increases by more than 500%, the majority of which is quickly degraded *post partum*.[5] As a result of elastin's impressive diversity of biological roles and exceptional mechanical properties, it is the best-characterized rubber-like elastomeric protein.

It is now possible to mimic the elastic properties and self-assembly of elastin using smaller recombinant polypeptides, which can be used to fabricate materials suitable for stress-strain measurements.[14,34] Both elastin and elastin-derived peptides self-aggregate upon heating to form an organized fibrillar structure in a process known as coacervation. Remarkable durability and intrinsic capacity for self-organization make elastin an ideal biomimetic model in the development of synthetic biomaterials.[14] Biomaterials composed of either elastin or elastin-derived peptides have desirable elastic mechanical properties: low stiffness, high resilience and high extensibility (refer to Table 1 for measurements).[1,14]

Elastin's mechanical properties are encoded in its amino acid sequence, which fulfills the essential requirements for rubber-like elasticity. The sequence of tropoelastin, the monomeric precursor of elastin, is composed of alternating cross-linking and hydrophobic domains. The covalent cross-linking of elastin monomers imparts strength and stability to the polymeric matrix, while the hydrophobic domains are thought to confer the propensities

Table 1. Elastomeric proteins—Sequence, structure and function

Protein	Biological Role	Sequence Features		Structural Properties	Elastic Mechanical Properties			
		Cross-Links	Motifs		Modulus	Resilience	Strength	Extensibility
Abductin	- The hinge ligament of the shells of bivalve molluscs is composed of abductin - It facilitates swimming - When compressed, abductin stores elastic energy, which when released is used to open the shell	- Nature of cross-links not completely characterized[71] - Possibly 3,3′ methylene bistyrosine[92] or disulfide cross-links[43,92] - May have cross-linking involving Tyr, Lys residues[71]	GGFGGM GGGX[71]	- PPII, β-turns, unordered[43]	4.7 MPa[9]	82-97%[9,27] 91% [9]		
ColP	- The elastic domains of ColP confer extensibility on the threads connecting mussels to surfaces, allowing them to extend without breaking	- Cross-links may be formed by histidines bound to transition metals[93] - DOPA lysines[43]	GPGGG[43] GXGPG, XGGPG, GGXPG (X = I,F or A)[12]	- No high-resolution structural data - Disordered based on X-ray fibre diffraction studies of intact byssal threads[73]	50 MPa[12] 15.6—26.4 MPa [74]	53%[12] 42-53%[74]	12.4- 34.6 MPa[74]	1.6 X [12] 0.8—1.97 X[74]

continued on next page

Table 1. Continued

Protein	Biological Role	Sequence Features		Structural Properties	Elastic Mechanical Properties			
		Cross-links	Motifs		Modulus	Resilience	Strength	Extensibility
Elastin	- Responsible for the elasticity of various vertebrate tissues (skin, aorta, lung alveoli, elastic ligaments, uterine tissue)[4]	- Desmosine, isodesmosine or lysinonorleucine cross-links, all formed from Lys residues[43,92]	VGVAPG[43,94] VGVPG[43,94] VPGG[43]	- The hydrophobic domains of elastin contain β-turns and PPII structure and are intrinsically disordered[16,17] - Alanine-rich cross-linking domains have significant α-helix content[95]	1.1 MPa[38]	90%[38]	2MPa[38]	1.5 X[38]
Flageliform Silk	- Capture spiral of orb-webs [63]	- Noncovalent cross-links[43]	GPGGSG-PGGY[43] GPGGX[65] GGX[65]	- Random coil/ unordered by CD[65] - Absence of β-sheet structure[96]	3 MPa[63] 1 MPa[96]	35%[63]	0.5 GPa[63] 95 MPa[96]	2.7 X[63] 4.7 X[96]

continued on next page

Table 1. Continued

Protein	Biological Role	Sequence Features		Structural Properties	Elastic Mechanical Properties			
		Cross-links	Motifs		Modulus	Resilience	Strength	Extensibility
Major Ampullate Silk	- Dragline silk and radial fibres of orb webs - Rubber-like properties have no known biological role	- Noncovalent cross-links composed of polyalanine intermolecular β–sheets	GPGXX[91]	- In the super contracted state, the elastic restoring force is only 14% due to internal energy. - Structure consists of alternating β-sheet crystalline domains and amorphous domains	10MPa[63] (10 GPa when dry)[1]	35%[63]	1.1 GPa[63]	0.3 X[63]
Resilin	- Energy storage protein in insects - Dragonfly wing hinge ligament[6] - Cicada tymbal (essential for sound production)[54] - Flea cuticle (stores energy for jumping)[7] - Tick cytoskeleton (essential for cyto-skeletal expansion during feeding)[55]	- No unique cross-linking domains[58] - Di-tyrosine and tri-tyrosine cross-links[43,58]	AQTPSS-QYGAP[60] GGRP-DSYGAP-GGGN[13,43] GYSGGR-PGGQ-DLG[43,58]	- PPII, β-turns, unordered[57]	2 MPa[1]	92%[1]	4 MPa[1]	1.9 X[1]

for self-aggregation and extensibility.[29] More than 80% of the sequence of the hydrophobic domains consists of proline, glycine, valine and alanine.[4] The hydrophobic domains of elastin have a pseudo-periodic, low complexity sequence, with repeat motifs PGVGVA, PGV and PGVGV.[16] There are several excellent reviews of elastin's biochemistry and structural properties.[4,29,35] Here, we briefly review what is currently known about the structural features of elastin's hydrophobic domains.

For several decades, the elastin field was plagued by controversy surrounding the structure of elastin and, correspondingly, the molecular mechanism of its elastic recoil. Models of elastin structure and function were simplistic and largely incompatible with each other.[29] Urry postulated that elastin's repetitive sequence must encode a perfectly repetitive structure: the β-spiral.[36] In this model, an ordered spiral consists of consecutive Type-II β-turns with PG motifs forming the corners of the turn. Elasticity was thought to arise from the "librational motions" of the β-spiral.[36] The β-spiral was also postulated as the structure of other elastomeric proteins, including wheat gluten, spider silk and resilin.[24,37] In contrast to the ordered view of the β-spiral model, Flory and Gosline put forth models of elastin as a random, rubber-like structure.[28,38] Thus, models of elastin structure ranged from ordered to predominantly random coil.

In support of rubber-like elasticity and a highly disordered structure, force-temperature measurements on elastin indicate that internal energy contributes between 10% and 26% to the elastic restoring force, with entropy being the dominant molecular driving force.[3,18,28,39] Consistent with these macroscopic measurements, solid-state NMR has provided significant insight, suggesting the absence of α-helix and β-sheet and a high degree of dynamic disorder.[17] In addition, ^{13}C NMR studies demonstrated that the hydrated state of elastin has significant structural mobility, which decreases as water is removed.[40] Taken together, the thermoelasticity and NMR evidence indicates that the polypeptide chains of elastin are highly mobile, and therefore possess high configurational entropy. These observations are inconsistent with models requiring conformationally-restricted structures, such as the β-spiral.[36]

However, the random network model of elastin is too simplistic to account for experimental data consistent with the presence of β-turns and polyproline II (PPII) structure. Circular dichroism (CD) and Fourier transform infrared spectroscopy (FTIR) have provided limited structural data suggesting a high degree of conformational flexibility, together with a measurable propensity to adopt β-turns and PPII conformations.[35,41,42] It should be noted that only qualitative interpretations of CD spectra of elastin are possible because the reference databases used by CD deconvolution programs consist primarily of globular proteins.[43] Given these limitations, it is not possible to obtain information about equilibrium populations of either PPII or β-turn structures using CD.

Similarly to many other intrinsically disordered proteins (IDPs), the insolubility, conformational heterogeneity and intrinsic flexibility of elastin have precluded the use of conventional high-resolution structural determination methods, including X-ray crystallography and solution NMR.[17] In contrast to experimental approaches, molecular dynamics (MD) simulations are not hindered by conformational disorder, and have therefore proven useful in obtaining atomic-level descriptions of the conformational ensembles of elastin-like peptides.[16,44-46] Molecular simulations provide information that can be used to characterize the ensemble of IDPs.[47] This is because MD simulations provide time trajectories containing information on the dynamics of all particles in the system. Due to limited computing power, MD simulation studies of elastin were until recently restricted to short time scales (nanoseconds)[48] or small oligopeptides (only eight residues).[46]

However, two recent studies from our laboratory have dramatically extended the scope of simulations of elastin-like peptides: it is now possible to reach the microsecond time scale for peptides of similar size to a hydrophobic domain (35 residues).[16,45,47] Using all-atom MD simulations with explicit water, we obtained a disordered conformational ensemble for the elastin-like peptide $(GVPGV)_7$.[45,47] In order to effectively sample many conformational states, we employed a novel enhanced sampling method.[45] The structure of elastin-like peptides is flexible and disordered, which indicates that the underlying energy landscape is defined by conformations that are very similar in energy, and these conformations exchange rapidly with one another. For this reason, it is relatively easy to obtain meaningful structural information and compute thermodynamic averages from molecular simulations.

The $(GVPGV)_7$ peptide populates a heterogeneous ensemble of conformations (Fig. 3). A selection of conformations from the simulation is shown, which represent a very small subset of the possible conformations of this peptide. Here, we illustrate the ability of MD simulations to provide an atomic-resolution description of a disordered state ensemble. The monomeric peptides (Fig. 3B) adopt collapsed, water swollen conformations reminiscent of the unfolded ensemble of globular protein domains.[49] The structures exhibit a highly flexible polypeptide backbone, with exchanging conformations and overall structural disorder. Although these structures contain no α-helix or extended β-sheet, they are not random coils. Ordered structure is observed predominantly in the form of PPII content and hydrogen-bonded turns, both of which are local. In agreement with recent studies of peptides modeling the unfolded state of proteins,[50] the PPII structure observed is not extensive but is instead confined to one or two consecutive residues. Consistent with spectroscopic studies, the structures populated by an elastin-like peptide in solution are disordered, but not random.

In addition, we show a snapshot from a simulation of an aggregate of eight $(GVPGV)_7$ peptides (Fig. 3A). Consistent with solid state NMR data indicating a lack of β-sheet structure in the aggregated state of elastin,[17] the aggregate of this elastin-like sequence is intrinsically disordered and highly hydrated. Retention of structural disorder in the aggregated state is analogous to the phenomenon of "fuzziness" in protein-protein interactions; the "bound" (aggregated) state of elastin retains both static and dynamic disorder in the same way that some IDPs remain disordered in complex with their binding partners.[51] While there was significant controversy surrounding the structural tendencies of elastin, it is now clear based on simulations, solid-state NMR data and CD spectra that elastin should be classified as an IDP.[4,47,52] The emerging consensus is a new model of elastin structure in which the hydrophobic domains form water-swollen, disordered aggregates characterized by an ensemble of many degenerate conformations devoid of any regular secondary structure. Local structure, in the form of turns and PPII, is present and is transiently populated as conformations rapidly interconvert on the nanosecond timescale. Further structural insight from our studies is reviewed below.

Resilin

Another rubber-like elastomeric protein with very similar mechanical properties to elastin is the insect cuticle protein, resilin. First discovered in the elastic tendon of dragonflies,[6] resilin is found in many arthropod species and is important to insect flight, locomotion and sound production.[6,7,53] The tymbal mechanism of the cicada utilizes resilin as an energy storage device. When the tymbal is compressed, elastic energy is stored in resilin and the subsequent release of this energy is accompanied by the cicada's characteristic sound.[54] Resilin in the cuticle of ticks facilitates the dramatic expansion of

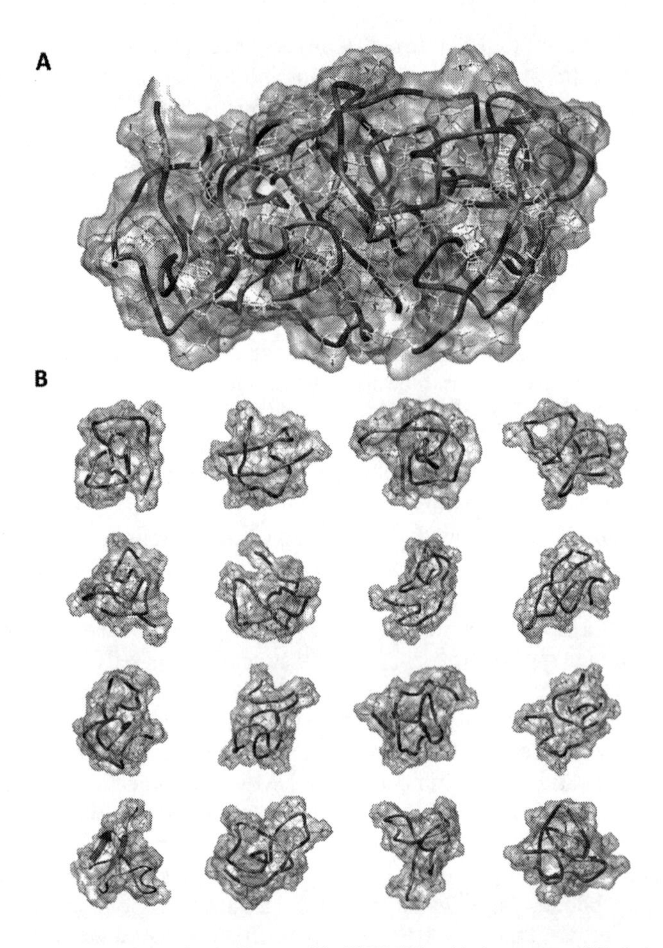

Figure 3. Configurations of the elastin-like peptide $(GVPGV)_7$. A) Aggregate of 8 chains; B) 16 unique configurations of a monomer. These configurations are only a small subset of the thousands of configurations generated by the simulations and are shown to illustrate the structural heterogeneity of this peptide. The peptide retains a water-swollen, disordered structure in both monomeric and aggregated states. There are, on average, 0.73 water molecules bound to each hydrogen-bonding backbone group.[16] For both the monomer and aggregate, MD simulations were conducted using the STDR method[45] in explicit water (which is not shown for clarity) for a total simulation time of 70 μs (monomer) and 110 μs (aggregate). Snapshots from 300 K are shown with a cartoon representation of the backbone, licorice representation of sidechains and a solvent-excluded surface (rendered using the visualization program UCSF Chimera[97]). The cartoon representation is coloured by residue (proline in yellow, glycine in purple and valine in orange). A color version of this figure is available online at www.landesbioscience.com/curie.

the cytoskeleton during feeding, a property not found in most other insects that shed their cytoskeleton before significant growth can occur.[55] Another biological role of resilin is as an energy storage device in the jumping mechanism of of fleas: muscle contraction alone is incompatible with the timescale of energy release (less than one millisecond) and the necessary power output.[7] Accordingly, the cuticle of fleas contains a resilin pad, the size of which correlates with jumping ability.[7]

Unlike elastin, which is challenging to isolate and purify,[4] the resilin pads and tendons in insects are isotropic and easily isolated, and therefore convenient for experimental characterization.[6] However, in order to produce sufficient amounts of resilin to manufacture biomaterials, several groups utilize recombinant expression systems.[13,56,57] A resilin-like protein (rec1 resilin) was cloned and expressed in *E. coli*.[13,58] Large quantities of soluble rec1 resilin were produced and cast into rods and strips. Importantly, synthetic rec1 resilin materials have the same resilience (90-92%) as elastic tendon isolated from dragonfly wing (92%). Rec1 resilin strips have a resilience of 97% in solution, dissipating only 3% of elastic energy as heat. Thus, resilin's resilience is greater than that of polybutadiene, a high resilience rubber, and is unmatched by any other elastomeric protein.[13] Furthermore, synthetic resilin can be stored in a dehydrated state and recover the same resilience upon rehydration.[13] Besides its high resilience, resilin also has a high extensibility. Synthetic resilin strips can be stretched to more than three times their original length without permanent deformation,[13] and elastic tendons composed of resilin can be compressed by a factor of one-third.[25]

In addition to these exceptional elastic properties, resilin has a high durability. It is deposited in the insect cuticle during the pupal stage[13] and remains in place throughout the adult lifetime. While insects generally have much shorter life spans than vertebrates, their resilin deposits have similar durability requirements as elastin. For example, the resilin deposit in the cicada's tymbal is compressed hundreds of millions of times in sound production, necessitating a high durability.[53] Taken together, resilin's high extensibility, high resilience, high durability and low stiffness make it ideal for a wide variety of biomaterials applications. To this aim, a resilin-like sequence was recently combined with a cell binding domain in a designed recombinant protein. The resulting biomaterial has a high extensibility (up to 200%) and is capable of cell adhesion. This study represents an important first step towards tissue engineering because fibroblast cells were able to adhere and proliferate on the extensible resilin-like scaffold.[56]

Based on their similar mechanical properties, it is not surprising that the sequences of elastin and resilin share a number of similar features. Both have a high content of proline and glycine and are highly repetitive (see Table 1 for specific repeat motifs).[16,57,58] However, unlike the predominantly nonpolar sequence of elastin, the sequence of resilin is depleted of large hydrophobic residues like valine and isoleucine and has a significantly enhanced content of hydrophilic residues.[57] Towards the development of a resilin-inspired biomaterial, Tamburro and coworkers recently identified PGGGN as a putative minimal repeat motif for resilin-like properties.[57] Peptides based on this motif were found to readily self-assemble into fibrillar structures.[57] In contrast to the regular alternance of cross-linking and hydrophobic domains in elastin, the sequence of resilin has no specific cross-linking domains; instead, tyrosine residues are interspersed throughout the elastic repeat motifs,[24] and form di- and tri-tyrosine cross-links. It is estimated that 20% of tyrosine residues are covalently cross-linked,[13] with a spacing of 40 to 60 residues between cross-links.[59] Thus, the sequence of resilin is compatible with the requirements for rubber-like elasticity, with cross-links separated by flexible polymeric chains.

Resilin was the first elastomeric protein to be identified as rubber-like on the basis of thermoelasticity experiments.[25] In agreement with an entropy-driven mechanism of elastic recoil, CD, NMR and Raman spectra of a resilin-like protein are all consistent with a heterogeneous and dynamic structure.[60] Vicinal coupling constants from NMR indicate an absence of either α-helix or β-sheet structure, while chemical shifts are consistent with "random coil" values.[60] Sequential and medium range NOEs indicate the presence of β-turn conformations, with PG, GG and PS motifs forming the corners of the turn.[57] Similarly, CD spectra are consistent with the presence of PPII structure and β-turns.[43] X-ray diffraction measurements indicate that resilin does not attain any significant ordered structure, even when stretched to three times its length or when dried.[59] Although the β-spiral structure has also been proposed for resilin,[58] all of the experimental evidence obtained to date is incompatible with this model. Instead, a common thread emerges: the polypeptide chain of resilin is highly flexible and intrinsically disordered.

Spider Silk

In order to construct their intricate webs, spiders utilize silks with diverse elastic mechanical properties. Using a complex spinning and extrusion process, spiders finely tune the elastic properties of the silk proteins (spidroins) produced in their abdomen.[61] The best-characterized spidroins belong to spiders from the family Araneoidea. Known for their orb-shaped webs, araneid spiders include the common garden spider, *Araneus diadematus,* and the golden orb-weaver, *Nephilia clavipes*. Araneid spiders produce seven unique types of silk, which are named after the specialized abdominal glands in which they are synthesized: (1) major ampullate silk is spun into fibres forming both the dragline and the radial threads of the web; (2) minor ampullate silk reinforces the dragline and web frame; (3) flagelliform silk forms the capture spiral; (4) aggregate silk is an aqueous 'glue' coating the capture spiral; (5) the 'cement-like' silk from the piriform gland is used to attach the web to a surface; (6) aciniform silk is used to restrain captured prey; and (7) tubuliform/cylindriform silk is used to protect the egg sac.[8,62,63]

While spiders make use of silk for a wide range of purposes, the most familiar use is in the construction of the spider web. A web is an incredibly efficient insect-catching device: a spider can produce a web covering nearly 1 m^2 using only 180 μg of protein.[8] Although the threads of the web are very thin, they exhibit toughness greater than the best synthetic materials, including Kevlar®. It is for this reason that the web does not break upon impact with an incoming insect, or when a trapped insect scrambles to get free.[8] Orb-web spiders construct their webs using a combination of two silks with complementary elastic mechanical properties.[64] The threads forming the radial frame of the web are stiff and extremely tough; they are composed of both major and minor ampullate silk. The capture spiral connecting the radial threads is sticky, highly extensible and easily stretched; it is composed of flagelliform silk, with stickiness and hydration provided by "glue" silk from the aggregate gland. Both major ampullate silk and flagelliform silk have low resilience (roughly 30%). This is essential for two reasons: (1) the elastic energy stored in the web by the impact of an incoming insect is dissipated as heat, preventing the insect from bouncing off the web and (2) the dissipation of elastic energy contributes to the overall toughness of the web, preventing the strands from breaking.[8]

Here, we are particularly interested in the two types of spider silk possessing rubber-like elasticity: major ampullate silk and flagelliform silk. Both major ampullate and flagelliform silk contract when immersed in water, but the effect is much more pronounced for major ampullate silk, and is called "supercontraction".[63] Thermoelastic measurements indicate that major ampullate silk exhibits rubber-like elasticity, but only in the supercontracted, hydrated state. The internal energy component of the elastic force is only 14%.[26] When hydrated, major ampullate silk has an elastic modulus three orders of magnitude smaller than in the dry state (10^7 Pa compared to 10^{10} Pa).[8] A biological function has not yet been identified for the supercontracted, hydrated state of major ampullate silk.[63] The structure of major ampullate silk is thought to consist of alternating crystalline and amorphous domains. The crystalline regions are short polyalanine stretches of 8 to 10 residues, which form β-sheets oriented parallel to the fibril axis.[63,65] Because the β-sheets are intermolecular, they effectively act as noncovalent cross-links,[63] while the conformational entropy of the amorphous domains results in elastic recoil.[26] As a result, the structure of major ampullate silk resembles a rubber with "crystalline inclusions".[8]

The mechanical properties of flagelliform silk are qualitatively similar to those of supercontracted major ampullate silk.[26] In contrast to elastin and resilin, flagelliform silk has a relatively low resilience, which is essential to its functional role: if it returned the stored elastic energy efficiently, insects that fly into a spider web would immediately bounce off.[63] In order to effectively trap insects in the web, flagelliform silk must be coated in a sticky mixture of hygroscopic peptides and glycoproteins, which are produced in the aggregate gland.[66] Flagelliform silk is a very promising biomaterial by virtue of its unusually high strength (0.5 GPa), which is approximately ten times that of any other rubber-like elastomer.[63] However, the cannibalistic and territorial nature of spiders precludes the possibility of directly harvesting sufficient quantities of silk to manufacture materials.[66] An additional complication arises from the importance of the effect of spinning on mechanical properties. Even if silk can be recombinantly expressed, mimics of the spider's spinnerets are essential to obtain strong and extensible fibres.

Compared to elastin and resilin, there is relatively little high-resolution structural information on flagelliform silk. A recent solid-state NMR study of the flagelliform-like sequence $(GPGGA)_6G$ demonstrated that the motif GPGG has a high propensity to form a β-turn.[67] In addition, ^{13}C chemical shifts and Raman spectra are consistent with a "random-coil" structure.[67,68] These results are inconsistent with the β-spiral structure, which was proposed for flagelliform silk because the sequence encodes repetitive pentapeptide motifs.[62] While cross-links are necessary to explain the remarkable strength of flagelliform silk, the residues involved in cross-linking have not yet been identified.[63]

Abductin

Abductin is an elastomeric protein that forms the hinge ligament of the bivalve mollusc shell.[27] In molluscs of the family *Pectinidae*, the shell opens and closes three times per second to facilitate swimming.[27] The name of abductin is derived from its biological role: the hinge composed of abductin acts as an abductor, an antagonist to the adductor muscle in the opening and closing motion of the shell.[9,69] The adductor muscle stores elastic energy in the abductin hinge by compressing it as the shell closes. The shell opens when the adductor muscle relaxes and the energy stored in the abductin hinge is released to oppose the force of the surrounding water.[9,70] In order to minimize the energy required by the mollusc for swimming, the energy stored in the abductin hinge must be recovered to open the shell. Thus, high resilience is essential to abductin's biological function.[9] Mechanical tests on abductin demonstrated its resilience to be between 82 and 96%, with significant variation between mollusc species.[27] Interestingly, an analysis of the amino acid composition of abductin from several species revealed that resilience is correlated with glycine content. In fact, the sequence of abductin is characterized by an unusually high glycine content (nearly 70% in fast-swimming mollusc species).[27] Glycine content is therefore a fundamental determinant of the rubber-like mechanical properties of abductin.

Since abductin's biological function demands a high resilience, it is not surprising that its amino acid sequence and its structural properties are similar to those of both elastin and resilin. Thermoelasticity measurements on abductin place it among the rubber-like elastomeric proteins, with a primarily entropy-driven elastic restoring force.[9] Like elastin, resilin, major ampullate silk and flagelliform silk, abductin exhibits rubber-like elasticity only when hydrated.[10] Accordingly, the abductin hinge ligament is composed of approximately 50% water.[9] The amino acid sequence of abductin is highly repetitive, with a consensus repeat motif GGFGGMGGGX.[71] Although the complete sequence of abductin from scallops has been determined, the residues involved in cross-linking have not been unambiguously identified.[71] CD and NMR spectra of abductin-like sequences are consistent with the presence of both PPII structure and β-turns,[43,70] but no high-resolution structural studies have been performed to date. Abductin's high resilience and compressibility make it an interesting biomaterial worthy of further structural and mechanical studies.

ColP from Byssal Threads

Marine mussels use byssal threads to attach themselves to solid substrates, such as rocks and harbour walls.[72,73] Since a secure attachment is vital to their survival,[74] byssal threads require both strength and extensibility. In addition, the low resilience of byssal threads effectively dissipates elastic energy like a damped spring, preventing the mussel from hitting the hard surface to which it is attached.[1,12] The mechanical properties of the byssal thread vary along its length:[72] the distal end (near the point of attachment to the surface) is stiff and strong, while the proximal end (near the shell) has a high extensibility and low elastic modulus.[12,72]

The byssal thread's continuum of mechanical properties is the result of a protein gradient along its length. The distal end is primarily composed of the protein ColD, whereas the proximal end is primarily composed of ColP, and the intermediate region contains a mixture of both ColP and ColD.[12,72] The protein ColP is itself a hybrid: it is the first known example of a block copolymer containing both elastic and collagen-like domains.[72] The collagen-like domain is predicted to adopt a triple helix structure, while the elastic domains are enriched in hydrophobic residues and contain many instances of the PG motif.[72] Based on sequence similarity between ColP's elastic domains and the repeat motifs of elastin and flagelliform silk, it has been proposed that ColP confers elasticity and extensibility on byssal threads.[72] As a consequence of combining these two types of sequences, proximal byssus has mechanical properties intermediate between those of elastin and collagen. Compared to collagen, it has higher toughness and extensibility, at the cost of reduced strength.[72] The extensibility of the proximal region is similar to that of elastin and resilin, with a significantly lower resilience and an elastic modulus that is an order of magnitude greater.[1,74] The hybrid nature of the byssal thread endows it with the remarkable strength and flexibility to maintain surface attachment against powerful tides.[73,75]

Very little is known about the structural properties of ColP. X-ray fibre diffraction studies of intact byssal threads indicate a gradual decrease in structural order along the length of the thread (in the distal to proximal direction). Diffraction patterns are consistent with the presence of ordered collagen-like structure in the distal region, gradually giving way to increased structural disorder in the proximal region.[73] While structural studies on intact byssal threads are a useful first step, higher-resolution structural studies, ideally on ColP or its elastic domains in isolation, are essential to elucidate the connection between structural disorder and elastic properties. Strictly speaking, ColP has not yet been demonstrated to be a rubber-like elastomer with thermoelasticity measurements, but here we group its elastic domains with the other rubber-like elastomeric proteins on the basis of high sequence similarity.

Other "Fuzzy" Elastomeric Proteins

Interestingly, some proteins that form rubber-like biomaterials in vitro do not require rubber-like elasticity to fulfill their biological role. The sequence signatures of disordered elastomeric proteins were recently found in a domain of the transcription factor Ultrabithorax (Ubx) from *Drosophila melanogaster*. The sequence of Ubx is enriched in glycine and contains multiple GGX and GXXP elastin-like motifs. In vivo, Ubx interacts directly with DNA to regulate transcription; it is not known to form aggregates as part of its biological function.[76,77] Remarkably, however, Ubx was recently found to self-assemble in vitro into elastic materials with several morphologies, including films, fibres and sheets.[77] Ubx 'ropes' were found to have an extensibility approximately one-third that of elastin. Similar to the rubber-like elastomeric proteins, Ubx materials are only extensible when hydrated and become brittle when desiccated. The identification of Ubx as an elastomeric protein suggests an interesting research direction: the identification of sequences with similar features to known elastomeric proteins as possible novel rubber-like elastomers.

In addition to looking for new elastomeric proteins through sequence similarity to known elastomeric proteins, it is also essential to investigate the protein constituents of extensible and soft biological tissues. It is likely that many more rubber-like elastomeric proteins exist that have not yet been discovered. For example, studies on octopus aorta revealed the presence of a rubber-like elastomeric protein.[78] The "octopus arterial elastomer" (OAE) performs the same biological role as elastin in vertebrate arteries. Like the other rubber-like elastomeric proteins, its elastic recoil is predominantly entropy-driven. However, the amino acid composition of OAE is very different than that of elastin, resilin or abductin. More than 33% of the sequence consists of charged residues, and there is significantly less proline and glycine. OAE is stiffer, less extensible and less resilient than elastin and abductin.[78] Although both elastin and OAE are responsible for the elastic recoil of the aorta, their sequence features are quite different. Thus, in developing a toolkit of elastomeric proteins for incorporation in biomaterials, it is valuable to study proteins from the myriad of elastic tissues that exist in nature. As a first step, we require information on the sequence determinants of rubber-like elasticity.

SEQUENCE FEATURES OF DISORDERED ELASTOMERIC PROTEINS

Low Sequence Complexity

The sequences of elastomeric proteins have very little sequence homology[19] and are instead characterized by a common 'style' of sequence. Elastomeric domains typically have low complexity sequences with repeat motifs (refer to examples in Table 1).[24] These repeat motifs often contain PG and GG dipeptides, which preferentially form β-turns.[4,79] For several decades, it was thought that the repetitive sequences of elastomeric proteins must encode a repetitive molecular structure (the β-spiral).[24] However, a wealth of experimental evidence indicates the absence of a well-defined or repetitive structure, which is corroborated by computational results. The observation that a repetitive sequence leads to a disordered structure is not unexpected, given the well-established connection between low sequence complexity and structural disorder.[80] In general, tandem repeats are more common in the sequences of IDPs when compared to all sequences in the Swiss-Prot database.[81] In this sense, elastomeric proteins are prototypical IDPs. Our studies have uncovered an important sequence feature that is common to all rubber-like elastomeric proteins, which we describe in the next section.

Proline and Glycine Control Self-Aggregation of Elastomeric Proteins

Rubber-like elastomeric proteins require a high level of structural disorder for entropy-driven elastic recoil. In addition, they must self-aggregate in order to form an elastomeric network. Their sequence must therefore preclude the possibility of forming well-structured protein aggregates. In particular, elastomers must avoid the formation of amyloid fibrils, which are characterized by a cross-β quaternary structure, with β-strands running perpendicular to the main axis of the fibril.[82,83] The deposition of amyloid fibrils is associated with more than forty tissue-degenerative pathologies, including Alzheimer's disease and Type II diabetes.[84,85] However, amyloid fibrils are not necessarily toxic,[83] and have even been found to have essential functional roles in both humans and bacteria.[86,87] Furthermore, it has been proposed that the amyloid fibril represents an inherent form of organization potentially accessible to all polypeptide chains under appropriate conditions.[85] Highly hydrophobic sequences, such as that of elastin, are, in principle, susceptible to forming amyloid. For example, exon 30 of human elastin forms amyloid-like fibrils when removed from the context of the full-length protein.[88] Elastin-like peptides with repeat motifs PGVGVA and PGVGV form biomaterials with mechanical properties similar to native elastin, but mutations of the tandem repeats to GGVGVA, GGVGV, or GVA promote the formation of amyloid fibrils under certain solution conditions.[16,34] Thus, it is important to understand how the hydrophobic domains of elastin, and indeed all self-associating elastomeric proteins, manage to avoid the amyloid fate.

To investigate how sequence modulates the ability of polypeptides to self-assemble into elastin-like or amyloid-like fibrils, we performed comparative molecular dynamics simulations of a model set of peptides based on elastin-like motifs PGV, GV, GVA and GGV, in both monomeric and aggregated states.[16] Elastin-like peptides are characterized by a higher hydration of the peptidic backbone and low peptide-peptide hydogen-bonding propensities. The opposite is true of amyloid-like peptides, which form the extended β-sheets that characterize amyloid fibrils. Moreover, the comparison of structural properties demonstrated that elastin-like and amyloidogenic peptides are separable in terms of backbone hydration and conformational disorder, and that these properties are modulated by proline and glycine. Why do elastin sequences combine the two extremes of backbone flexibility? Proline and glycine conspire to keep the backbone disordered and hydrated, but for opposite reasons: proline, because it is too constrained to form secondary structure, and glycine, because in water it is too flexible and entropically-disinclined to do so. Proline is the primary determinant: its conformationally-restricted main chain induces a significant propensity for PPII structure and a reduced ability to form β-sheet. Both proline and glycine favour conformational disorder of the backbone over the formation of self-interactions.[16]

The generalization of this finding to the structure and function of elastomeric proteins was confirmed by our observation of an approximate threshold in combined proline and glycine content separating the composition of known amyloidogenic sequences from those of known elastomeric proteins (see the PG diagram, Fig. 4). Remarkably, the compliance with a PG composition threshold is not limited to the hydrophobic domains

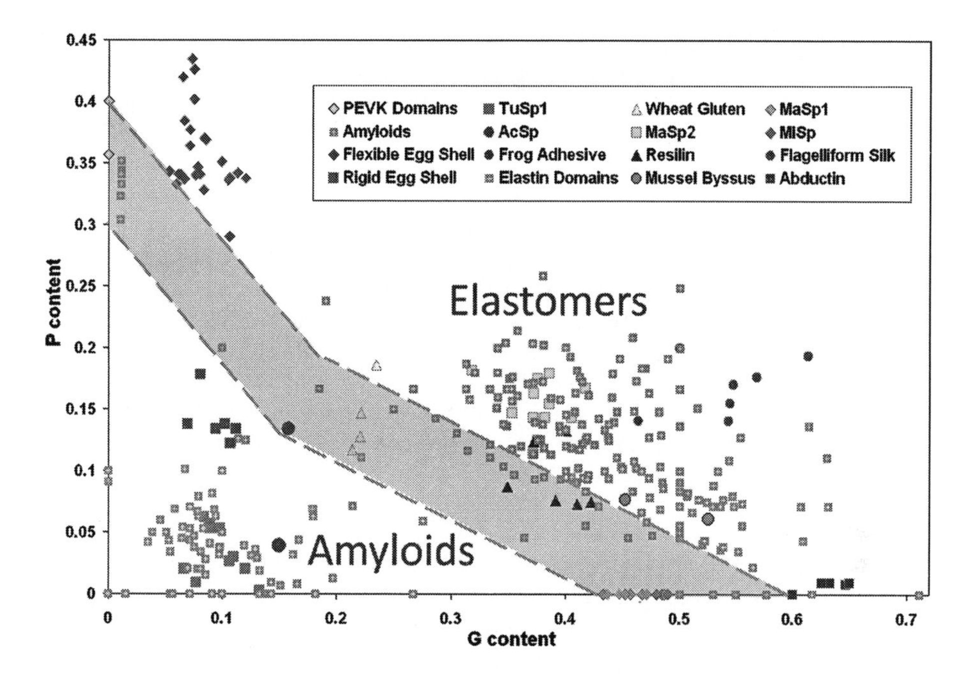

Figure 4. The PG Diagram. Elastomeric proteins from diverse organisms obey a simple design principle: their combined fractional content of proline and glycine is above a defined threshold. In contrast, amyloidogenic proteins have combined proline and glycine compositions below this threshold. Separating elastomeric and amyloidogenic proteins is a coexistence region (shaded in gray) where both types of protein self-aggregation are possible. Adapted from Rauscher et al, Structure 2006; 14(11):1667-1676;[16] ©2006 with permission from Elsevier. A color version of this figure is available online at www.landesbioscience.com/curie.

of elastin, but is also observed for the elastic domains of ColP, abductin, resilin and spider silk. The fact that approximately two glycines are equivalent to one proline at this threshold confirms the role of proline as the primary determinant of elastin's properties. The transition in composition space between elastomers and amyloids does not appear to be an abrupt one. Rather, a coexistence region includes sequences that are either elastomeric, amyloidogenic, or both.[16] Additional factors, such as the compositions of other residues and solution conditions, are expected to contribute to the modulation of protein aggregation tendencies.[84]

In support of the role of proline as the primary determinant of elastomeric properties, proline content is the main sequence difference separating major ampullate silks with or without rubber-like elasticity.[89,90] The major ampullate silks of orb-weavers *Araneus diadematus* and *Nephilia clavipes* have nearly the same glycine content (40% and 45%, respectively) but significantly different proline content (16% and 3.5%, respectively). The composition of *Araneus* silk places it well above the PG composition threshold, while *Nephilia* silk is near the lower boundary of the coexistence region. The differences in proline content between these two species result in opposite mechanisms of elastic recoil: *Araneus* silk has entropically-driven elastic recoil, while the elastic recoil of *Nephilia* silk is almost entirely due to internal energy. On the basis of both thermoelasticity and birefringence measurements, it is thought that the molecular structure of *Araneus* silk is highly disordered, while that of *Nephilia* silk contains energetically-stable secondary structure.[89,90] In addition, based on a study of major ampullate silks from a wide range of species, increased proline content is strongly correlated with decreased stiffness, increased extensibility and increased capacity to shrink (supercontract).[91] Thus, proline content is an essential factor in determining the thermoelastic and mechanical properties of spider silk.

In support of the crucial role of glycine content in elasticity, Dicko et al observed that increasing glycine content correlates with increasing structural disorder for the various silks produced by *Nephilia edulis*.[61] In this study, structural disorder was quantified by a folding index, defined as the ratio of CD ellipticities at 220 nm and 200 nm; the folding index is a measure of the ratio of folded to unfolded structures. Interestingly, glycine-rich silks, such as major ampullate and flagelliform silks, have a significantly lower folding index than glycine-poor silks, such as aciniform, piriform and cylindriform.[61] These results are consistent with the view that increasing glycine content increases structural disorder, and therefore is correlated with the onset of elastomeric properties.

UNIFIED MODEL OF RUBBER-LIKE ELASTOMERIC STRUCTURE AND FUNCTION

The PG diagram (Fig. 4) points to a direct relationship between amino acid composition, conformational disorder and elastomeric properties of self-assembling elastomeric proteins. Significantly, the incompatibility of amyloid and elastomeric protein organization suggests that avoiding the formation of a water-excluding core involving extensive self-interactions is not only a fundamental requirement, but may very well constitute the single most essential design principle of self-assembling elastomeric proteins. The functional state of an elastomer may thus be described as a water-swollen, disordered aggregate characterized by an ensemble of many degenerate conformations (Fig. 3) that cannot form an ordered structure and are incompatible with the amyloid state. In this loosely-aggregated, hydrated state, the polypeptide chains can readily extend under strain.

This analysis leads to a simple, unified model of elastomeric structure and function. Like the native state of globular proteins, the structure of amyloids is characterized by a water-excluding core and extensive backbone self-interactions. In contrast, the unfolded state of proteins is a large ensemble of relatively disordered conformational states similar to that depicted in Figure 3B. At sufficient concentrations, unfolded proteins are prone to self-aggregation, which often results in the formation of amyloid fibrils. The PG diagram suggests that elastomeric proteins are designed to avoid both folding and amyloid formation, and that chain entropy and hydration play a central role in their function, consistent with rubber-like elasticity. Two major entropic forces are at play in the folding and aggregation of elastomeric proteins: polypeptide chain entropy and hydrophobicity. Chain entropy opposes both folding and full extension of the polypeptide chain since both events dramatically decrease the number of accessible conformations. Hydrophobic forces drive the emergence of collapsed states of polypeptide chains. Elastomeric chains remain hydrated and disordered even after aggregation because their polypeptide backbone is inherently unable to form extensive self-interactions. The relative contributions of chain entropy and entropy due to hydrophobic packing are likely to be highly sequence-dependent. In particular, both resilin and elastin exhibit very similar elastic properties, but have very different sequence hydrophobicities.

CONCLUSION AND PERSPECTIVES

There is an emerging consensus that rubber-like elastomeric proteins are intrinsically disordered, and therefore exhibit entropy-driven elastic recoil. Although a fundamental requirement for elastomeric domains is to remain disordered even when aggregated, they are not "random coils". Spectroscopic evidence and computational studies both point to the presence of significant amounts of transiently populated hydrogen-bonded turns and PPII structure. Importantly, the work reviewed above shows that the study of rubber-like elastomeric proteins benefits greatly from a synergy of theoretical and experimental approaches. Molecular simulations offer high-resolution structural information that is complementary to data obtained using spectroscopic approaches, including CD, FTIR and solid-state NMR. However, despite significant progress in the structural characterization of disordered states, there is at present very little insight into how the global effect of chain entropy relates to the fine balance of microscopic properties resulting in the distribution of conformations of the polypeptide chain and various side chains, their hydration, aggregation and extension. Such detailed understanding is required to explain why the mechanical properties of rubber-like elastomers exhibit significant variation. Understanding the structural properties of rubber-like elastomeric proteins is a necessary prerequisite to their effective use in biomaterials applications, and furthermore, to the rational design of novel elastomeric proteins.

Thus far, the majority of detailed studies on elastomeric proteins have focussed on elastin, and more recently, on spider silks. This is likely because biological science is driven to achieve advances that directly impact human health, which often narrows the focus of investigations to proteins related to specific human diseases. While there are advantages to such a focussed approach, namely, the wealth of structural data on elastin and elastin-derived peptides, we have demonstrated in this chapter that there are many similarities between elastin and the other rubber-like elastomeric proteins, all of which have tremendous potential in biomaterials development. For the most part, biomaterials scientists have not yet exploited the emerging knowledge of the myriad of elastomeric proteins that are adapted to various functional roles in nature. Thus, it is essential to obtain more detailed structural and mechanical characterization of resilin, abductin, ColP and other rubber-like elastomeric proteins. Simultaneous to fundamental structural investigations, hypotheses regarding essential sequence features can be tested by recombinantly producing elastomeric materials. These studies will provide complementary information to our knowledge of elastin and spider silks.

Thus, as a next step, our toolkit of rubber-like elastomers needs to be dramatically expanded. It is likely that many more rubber-like elastomeric proteins can be discovered by characterizing the proteins responsible for elastic recoil in biological tissues, which is the approach that led to the discovery of the octopus arterial elastomer. Once the essential sequence features of rubber-like elastomers have been identified, we can search genomic databases based on sequence similarity. It is likely other proteins, like Ubx, self-assemble to form elastomeric materials even though elastic properties have no known relevance to their role in vivo.

Many of the elastomeric biomaterials reviewed in this chapter exhibit a common design theme: the combination of elastomers with diverse mechanical properties in hybrid materials tuned to fulfill a particular biological role. Combinations of elastomers are found in byssal threads, which unite collagen-like and elastin-like sequences; the spider orb-web, which combines stiff and strong major ampullate silk with sticky and extensible flagelliform silk; and mammalian arterial walls, which achieve resilience and strength with a combination of elastin and collagen fibrils. In addition to designs incorporating elastomeric domains with differing mechanical properties, it is also possible to incorporate other biologically-active domains, such as those facilitating cell adhesion.[56] Furthermore, even the sequence of elastin exhibits duality in its alternating cross-linking and hydrophobic domains. Only after we understand the sequence determinants of elasticity for elastomers in isolation will it be possible to effectively combine different elastomers in hybrid materials, which represent a clear next step in biomimetic materials.

Through sequence, structure and mechanical studies of rubber-like elastomeric proteins, recent advances have led to the development of a framework for the rational design of self-assembling biomaterials. These studies will advance the development of biomimetic materials for a variety of applications, including vascular grafts, tissue replacements and scaffolds for tissue regeneration.

ACKNOWLEDGEMENTS

We are grateful to Dr. Fred Keeley and Dr. Lisa Muiznieks for useful discussions. We thank the Shared Hierarchical Academic Research Computing Network (SHARCNET) for a generous allocation of CPU resources. Computations were also performed on the general purpose cluster (GPC) supercomputer at the SciNet HPC Consortium. SciNet is funded by: the Canada Foundation for Innovation under the auspices of Compute Canada; the Government of Ontario; Ontario Research Fund—Research Excellence; and the University of Toronto. We gratefully acknowledge the Canadian Institutes of Health Research (Grant No. MOP84496) for support. S.R. is funded by a Canada Graduate Scholarship from the Natural Sciences and Engineering Research Council (NSERC) and the Research Training Center at the Hospital for Sick Children. R.P. is a CRCP chairholder.

REFERENCES

1. Gosline J, Lillie M, Carrington E et al. Elastic proteins: biological roles and mechanical properties. Philos Trans R Soc Lond Ser B-Biol Sci 2002; 357(1418):121-132.
2. Mark JE. Rubber Elasticity. J Chem Educ 1981; 58(11):898-903.
3. Mark JE, Erman B. Rubberlike elasticity: A molecular primer. Cambridge: Cambridge University Press; 2007.
4. Muiznieks LD, Weiss AS, Keeley FW. Structural disorder and dynamics of elastin. Biochem Cell Biol 2010; 88(2):239-250.
5. Woessner JF, Brewer TH. Formation and breakdown of collagen and elastin in human uterus during pregnancy and postpartum involution. Biochem J 1963; 89(1):75-82.
6. Weis-Fogh T. A rubber-like protein in insect cuticle. J Exp Biol 1960; 37(4):889-907.
7. Bennet-Clark HC, Lucey ECA. The jump of a flea: A study of the energetics and a model of the mechanism. J Exp Biol 1967; 47(1):59-76.
8. Gosline JM, Demont ME, Denny MW. The structure and properties of spider silk. Endeavour 1986; 10(1):37-43.
9. Alexander RM. Rubber-like properties of the inner hinge-ligament of Pectinidae. J Exp Biol 1966; 44(1):119-130.
10. Kelly RE, Rice RV. Abductin: A rubber-like protein from the internal triangular hinge ligament of Pecten. Science 1967; 155(3759):208-210.
11. Denny M, Miller L. Jet propulsion in the cold: mechanics of swimming in the Antarctic scallop Adamussium colbecki. J Exp Biol 2006; 209(22):4503-4514.
12. Waite JH, Vaccaro E, Sun CJ et al. Elastomeric gradients: a hedge against stress concentration in marine holdfasts? Philos Trans R Soc Lond Ser B-Biol Sci 2002; 357(1418):143-153.

13. Elvin CM, Carr AG, Huson MG et al. Synthesis and properties of crosslinked recombinant proresilin. Nature 2005; 437(7061):999-1002.
14. Bellingham CM, Lillie MA, Gosline JM et al. Recombinant human elastin polypeptides self-assemble into biomaterials with elastin-like properties. Biopolymers 2003; 70(4):445-455.
15. Bella J, Eaton M, Brodsky B et al. Crystal and molecular structure of a collagen-like peptide at 1.9 Å resolution. Science 1994; 266(5182):75-81.
16. Rauscher S, Baud S, Miao M et al. Proline and glycine control protein self-organization into elastomeric or amyloid fibrils. Structure 2006; 14(11):1667-1676.
17. Pometun MS, Chekmenev EY, Wittebort RJ. Quantitative observation of backbone disorder in native elastin. J Biol Chem 2004; 279(9):7982-7987.
18. Andrady AL, Mark JE. Thermoelasticity of swollen elastin networks at constant composition. Biopolymers 1980; 19(4):849-855.
19. Shewry PR, Tatham AS, Bailey AJ eds. Elastomeric proteins: Structures, biomechanical properties and biological roles. Cambridge: Cambridge University Press; 2002.
20. Alexander RM. Animal Mechanics. Worcester: Macmillan and Co Ltd; 1983.
21. Miserez A, Wasko SS, Carpenter CF et al. Non-entropic and reversible long-range deformation of an encapsulating bioelastomer. Nature Materials 2009; 8(11):910-916.
22. Tatham AS, Shewry PR. Elastomeric proteins: biological roles, structures and mechanisms. Trends Biochem Sci 2000; 25(11):567-571.
23. Mark JE. Molecular aspects of rubberlike elasticity. Angew Makromol Chem 1992; 202:1-30.
24. Tatham AS, Shewry PR. Comparative structures and properties of elastic proteins. Philos Trans R Soc Lond Ser B-Biol Sci 2002; 357(1418):229-234.
25. Weis-Fogh T. Thermodynamic properties of resilin, a rubber-like protein. J Mol Biol 1961; 3(5):520-531.
26. Gosline JM, Denny MW, DeMont ME. Spider silk as rubber. Nature 1984; 309(5968):551-552.
27. Kahler GA, Fisher FM, Sass RL. The chemical composition and mechanical properties of the hinge ligament in bivalve molluscs. Biol Bull 1976; 151(1):161-181.
28. Hoeve CAJ, Flory PJ. The elastic properties of elastin. Biopolymers 1974; 13(4):677-686.
29. Vrhovski B, Weiss AS. Biochemistry of tropoelastin. Eur J Biochem 1998; 258(1):1-18.
30. Starcher BC. Determination of the elastin content of tissues by measuring desmosine and isodesmosine. Anal Biochem 1977; 79(1-2):11-15.
31. Gibbons CA, Shadwick RE. Circulatory mechanics in the toad Bufo marinus I. Structure and mechanical design of the aorta. J Exp Biol 1991; 158:275-289.
32. Shadwick RE. Mechanical design in arteries. J Exp Biol 1999; 202(23):3305-3313.
33. Starcher B, Percival S. Elastin turnover in the rat uterus. Connect Tissue Res 1985; 13(3):207-215.
34. Miao M, Bellingham CM, Stahl RJ et al. Sequence and structure determinants for the self-aggregation of recombinant polypeptides modeled after human elastin. J Biol Chem 2003; 278(49):48553-48562.
35. Tamburro AM, Bochicchio B, Pepe A. Dissection of human tropoelastin: Exon-by-exon chemical synthesis and related conformational studies. Biochemistry 2003; 42(45):13347-13362.
36. Urry DW, Hugel T, Seitz M et al. Elastin: A representative ideal protein elastomer. Philos Trans R Soc Lond Ser B-Biol Sci 2002; 357(1418):169-184.
37. Becker N, Oroudjev E, Mutz S et al. Molecular nanosprings in spider capture-silk threads. Nature Materials 2003; 2(4):278-283.
38. Aaron BB, Gosline JM. Elastin as a random-network elastomer: A mechanical and optical analysis of single elastin fibers. Biopolymers 1981; 20(6):1247-1260.
39. Dorrington KL, McCrum NG. Elastin as a rubber. Biopolymers 1977; 16(6):1201-1222.
40. Perry A, Stypa MP, Tenn BK et al. Solid-state ^{13}C NMR reveals effects of temperature and hydration on elastin. Biophys J 2002; 82(2):1086-1095.
41. Muiznieks LD, Jensen SA, Weiss AS. Structural changes and facilitated association of tropoelastin. Arch Biochem Biophys 2003; 410(2):317-323.
42. Bochicchio B, Floquet N, Pepe A et al. Dissection of human tropoelastin: Solution structure, dynamics and self-assembly of the exon 5 peptide. Chem Eur J 2004; 10(13):3166-3176.
43. Bochicchio B, Pepe A, Tamburro AM. Investigating by CD the molecular mechanism of elasticity of elastomeric proteins. Chirality 2008; 20(9):985-994.
44. Glaves R, Baer M, Schreiner E et al. Conformational dynamics of minimal elastin-like polypeptides: The role of proline revealed by molecular dynamics and nuclear magnetic resonance. ChemPhysChem 2008; 9(18):2759-2765.
45. Rauscher S, Neale C, Pomès R. Simulated tempering distributed replica sampling, virtual replica exchange and other generalized-ensemble methods for conformational sampling. J Chem Theor Comput 2009; 5(10):2640-2662.

46. Baer M, Schreiner E, Kohlmeyer A et al. Inverse temperature transition of a biomimetic elastin model: Reactive flux analysis of folding/unfolding and its coupling to solvent dielectric relaxation. J Phys Chem B 2006; 110(8):3576-3587.
47. Rauscher S, Pomès R. Molecular simulations of protein disorder. Biochem Cell Biol 2010; 88(2):269-290.
48. Li B, Alonso DOV, Daggett V. The molecular basis for the inverse temperature transition of elastin. J Mol Biol 2001; 305(3):581-592.
49. Vendruscolo M, Dobson CM. Towards complete descriptions of the free energy landscape of proteins. Phil Trans R Soc Lond A 2005; 363:433-452.
50. Zagrovic B, Lipfert J, Sorin EJ et al. Unusual compactness of a polyproline type II structure. Proc Natl Acad Sci USA 2005; 102:11698-11703.
51. Tompa P. Structural disorder in amyloid fibrils: its implication in dynamic interactions of proteins. FEBS J 2009; 276(19):5406-5415.
52. Dyksterhuis LB, Carter EA, Mithieux SM et al. Tropoelastin as a thermodynamically unfolded premolten globule protein: The effect of trimethylamine N-oxide on structure and coacervation. Arch Biochem Biophys 2009; 487(2):79-84.
53. Bennet-Clark H. The first description of resilin. J Exp Biol 2007; 210(22):3879-3881.
54. Bennet-Clark HC. Tymbal mechanics and the control of song frequency in the cicada Cyclochila australasiae. J Exp Biol 1997; 200(11):1681-1694.
55. Dillinger SCG, Kesel AB. Changes in the structure of the cuticle of Ixodes ricinus L. 1758 (Acari, Ixodidae) during feeding. Arthropod Structure and Development 2002; 31(2):95-101.
56. Charati MB, Ifkovits JL, Burdick JA et al. Hydrophilic elastomeric biomaterials based on resilin-like polypeptides. Soft Matter 2009; 5(18):3412-3416.
57. Tamburro AM, Panariello S, Santopietro V et al. Molecular and supramolecular structural studies on significant repetitive sequences of resilin. ChemBioChem 2010; 11(1):83-93.
58. Ardell DH, Andersen SO. Tentative identification of a resilin gene in Drosophila melanogaster. Insect Biochem Mol Biol 2001; 31(10):965-970.
59. Elliott GF, Huxley AF, Weis-Fogh T. On the structure of resilin. J Mol Biol 1965; 13(3):791-795.
60. Nairn KM, Lyons RE, Mulder RJ et al. A synthetic resilin is largely unstructured. Biophys J 2008; 95(7):3358-3365.
61. Dicko C, Porter D, Bond J et al. Structural disorder in silk proteins reveals the emergence of elastomericity. Biomacromolecules 2008; 9(1):216-221.
62. Hayashi CY, Lewis RV. Spider flagelliform silk: lessons in protein design, gene structure and molecular evolution. Bioessays 2001; 23(8):750-756.
63. Gosline JM, Guerette PA, Ortlepp CS et al. The mechanical design of spider silks: From fibroin sequence to mechanical function. J Exp Biol 1999; 202(23):3295-3303.
64. Vollrath F, Porter D. Silks as ancient models for modern polymers. Polymer 2009; 50(24):5623-5632.
65. Teulé F, Furin WA, Cooper AR et al. Modifications of spider silk sequences in an attempt to control the mechanical properties of the synthetic fibers. J Mater Sci 2007; 42(21):8974-8985.
66. Vendrely C, Scheibel T. Biotechnological production of spider-silk proteins enables new applications. Macromol Biosci 2007; 7(4):401-409.
67. Ohgo K, Kawase T, Ashida J et al. Solid-state NMR analysis of a peptide (Gly-Pro-Gly-Gly-Ala)$_6$-Gly derived from a flagelliform silk sequence of Nephila clavipes. Biomacromolecules 2006; 7(4):1210-1214.
68. Rousseau ME, Lefèvre T, Pézolet M. Conformation and Orientation of Proteins in Various Types of Silk Fibers Produced by Nephila clavipes Spiders. Biomacromolecules 2009; 10(10):2945-2953.
69. Vogel S. Animal locomotion—Squirt smugly, scallop! Nature 1997; 385(6611):21-22.
70. Bochicchio B, Jimenez-Oronoz F, Pepe A et al. Synthesis of and structural studies on repeating sequences of abductin. Macromol Biosci 2005; 5(6):502-511.
71. Cao QP, Wang YJ, Bayley H. Sequence of abductin, the molluscan 'rubber' protein. Curr Biol 1997; 7(11):R677-R678.
72. Coyne KJ, Qin XX, Waite JH. Extensible collagen in mussel byssus: A natural block copolymer. Science 1997; 277(5333):1830-1832.
73. Hagenau A, Scheidt HA, Serpell L et al. Structural Analysis of Proteinaceous Components in Byssal Threads of the Mussel Mytilus galloprovincialis. Macromol Biosci 2009; 9(2):162-168.
74. Brazee SL, Carrington E. Interspecific comparison of the mechanical properties of mussel byssus. Biol Bull 2006; 211(3):263-274.
75. Deming TJ. Mussel byssus and biomolecular materials. Curr Opin Chem Biol 1999; 3(1):100-105.
76. Bondos SE. Roles for intrinsic disorder and fuzziness in generating context-specific function in ultrabithorax, a hox transcription factor. In: Fuxreiter M, Tompa P, eds. Fuzziness: Structural Disorder in Protein Complexes. Austin/New York: Landes Bioscience/Springer Science+Business Media, 2010:86-105.
77. Greer AM, Huang Z, Oriakhi A et al. The Drosophila transcription factor ultrabithorax self-assembles into protein-based biomaterials with multiple morphologies. Biomacromolecules 2009; 10(4):829-837.

78. Shadwick RE, Gosline JM. Physical and chemical properties of rubber-like elastic fibers from the octopus aorta. J Exp Biol 1985; 114:239-257.
79. Kim W, Conticello VP. Protein engineering methods for investigation of structure-function relationships in protein-based elastomeric materials. Polymer Reviews 2007; 47(1):93-119.
80. Romero P, Obradovic Z, Li XH et al. Sequence complexity of disordered protein. Proteins 2001; 42(1):38-48.
81. Tompa P. Intrinsically unstructured proteins evolve by repeat expansion. Bioessays 2003; 25(9):847-855.
82. Petkova AT, Ishii Y, Balbach JJ et al. A structural model for Alzheimer's beta-amyloid fibrils based on experimental constraints from solid state NMR. Proc Natl Acad Sci USA 2002; 99(26):16742-16747.
83. Fowler DM, Koulov AV, Balch WE et al. Functional amyloid—from bacteria to humans. Trends Biochem Sci 2007; 32(5):217-224.
84. Monsellier E, Chiti F. Prevention of amyloid-like aggregation as a driving force of protein evolution. EMBO Reports 2007; 8(8):737-742.
85. Dobson CM. Protein folding and misfolding. Nature 2003; 426(6968):884-890.
86. Watt B, van Niel G, Fowler DM et al. N-terminal domains elicit formation of functional Pmel17 amyloid fibrils. J Biol Chem 2009; 284(51):35543-35555.
87. Wang X, Zhou Y, Ren JJ et al. Gatekeeper residues in the major curlin subunit modulate bacterial amyloid fiber biogenesis. Proc Natl Acad Sci USA 2010; 107(1):163-168.
88. Tamburro AM, Pepe A, Bochicchio B et al. Supramolecular amyloid-like assembly of the polypeptide sequence coded by exon 30 of human tropoelastin. J Biol Chem 2005; 280(4):2682-2690.
89. Savage KN, Gosline JM. The role of proline in the elastic mechanism of hydrated spider silks. J Exp Biol 2008; 211(12):1948-1957.
90. Savage KN, Gosline JM. The effect of proline on the network structure of major ampullate silks as inferred from their mechanical and optical properties. J Exp Biol 2008; 211(12):1937-1947.
91. Liu Y, Sponner A, Porter D et al. Proline and processing of spider silks. Biomacromolecules 2008; 9(1):116-121.
92. Andersen SO. Isolation of a new type of cross link from hinge ligament protein of molluscs. Nature 1967; 216(5119):1029-1030.
93. Waite JH, Lichtenegger HC, Stucky GD et al. Exploring molecular and mechanical gradients in structural bioscaffolds. Biochemistry 2004; 43(24):7653-7662.
94. Chung MIS, Miao M, Stahl RJ et al. Sequences and domain structures of mammalian, avian, amphibian and teleost tropoelastins: Clues to the evolutionary history of elastins. Matrix Biol 2006; 25(8):492-504.
95. Djajamuliadi J, Kagawa TF, Ohgo K et al. Insights into a putative hinge region in elastin using molecular dynamics simulations. Matrix Biol 2009; 28(2):92-100.
96. Blackledge TA, Hayashi CY. Silken toolkits: biomechanics of silk fibers spun by the orb web spider Argiope argentata (Fabricius 1775). J Exp Biol 2006; 209(13):2452-2461.
97. Pettersen EF, Goddard TD, Huang CC et al. UCSF Chimera—a visualization system for exploratory research and analysis. J Comput Chem 2004; 25(13):1605-1612.

CHAPTER 11

"FUZZINESS" IN THE CELLULAR INTERACTOME: A Historical Perspective

G. Rickey Welch
Department of Biological Sciences and Department of History, University of Maryland, Baltimore, Maryland, USA
Email: welch@umbc.edu

Abstract: Some historical background is given for appreciating the impact of the empirical construct known as the cellular protein-protein interactome, which is a seemingly de novo entity that has arisen of late within the context of postgenomic systems biology. The approach here builds on a generalized principle of "fuzziness" in protein behavior, proposed by Tompa and Fuxreiter.[1] Recent controversies in the analysis and interpretation of the interactome studies are rationalized historically under the auspices of this concept. There is an extensive literature on protein-protein interactions, dating to the mid-1900s, which may help clarify the "fuzziness" in the interactome picture and, also, provide a basis for understanding the physiological importance of protein-protein interactions in vivo.

INTRODUCTION: PROTEINS IN MOTION

Fuzzy: "Not firm or sound in substance; frayed into loose fibres; blurred; indistinct"
—Oxford English Dictionary

There is no better descriptor of the physical aura of proteins than the oh-so-noble expression, "fuzzy"! Under defined laboratory conditions, proteinaceous macromolecules display a rich diversity of fluctuational motions, spanning a wide range of timescales, about some average "structure." The significance of such behavior in our understanding of the functionality of proteins led to the rise of the subfield of biophysical chemistry known as "protein dynamics" in the latter part of the 20th century. As an indication of the timeliness today, *Science*[2] devoted a special issue to the subject of protein dynamics

Fuzziness: Structural Disorder in Protein Complexes, edited by Monika Fuxreiter and Peter Tompa.

and it is featured recently in an "Insight" supplement on "Proteins to Proteomes" in *Nature*[2] (for reviews of the older literature see refs. 4-6). Vinson[2] described the character of the protein-dynamical picture aptly, as follows: "The view that has emerged is that of an intricate ballet. Individual proteins are in constant motion, sampling an ensemble of different conformations and perhaps changing interaction partners as they play their part in a particular biological process. How do these dynamics affect function? The conformational space that a protein can explore can be described by an energy landscape, in which different conformations are populated based on their energies and rates of interconversion are dependent on the energy barriers between states. The landscape and thus the relative populations of conformational states, can be modulated, for example, by interactions with other proteins or by covalent modifications such as phosphorylation." Considering the complexity of such internal motions, across so many timescales and internal degrees of freedom, it is a wonder that proteins manifest such defined properties as ligand binding, enzyme catalysis, etc. From a physiological viewpoint, it stands to reason that the thermal sampling of conformational substates within proteins cannot be a completely random walk; that is, there must be constraint on the ergodic trajectories in the phase space of conformational dynamics.[7] As discussed of late by Henzler-Wildman and Kern,[3] "Because biological function is the property selected by evolution, [the] conformational substates sampled by a protein and the pathways between them, are not random but rather a result of the evolutionary selection of states that are needed for protein function... In other words, the dynamic landscape is an intrinsic property (or 'personality') of a protein and is encoded in its fold." The idea that protein fluctuations entail such a "personality" was proposed long ago,[5,7] and recent studies[8] have lended support thereto. Quite simply, proteins act as "energy funnels."

Thus, it has come to be acknowledged that functionality, as well as adaptability and evolvability, in proteins is linked to protein dynamics.[9] Various changes in the traditional, static view of protein structure-and-function have ensued. For example, it is now recognized that there are classes of proteins with "designed" disordered states—what have been dubbed "intrinsically unstructured proteins".[10,11] Importantly, Tompa and Fuxreiter[1] have extended this notion of structural disorder and polymorphism to the description of interacting protein systems—in the process, introducing (what they call) the principle of "fuzziness" into vernacular of protein biochemistry. Here, I wish to give some historical background to the general idea of "fuzziness" in our understanding of enzyme-enzyme interactions in living cells—within the context of the postgenomic concept of the *interactome*.

THE SOCIAL LIFE OF PROTEINS IN THE CELL

Most proteins are by nature gregarious entities; they prefer to be with their own kind. One of the oldest tricks-of-the-trade in laboratory biochemistry is the necessity, in many cases, for the addition of a "neutral" protein (e.g., bovine serum albumin) to a purified enzyme preparation in order to stabilize or enhance catalytic activity. Sometimes artificial polymeric agents (e.g., polyethylene glycol) will do the trick. We now know that these in vitro effects are indicative of the reality of dense, heterogeneous microenvironments in vivo, wherein most individual enzymes operate in close proximity (or are bound) to protein (or lipoprotein) arrays.[12] Protein-protein and protein-cytomatrix associations in living cells pervade the multienzyme systems in most metabolic pathways, as well as such processes as signal transduction, cell trafficking, DNA replication, transcription and translation.[13-15] There is a vast body of work in this area, incorporating a range of experimental approaches, dating from the mid 1900s. The physiological function(s) of organized enzyme states has been widely examined.[16] The social life of proteins (particularly enzymes) in situ is a far cry from that observed in dilute aqueous solution in vitro.

Interest (and, one might argue, awareness) of the importance of enzyme organization in biochemistry waned in the latter part of the 20th century, with the coming of the genomic age. At the turn of the new millennium (and the rise of the postgenomic era), large-scale, high-throughput experimental methods began to generate massive amounts of data relating to all facets of cellular operation—via an expanding hierarchy of empirical portals known as "-omes" that reflect the defined levels of complexity. Indication of the widespread existence of protein-protein interactions in living cells has rapidly mounted in this period, relying heavily on such analytical detection tools as the yeast two-hybrid technique and affinity purification (for a review of the methods, see ref. 17). Whole cellular interactomes (with associated protein-protein interaction maps) have been published for a variety of organisms. The attempt to integrate the "-omic" information into a holistic picture of cellular function today falls under the rubric of "systems biology." The newfound attention to the role of supramolecular organization in the cell features prominently in this new field. As remarked in a recent *Nature* editorial,[18] "How are innumerable protein functions integrated so that a living cell interacts coherently with its environment? This question is central to an emerging science of biological information processing—systems biology."

Study of the cellular interactome—both experimentally and theoretically—is now a major research enterprise, as observed by the dramatic rise of publications and scientific conferences on the subject in the last decade. Analogies and metaphors from various fields of knowledge abound, in the struggle to make sense of this observational construct. Sociology is a particularly evocative discipline of interest lately. In a "News Feature" commentary in *Nature* entitled "Proteomics: The Society of Proteins," Abbott[19] pointed to "an emerging biological concept, that proteins do not work alone." Recently, Robinson, Sali and Baumeister[20] highlighted this consideration of protein-protein interactions in a treatise labeled as "The Molecular Sociology of the Cell." The application of concepts from sociology to the description of protein-protein interactions is, in fact, well known in the older, pregenomic-era literature on cellular organization (for review see refs. 21, 22). The conception of (what have been termed) "social sites" on globular protein surfaces[23] has even led to the proposal for the formal establishment of a fifth ("quinary") level of protein structure, in order to include the assortment of heterogeneous interactions that relate to the organized life of most proteins in vivo.[24]

It was appreciated early in the study of interacting protein systems that the functional association of the macromolecular partners has been achieved via highly specific binding sites, where the aggregation process is often subject to the subtleties of regulatory and microenvironmental effects in vivo.[16,21] Looking at the surface characteristics of typical globular proteins, Anderson[23,25] argued from a fundamental immunological perspective of "self-recognition" that the protein-protein interaction patterns must be limited and sharply defined in functionality. Such conditions involve a great degree of conservatism in the evolutionary landscape of specifically-integrated cellular proteins.[24] This early impression has become reified in the observed network properties of the newly-emerging interactome mappings.[26]

From the cytosociological standpoint, today's interactome studies would seem to represent prima facie the fulfillment (or perhaps the culmination) of an older chronological train of thought on the organizational character of protein molecules in the living cell. However, there are conceptual gaps, between this historical legacy and the empirical approach presently being taken to understand protein-protein interactions in vivo, which demand resolution.

THE INTERACTOME: WHAT IS REAL AND NOT REAL?

Critical questions regarding the characterization and interpretation of the cellular protein–protein interactome have been raised by a growing number of observers. Such concerns, for my purpose, are well illustrated in a recent string of communications in *Trends in Biochemical Sciences* (summarized in ref. 27). Much of the attention centers on the nature of the detection methods. Mackay, Sunde, Lowry, et al[28] claimed that many reports of protein interactions are founded on "insufficient data," and that it is becoming "dangerously acceptable" to conclude that proteins interact based on questionable results and limited methodologies. Chatr-aryamontri et al[29] defended the importance of extant protein-interaction databases (e.g., the Molecular INTeraction database [MINT], http://mint.bio.uniroma2.it/mint/Welcome.do) in revealing biologically relevant information, while affirming "that no single experimental approach has maximum sensitivity (i.e., no false negative) and specificity (i.e., no false positive) and that confidence can only be built on the integration of orthogonal experimental evidence." Mackay et al[30] further warned that the common experimental detection methods are carried out with cell lysates in an "uncompartmentalized soup," in a molecular environment that is rife with "promiscuous stickiness." On another note, Wilkins and Kummerfeld[31] questioned the static picture of the interactome that has emerged from published protein-interaction network models, with regards to the possible physiological variability in the properties of the protein constituents of such networks in vivo. In this same sequence of correspondences, Tompa and Fuxreiter[1] added their aforementioned argument that the occurrence of dynamic structural disorder in protein complexes must play a role in depictions of the interactome.

Let us consider the question of the referential context for the large-scale, high-throughput interactome experiments and the interpretation of ensuing results. Without giving an exhaustive list of the manifold publications in this area, it is fair to say that many of the summary reports of cellular interactomes cite the commentary paper by Alberts,[32] entitled "The cell as a collection of protein machines: preparing the next generation of molecular biologists," as the point of reference for the biological importance of the interactome concept. This banner article highlights the properties of large protein complexes involved in such functions as the cell cycle, RNA processing and DNA replication and transcription. The fact that many of the aggregates in such processes are isolable in stable form might be leading to a false sense that protein-protein interactions represent a simple binary occurrence in living cells and that the associations can be readily (and universally) identified experimentally by the detection methods that are currently used in the interactome determinations. This simplistic (mis)perception would seem to be evident, for example, in the quest for so-called "gold standard" methodologies for ascertaining the reliability of the protein interactions (for example, see refs. 28, 29). Jansen and Gerstein[33] cast this mission in the view that "there is a degree of uncertainty related to the ultimate goal of functional genomics," owing to the circumstance that "the concept of 'protein function' is rather 'fuzzy' because it is often based on whimsical terms or contradictory nomenclature." Curiously, those authors further asserted that "unlike protein function, protein–protein interactions are relatively clearly defined." In truth, it is "fuzziness" in the latter notion, historically, that belies the validity of today's interactome network models!

Although one might debate the singular motivational influence that Alberts's seminal paper[32] has had on interactome studies, the fact is that attention has focused heavily on such processes as cell signaling, the cell cycle, membrane trafficking and DNA replication and transcription (e.g., see http://www.reactome.org). These systems are, without doubt, of great biological (and biomedical) importance. Notwithstanding, as discussed above, there is a vast literature on the subject of macromolecular interactions that long predates the "-ome/-omic" era. Indeed, multienzyme complexes are in great abundance in the metabolic pathways of the living cell. Moreover, the role of these aggregates as "molecular machines" has long been recognized (for review see ref. 34). Notably, these supramolecular systems are defined by a wide range of binding strengths, engendering a large and diverse assortment of weak-to-stable functional aggregates.[12,13] The interaction modality itself is often modulated in living cells via the protein-conformational effects of regulatory ligands and microenvironmental factors in situ.[35] These macromolecular assemblages are known for their sensitivity to extraction conditions and artifacts of the isolation procedure (both false negatives and false positives) are a notorious characteristic of the in vitro analysis of these systems.[12-16] Accordingly, the caveats discussed in such writings as references 1, 27-31 represent noteworthy concerns that are supported by a long, established record of observations on protein complexes.

The large-scale interactome experiments aside, the ensuing curation efforts for such interactomic databases as MINT[29] have produced an impressively large (and growing) number of protein-protein interactions, spanning many organisms and canvassing a widening array of detection methods.[17] There is every reason to presume that, as a long-term goal, the consonance of the postgenomic protein-interaction datasets with the historical record of published studies on specific proteins is (or will be) of direct interest to those laboring in the curation activity. Continued mining of the substantial data on protein interactions from the pregenomic past (accessible, for example, through such portals as ref. 16) will most certainly help to advance this occupation. In the meantime, those of us on the outside looking in should be patient and optimistic as to the future successes of this vital curation work. Perkel[36] has summarized the status of the interactome situation rather fittingly as follows, "Many researchers ask, 'If the studies are so comprehensive, then why isn't my interaction there?' Well, for starters, it's probably because investigators haven't gotten to it yet. Also, interactome mapping is still a primitive science. For all their colorful nodes and edges, these maps could well warn, 'Here, there be dragons.' That's not to say they are without value... By revealing potential functional linkages, they suggest molecular explanations to biological phenomena and provide avenues for follow-up. The information is useful when viewed with an informed mind. Many of the principals involved in these projects compare the state of the field today with DNA sequencing technology in the late 1970s: immature, error-prone, yet filled with promise."

CONCLUSION: THE INTERACTOME AND PHYSIOLOGY

A fundamental empirical problem in the large-scale interactome studies at this point, as stated by Legrain, Wojcik and Gauthier,[37] is that the protein-interaction maps reflect "technology-driven experiments rather than hypothesis-driven experiments." So much of the work in systems biology these days is devoted to advancement in the analytical technology per se and to the manipulation of incomprehensibly large datasets. As eloquently discussed by Kell and Oliver[38] scientific knowledge discovery in the postgenomic era requires both hypothesis-driven and technology-led approaches (what the authors symbolize as "Ideas" and "Data," respectively) and, moreover, demands that they be "complementary and iterative partners." It is apparent from some of the more recent interactome studies that the physiological "Ideas" are, indeed, operating iteratively with the "Data"—as evidenced, for example, in the extension of interactome analyses to in vivo conditions.[39] Moreover, such in situ techniques as electron tomography[17,20] offer great promise. The historical literature base on protein interactions[16,40] provides a wealth of "Ideas" as to the physiological rationale for the organizational state, in addition to a large (and untapped) database on specific systems. The myopia of the data deluge in today's systems biology has created a blinding sense of the Present without an appreciation of the Past. A studied reflection of the historical ideas leads convincingly to the conclusion that systems biology only has meaning when it connects to physiology.[41]

Finally, it must be borne in mind that the protein-interaction mapping for a given cell-type—even when obtained under the most reliable of experimental conditions—represents a relative and variable thing. As trite as it might be to say, the interactome (like all "-omes" downstream from the genome) is a dynamic entity, the structure-and-function of which is constantly subject to intra- and extracellular physiological influences. Pregenomic systems biology has made it all too evident that the interactome is an observationally "fuzzy," yet seductively attractive, object of curiosity.

ACKNOWLEDGEMENTS

This chapter is dedicated to the memory of my friend and colleague, Béla Somogyi, who introduced me to the importance of protein motions in biochemistry.

REFERENCES

1. Tompa P, Fuxreiter M. Fuzzy complexes: Polymorphism and structural disorder in protein-protein interaction. Trends Biochem Sci 2008; 33:2-8.
2. Vinson V. Proteins in motion: Introduction. Science 2009; 324(5924):197.
3. Henzler-Wildman K, Kern D. Dynamic personalities of proteins. Nature 2007; 450(7172):964-72.
4. McCammon JA, Harvey SC. Dynamics of Proteins and Nucleic Acids. Cambridge: Cambridge University Press, 1987.
5. Welch GR, Somogyi B, Damjanovich S. The role of protein fluctuations in enzyme action. Prog Biophys Mol Biol 1982; 39:109-146.
6. Welch GR, ed. The Fluctuating Enzyme. New York: Wiley, 1986.
7. Somogyi B, Welch GR, Damjanovich S. The protein-dynamical basis of energy transduction in enzymes. Biochim Biophys Acta (Reviews on Bioenergetics) 1984; 768:81-112.
8. Swint-Kruse L, Fisher HF. Enzymatic reaction sequences as coupled multiple traces on a multidimensional landscape. Trends Biochem Sci 2008; 33:104-112.
9. Tokuriki N, Tawfik DS. Protein dynamism and evolvability. Science 2009; 324:203-207.

10. Wright PE, Dyson HJ. Intrinsically unstructured proteins: Re-assessing the protein structure-function paradigm. J Mol Biol 1999; 293:321-331.
11. Tompa P. Intrinsically unstructured proteins. Trends Biochem Sci 2002; 27:527-533.
12. Westerhoff HV, Welch GR. Enzyme organization and the direction of metabolic flow. In: Stadtman ER, Chock PB, eds. From Metabolite, to Metabolism, to Metabolon (Current Topics in Cellular Regulation, Vol. 33). New York: Academic Press, 1992:361-390.
13. Ovádi J. Cell Architecture and Metabolic Channeling. Austin: Landes Bioscience 1995.
14. Brindle K, ed. Enzymology In Vivo (Advances in Molecular and Cell Biology, Vol. 11). London: JAI Press, 1995.
15. Srere P. Complexes of sequential metabolic enzymes. Annu Rev Biochem 1987; 56:21-56.
16. Srere P. Macromolecular interactions: Tracing the roots. Trends Biochem Sci 2000; 25:150-153.
17. Aloy P, Russell RB. Structural systems biology: Modelling protein interactions. Nature Rev Mol Cell Biol 2006; 7:188-197.
18. Chouard T, Finkelstein J. Proteins to proteomes. Nature 2007; 450:963.
19. Abbott A. Proteomics: The society of proteins. Nature 2002; 417:894-896.
20. Robinson CV, Sali A, Baumeister W. The molecular sociology of the cell. Nature 2007; 450:973-982.
21. Welch GR. On the role of organized multienzyme systems in cellular metabolism: A general synthesis. Prog Biophys Mol Biol 1977; 32:103-191.
22. Welch GR, Keleti T. On the "cytosociology" of enzyme action in vivo: A novel thermodynamic correlate of biological evolution. J Theor Biol 1981; 93:701-735.
23. Anderson NG. Interactive macromolecular sites I. Basic theory. J Theor Biol 1976; 60:401-412.
24. McConkey EH. Molecular evolution, intracellular organization and the quinary structure of proteins. Proc Natl Acad Sci USA 1982; 79:3236-3240.
25. Anderson NG. Interactive macromolecular sites II. Role in prebiotic macromolecular selection and early cellular evolution. J Theor Biol 1976; 60:413-419.
26. Barabási A-L, Oltvai ZN. Network biology: Understanding the cell's functional organization. Nature Rev Genetics 2004; 5:101-113.
27. Welch GR. The "fuzzy" interactome. Trends Biochem Sci 2009; 34:1-2.
28. Mackay JP, Sunde M, Lowry JA et al. Protein interactions: Is seeing believing? Trends Biochem Sci 2007; 32:530-531.
29. Chatr-aryamontri A, Ceol A, Licata L et al. Protein interactions: Integration leads to belief. Trends Biochem Sci 2008; 33:241-242.
30. Mackay JP, Sunde M, Lowry JA et al. Response to Chatr-aryamontri et al: Protein interactions: To believe or not to believe? Trends Biochem Sci 2008; 33:242-243.
31. Wilkins MR, Kummerfeld SK. Sticking together? Falling apart? Exploring the dynamics of the interactome. Trends Biochem Sci 2008; 33:195-200.
32. Alberts B. The cell as a collection of protein machines: Preparing the next generation of molecular biologists. Cell 1998; 92:291-294.
33. Jansen R, Gerstein M. Analyzing protein function on a genomic scale: The importance of gold-standard positives and negatives for network prediction. Curr Opin Microbiol 2004; 7:535-545.
34. Welch GR, Kell DB. Not just catalysts—molecular machines in bioenergetics. In: Welch GR, ed. The Fluctuating Enzyme. New York: Wiley, 1986:451-492.
35. Welch GR. The organization of metabolic pathways in vivo. In: Bittar EE, Bittar N, eds. Cell Chemistry and Physiology (Principles of Medical Biology, Vol. 4). London: JAI Press, 1995:77-92.
36. Perkel JM. Validating the interactome. Scientist 2004; 18:18.
37. Legrain P, Wojcik J, Gauthier J-M. Protein-protein interaction maps: A lead towards cellular functions. Trends Genet 2001; 17:346-352.
38. Kell DB, Oliver SG. Here is the evidence, now what is the hypothesis? The complementary roles of inductive- and hypothesis-driven science in the postgenomic era. BioEssays 2004; 26:99-105.
39. Tarassov K, Messier V, Landry CR et al. An in vivo map of the yeast protein interactome. Science 2008; 320:1465-1470.
40. Clegg JS, Kell DB, Knull H et al. Macromolecular interactions: Tracing the roots. Trends Biochem Sci 2001; 26:91.
41. Welch GR. Physiology, physiomics and biophysics: A matter of words. Prog Biophys Mol Biol 2009; 100:4-17.

INDEX

D

E

F

G

H

I

L

M

N

O

P

Q

R

S

T

U

V

X